Negligence in Building Law

Cases and Commentary

Negligence in Building Law
Cases and Commentary

Jon Holyoak
Department of Law
Leicester University

OXFORD
BLACKWELL SCIENTIFIC PUBLICATIONS
LONDON EDINBURGH BOSTON
MELBOURNE PARIS BERLIN VIENNA

© Jon Holyoak 1992

Blackwell Scientific Publications
Editorial Offices:
Osney Mead, Oxford OX2 0EL
25 John Street, London WC1N 2BL
23 Ainslie Place, Edinburgh EH3 6AJ
3 Cambridge Center, Cambridge,
  Massachusetts 02142, USA
54 University Street, Carlton
  Victoria 3053, Australia

Other Editorial Offices:
Librairie Arnette SA
2, rue Casimir-Delavigne
75006 Paris
France

Blackwell Wissenschafts-Verlag
Meinekestrasse 4
D-1000 Berlin 15
Germany

Blackwell MZV
Feldgasse 13
A-1238 Wien
Austria

First published 1992

Set by DP Photosetting, Aylesbury, Bucks
Printed and bound in Great Britain by
Hartnolls Ltd, Bodmin, Cornwall

DISTRIBUTORS

Marston Book Services Ltd
PO Box 87
Oxford OX2 0DT
(Orders: Tel: 0865 791155
        Fax: 0865 791927
        Telex: 837515)

USA
Blackwell Scientific Publications, Inc.
3 Cambridge Center
Cambridge, MA 02142
(*Orders:* Tel: 800 759-6102
           617 225-0401)

Canada
Oxford University Press
70 Wynford Drive
Don Mills
Ontario M3C 1J9
(*Orders:* Tel: 416 441-2941)

Australia
Blackwell Scientific Publications
(Australia) Pty Ltd
54 University Street
Carlton, Victoria 3053
(*Orders:* Tel: 03 347-0300)

British Library
Cataloguing in Publication Data

A catalogue record for this book
is available from the British Library.

ISBN 0–632–02078–4

Library of Congress
Cataloging in Publication Data

Holyoak, Jon.
    Negligence in building law : cases and
commentary / Jon Holyoak. p.   cm.
    Includes index.
    ISBN 0-632-02078-4
    1. Liability for building accidents – Great
Britain – Cases. 2. Construction industry –
Law and legislation – Great Britain –
Cases. 3. Negligence – Great Britain –
Cases.   I. Title.
KD1980.B84H65   1992
346.4103'2—dc20
[344.10632]   92-5670
                                     CIP

Contents

Preface

Liability for negligence has become a matter of great concern to all professions in recent years. Equally the subject has itself changed, at times dramatically, in recent years. The aim of this book is to explain the workings of the law of negligence as it affects property and construction professionals and it represents a first attempt to assess the overall picture after the seminal 1990 decision in *Murphy* v. *Brentwood District Council*.

The first part of the book provides an overview of the general principles of negligence. The second part considers common issues in negligence cases where matters are complicated by the presence of contractual arrangements, as would be typical in a construction liability case. The third and major section of the book reviews the current liability picture for each group of property and construction professionals.

The 'casebook' format, blending full commentary with extensive quotation from the cases and other materials, is an ideal format for showing both the position the law has reached and the way in which that position is reached. It is hoped that the book will be appreciated by a wide range of readers. Its prime target is the professionals for whom these issues are part of daily life – surveyors and construction professionals, who need to understand the potential impact liability in negligence can have, their insurers, and lawyers, faced with potentially difficult and complex cases and who must reassess the picture after *Murphy*. Teachers and students of construction law should also find it useful.

I am happy to record my thanks to all who have helped me to produce this book. The greater part of the writing was done while on study leave at the Law Faculty of the National University of Singapore. I am grateful for the facilities provided and the opportunity to work on research without the distractions that Leicester insists on providing. I am grateful to all who have helped with secretarial work on the book from the Faculty of Law at Leicester's admirable office staff, particularly Heather Langman (sadly no longer at Leicester), Andrea Brew and Wendy Chantrell (who happily still are). Patrick Leece provided invaluable support in the final compilation of the book, which is much appreciated.

Blackwells, in the person of Julia Burden, have been an agreeable company for which to write; their assistance in production and encouragement not to be too late in producing a manuscript have been both welcome and necessary.

The idea of this book was Douglas Wood's. Although he did much work on the project, circumstances conspired to prevent him completing it. I am happy to acknowledge the effort he put in to this project and that I have had free access to his work. It is only right, though, that I should make clear that the responsibility for format and content, including such errors as there may be, is mine.

Jon Holyoak
Leicester
January 1992

Acknowledgements

It is a pleasure to acknowledge permission received from the publishers of the various Law Reports which I have used in this book. Thanks are due to the Incorporated Council for Law Reporting for England and Wales (the Official Law Reports); the Estates Gazette; the Longman Group UK (Building Law Reports); Sweet and Maxwell (Construction Law Journal); Tolleys (Professional Negligence); Butterworths (All England Law Reports, Construction Law Review and LEXIS); and Lloyds Legal Publishing Division (Lloyds Reports).

PART A
The Tort of Negligence

It makes good moral sense to penalize careless behaviour, especially careless behaviour that causes damage. Whether to punish the wrong-doer, deter his successors or to compensate his luckless victim (and each of these has been suggested as a justification) it seems right to most people that the wrongdoer should compensate the victim.

Such a broad approach is not, however, that taken by the law. It is only since 1932 that the law has recognized that careless conduct exists as a general category to which common principles should apply. Within the law as it has evolved, however, there has been a continual presumption that simple 'negligence' or 'carelessness' is too broad a category to form a basis for meaningful legal discussion. Instead the tort of negligence has become sub-divided and compartmentalized, and in order to decide whether a defendant has been careless three distinct stages have to be examined.

(1) *Duty of care*
Are the parties in a relationship sufficiently close that the tort of negligence should be applicable between them? This is an area of great current topicality of particular importance to construction disputes.

(2) *Breach of duty*
Has the defendant fallen below the standard of care that the law expects from him? This question is the practical issue at the heart of all negligence litigation.

(3) *Damage*
Is there damage? Is the damage consequent upon the breach of a duty of care?

Each of these elements will be examined in turn in successive chapters.

Chapter 1

The Duty of Care

The idea of a duty of care has been with us in its modern form for sixty years. And it is within the last fifteen of those years that the concept has, first of all, extended itself so as to have a highly significant impact on building litigation and practice, and more recently adjusted itself so as to adapt to contemporary conditions, and lessen, but not eliminate, the burden of negligence liability placed on the construction industry.

This chapter will trace its way through these developments, but will start at the beginning of the story, with a case that at first sight seems a long way from the world of major construction projects, but which lays the foundations for discussion of what the duty of care is, and why the concept has been created.

Duty of care – a starting point

DONOGHUE (or McALISTER) v. STEVENSON

[1932] AC 562   House of Lords

The facts assumed for the purposes of a preliminary ruling as to whether a duty of care existed were that the plaintiff and her friend went out on the town one day, to Minchella's cafe in Paisley, Scotland. The friend purchased for the plaintiff a bottle of the defendant's ginger beer. Coming to the end of the bottle, she opened it to finish off the contents, and the remaining ginger beer fell into her glass, accompanied by a decomposing dead snail. Gastro-enteritis was alleged to ensue.

No precedent existed for a claim in negligence by a consumer against a manufacturer, though many cases had created specific liabilities within a narrow framework based on negligence. It was held that a duty of care could exist in such circumstances.

LORD ATKIN (at p. 579):

> It is remarkable how difficult it is to find in the English authorities statements of general application defining the relations between parties that give rise to the duty. The courts are concerned with the particular relations which come before them in actual litigation, and it is sufficient to say whether the duty exists in these circumstances. The result is that the courts have been engaged upon an elaborate classification of duties as they exist in respect of property, whether real or personal, with further divisions as to ownership, occupation or control, and distinctions based on the particular relations of the one side or the other, whether manufacturer, salesman or landlord, customer, tenant, stranger, and so on. In this way it can be ascertained at any time whether the law recognizes a duty, but only where the case can be referred to some particular species which has been examined and classified. And yet the duty which is common to all the cases where liability is established must logically be based upon some element common to the cases where it is found to exist. To seek a complete logical definition of the general principle is probably to go beyond the function of the judge, for the more general the definition the more likely it is to omit essentials or to introduce non-essentials.
>
> At present I content myself with pointing out that in English law there must be, and is, some general conception of relations giving rise to a duty of care, of which the particular cases found in the books are but instances. The liability for negligence, whether you style it such or treat it as in other systems as a species of 'culpa', is no doubt based upon a general public sentiment of moral wrongdoing for which the offender must pay. But acts or omissions which any moral code would censure cannot in a practical world be treated so as to give a right to every person injured by them to demand relief. In this way rules of law arise which limit the range of complainants and the extent of their remedy.

This section becomes particularly significant later on. Note for now that the concept of duty of care is a restrictive one, designed to limit the number of negligence cases, not to expand them.

What then is the nature of this limiting device, the duty of care?

LORD ATKIN (continued, at p. 580):

> The rule that you are to love your neighbour becomes in law, you must not injure your neighbour; and the lawyer's question, 'Who is my neighbour?' receives a restricted reply.
>
> You must take reasonable care to avoid acts or omissions which you can reasonably foresee would be likely to injure your neighbour. Who, then, in law is my neighbour? The answer seems to be – persons who are so closely and directly affected by my act that I ought reasonably to have them in contemplation as being so affected when I am directing my mind to the acts or omissions which are called in question.

Two things are significant in this passage, one of the few known to almost all law students. First is the form of the restriction placed on the duty concept. It is a restriction based on proximity between the parties. Plaintiff and defendant must be in a relationship such that the actions of one closely and directly affect the other. This need not be geographical; Donoghue and Stevenson, it can be assumed, never met. Nevertheless it is clear that a mistake by Stevenson in manufacture of the drink would immediately affect Donoghue. So it was held, by a 3–2 majority, that a duty of care did exist between manufacturer and ultimate consumer.

LORD MACMILLAN (at p. 618)

> The law takes no cognizance of carelessness in the abstract. It concerns itself with carelessness only where there is a duty to take care and where failure in that duty has caused damage. In such circumstances carelessness assumes the legal quality of negligence and entails the consequences in law of negligence. What, then, are the circumstances which give rise to this duty to take care? In the daily contacts of social and business life human beings are thrown into, or place themselves in, an infinite variety of relations with their fellows; and the law can refer only to the standards of the reasonable man in order to determine whether any particular relation gives rise to a duty to take care as between those who stand in that relation to each other. The grounds of action may be as various and manifold as human errancy; and the conception of legal responsibility may develop in adaptation to altering social conditions and standards. The criterion of judgment must adjust and adapt itself to the changing circumstances of life. The categories of negligence are never closed. The cardinal principle of liability is that the party complained of should owe to the party complaining a duty to take care, and that the party complaining should be able to prove that he has suffered damage in consequence of a breach of that duty. Where there is room for diversity of view, it is in determining what circumstances will establish such a relationship between the parties as to give rise, on the one side, to a duty to take care, and on the other side to a right to have care taken.

'The categories of negligence are never closed.' A famous threat indeed, that has at times seemed justified not least in the headlong expansion of the tort into the construction industries in the 1970s and 1980s. But note also, firstly, that here at the very heart of duty of care is a link between the imposition (or otherwise) of a duty of care and the views of society i.e. 'the changing circumstances of life'. Note, secondly, the linking together of duty, breach and damage, which will be extended here over three chapters.

As we have noted, the decision in favour of a duty of care, was by a bare majority. The arguments of the minority in dissent were partly based on the implicit attack on the primacy of contract. Since Donoghue had been bought the drink, she had not paid for it and therefore had no contractual rights. (Legal history may have been very different if it had been Donoghue's round.) Also, the remorseless logic that could flow from a finding in favour of Donoghue concerned the Law Lords in the minority.

LORD BUCKMASTER (at p. 577):

> The principle contended for must be this: that the manufacturer, or indeed the repairer, of any article, apart entirely from contract, owes a duty to any person by whom the article is lawfully used to see that it has been carefully constructed. All rights in contract must be excluded from consideration of this principle; such contractual rights as may exist in successive steps from the original manufacturer down to the ultimate purchaser are ex hypothesi immaterial. Nor can the doctrine be confined to cases where inspection is difficult or impossible to introduce. This conception is simply to misapply to tort doctrine applicable to sale and purchase.
>
> The principle of tort lies completely outside the region where such considerations apply, and the duty, if it exists, must extend to every person who, in lawful circumstances, uses the article made. There can be no special duty attaching to the manufacture of food apart from that implied by contract or imposed by statute. If such a duty exists, it seems to me it must cover the construction of every article, and I cannot see any reason why it should not apply to the construction of a house. If one step, why not fifty? Yet if a house be, as it sometimes is, negligently built, and in consequence of that negligence the ceiling falls and injures the occupier or any one else, no action against the builder exists according to the English law, although I believe a right did exist according to the laws of Babylon.

It is the loose connection between liability in contract and that in negligence that lies at the heart of *Donoghue* v. *Stevenson*. It was necessary for the plaintiff to use negligence because the rules of privity of contract (i.e. no payment = no rights) stopped her using any contractual remedies. Her friend had no useful rights in contract, having suffered no damage of substance, and thus being confined to a claim against the cafe owner, for the price of the ginger beer.

It is this same connection between contract and tort that explains the relevance of this decision to construction law. Just as, in a pub, not everyone's drink has been purchased directly from its manufacturer, so likewise most people's houses and factories have not been purchased directly from the builder. Complex chains of relationships might evolve. Compare:

Manufacturer → Retailer → Purchaser → Ultimate Consumer,

as a model of a *Donoghue*-type product liability case, with

Builder → First Purchaser → Second Purchaser → Lessee

as a simple example of the type of premises liability case with which we shall be concerned. It can be seen that the same problem arises in each of a string of contracts which bind each party in the chain to the one either side, but do not, in contract law, give any more general connection between the parties to the chain as a whole.

Duty of care – into maturity

Donoghue was quickly and widely accepted as good authority, but it took much longer for its full implications to become clear. It was in the 1970s that decisive steps were taken that would mean the tort of negligence would enter the field of construction law.

HOME OFFICE v. DORSET YACHT CO.

[1970] AC 1004 House of Lords

Poor supervision of Borstal boys on an island in Poole Harbour meant that they were able to attempt an escape. The action was brought by the owner of a motor yacht damaged in this attempt against the Home Office, as the operators of the Borstal system. It was held that the Home Office were under a duty of care, in spite of their argument that no authority existed for the imposition of such a duty.

LORD REID at (p. 1026):

> This would at one time have been a strong argument. About the beginning of this century most eminent lawyers thought that there were a number of separate torts involving negligence, each with its own rules, and they were most unwilling to add more. They were of course aware from a number of leading cases that in the past the courts had from time to time recognised new duties and new grounds of action. But the heroic age was over; it was time to cultivate certainty and security in the law: the categories of negligence were virtually closed. The Attorney-General invited us to return to those halcyon days, but, attractive though it may be, I cannot accede to his invitation.
>
> In later years there has been a steady trend towards regarding the law of negligence as depending on principle so that, when a new point emerges, one should ask not whether it is covered by authority but whether recognized principles apply to it. *Donoghue* v. *Stevenson* [1932] AC 562 may be regarded as a milestone, and the well-known passage in Lord Atkin's speech should I think be regarded as a statement of principle. It is not to be treated as if it were a statutory definition. It will require qualification in new circumstances. But

I think that the time has come when we can and should say that it ought to apply unless there is some justification or valid explanation for its exclusion.

This untrammelled the duty of care from being a narrow precedent to becoming instead a major and fundamental principle of English law, therefore likely to be much more difficult to restrict or control merely because it was to be used in a novel factual setting.

LORD REID (continued, at p. 1032):

> Finally I must deal with public policy. It is argued that it would be contrary to public policy to hold the Home Office or its officers liable to a member of the public for this carelessness – or, indeed, any failure of duty on their part. The basic question is: who shall bear the loss caused by that carelessness – the innocent respondents or the Home Office, who are vicariously liable for the conduct of their careless officers? I do not think that the argument for the Home Office can be put better than it was put by the Court of Appeals of New York in *Williams* v. *State of New York* (1955) 127 NE 2d 545, 550:
>
>> '. . . public policy also requires that the State be not held liable. To hold otherwise would impose a heavy responsibility upon the State, or dissuade the wardens and principal keepers of our prison system from continued experimentation with "minimum security" work details – which provide a means for encouraging better-risk prisoners to exercise their senses of responsibility and honor and so prepare themselves for their eventual return to society. Since 1917, the legislature has expressly provided for out-of-prison work, Correction Law § 182, and its intention should be respected without fostering the reluctance of prison officials to assign eligible men to minimum security work, lest they thereby give rise to costly claims against the State, or indeed inducing the State itself to terminate this "salutary procedure" looking towards rehabilitation.'
>
> It may be that public servants of the State of New York are so apprehensive, easily dissuaded from doing their duty and intent on preserving public funds from costly claims that they could be influenced in this way. But my experience leads me to believe that Her Majesty's servants are made of sterner stuff. So I have no hesitation in rejecting this argument. I can see no good ground in public policy for giving this immunity to a government department. I would dismiss this appeal.

The importance of this is not so much in its result, as in the easy invocation of issues of policy. Even more flexibility is placed on offer. Lord Reid accepts that policy is an integral part of the duty question in finding that the plaintiffs succeeded in the claim.

These two elements of *Home Office* paved the way for the next logical step, one which meant that construction lawyers and other professionals could no longer ignore negligence, taken in a case that has dominated premises liability litigation from then on to recent times.

ANNS v. LONDON BOROUGH OF MERTON

[1978] AC 728   House of Lords

Mr and Mrs Anns leased a flat in Wimbledon from the building owners of the block. Eight years later, the flat began to crack and slope, due to allegedly inadequate foundations. No contract claim could be brought, due to the limitation of actions rules (see Chapter 4, below). A negligence claim was made instead, against the builder, who prudently settled the claim prior to the trial, and also against the local authority, for its allegedly negligent failure to use its powers of inspection of the foundations of new buildings so as to prevent the damage to the Anns' flat. It was held that a duty of care could exist. A key question was what should be the proper approach to the duty of care question in a novel claim such as this.

LORD WILBERFORCE (at p. 751):

> Through the trilogy of cases in this House – *Donoghue* v. *Stevenson* [1932] AC 562, *Hedley Byrne & Co. Ltd* v. *Heller and Partners Ltd* [1964] AC 465, and *Dorset Yacht Co. Ltd* v. *Home Office* [1970] AC 1004 – the position has now been reached that in order to establish that a duty of care arises in a particular situation, it is not necessary to bring the facts of that situation within those of previous situations in which a duty of care has been held to exist. Rather the question has to be approached in two stages. First one has to ask whether, as between the alleged wrongdoer and the person who has suffered damage there is a sufficient relationship of proximity or neighbourhood such that, in the reasonable contemplation of the former, carelessness on his part may be likely to cause damage to the latter – in which case a prima facie duty of care arises. Secondly, if the first question is answered affirmatively, it is necessary to consider whether there are any considerations which ought to negative, or to reduce or limit the scope of the duty or the class of person to whom it is owed or the damages to which a breach of it may give rise.

This passage came to attain notoriety in the years that followed. Criticism of *Anns* has, as we will see, mounted to such a peak that the case was finally overruled by the House of Lords in 1990. However, to ignore the case because of this would be misleading, for several reasons:

(1) *Anns* was the first House of Lords decision to bring the tort of negligence centre-stage into the construction world;

(2)   *Anns* was not without logic. *Home Office* confirmed that the duty test was a question of principle; *Anns* simply confirmed that, like all good principles, it should apply across the board;

(3)   *Anns* was not saying that a duty of care was invariable. The creation of a duty in any case of sufficient proximity was only a presumption, and Lord Wilberforce made it clear that long-established exceptions to mainstream duty, such as cases involving economic loss, would remain, and the availability of 'any considerations' to negate a duty clearly allowed a fair amount of flexibility for the judges;

(4)   In any event, *Anns* at least provided answers to general questions about duty of care to which, after its summary abolition, it is now hard to find any clear answer.

However, it may be argued that the problem lay, not so much with *Anns* itself, as with the failure of subsequent courts to appreciate that this was a two-stage, not a one-stage (pro-duty), formula. *Anns* roared on, with the application of the case reaching a high-point (or low-point, depending on view) with its arrival into the delicate area of economic, rather than physical loss, in *Junior Books* v. *Veitchi & Co.* [1983] AC 520 (on this case, and economic loss generally, see pp. 22 below).

Duty of care – into decline

Slowly, the courts, from the mid-1980s onward, began to try to tame the beast that duty of care had become, but different cases tried different methods and gave different reasons for this nonetheless constant trend, in cases both within and outside the building profession.

LEIGH AND SILLIVAN LTD v. ALIAKMON SHIPPING CO.

[1986] AC 785   House of Lords

Cargo on a ship was damaged during a voyage. The purchasers of the cargo were committed to the contract, but were not yet the legal owners of the cargo at the time of the damage. It was held that no duty of care was owed.

LORD BRANDON (at p. 809):

> My Lords, there is a long line of authority for a principle of law that, in order to enable a person to claim in negligence for loss caused to him by reason of loss of or damage to property, he must have had either the legal ownership of

or a possessory title to the property concerned at the time when the loss or damage occurred, and it is not enough for him to have only had contractual rights in relation to such property which have been adversely affected by the loss of or damage to it.

At p. 813, Lord Brandon summarizes counsel's argument on behalf of the cargo owners.

[Counsel for Leigh and Sillivan Ltd] submitted that the proper way for your Lordships to approach the present case was to ask and answer the two questions set out by Lord Wilberforce. He said that the answer to the first question must be that there was, as between the shipowners and the buyers, a sufficient relationship of proximity or neighbourhood such that, in the reasonable contemplation of the former, want of care on their part might be likely to cause damage in the form of pecuniary loss to the latter, so that a prima facie duty of care arises. With regard to the second question, relating to considerations which ought to limit the scope of the duty, he conceded that it would be unjust to the shipowners to be liable to the buyers in tort for negligence without reference to the terms of the bills of lading under which the shipowner carried the goods; and he sought to find a legal rationale for the qualification of the duty of care by reference to those terms on the basis that those were the terms of the bailment of the goods by the sellers to the shipowners to which the buyers had, by entering into a c. and f. contract with the sellers, impliedly consented.

At p. 815, Lord Brandon explained why he did not accept this view.

I now return to consider [counsel for Leigh and Sillavan Ltd] submissions based on what Lord Wilberforce said in *Anns'* case [1978] AC 728. There are two preliminary observations which I think that it is necessary to make with regard to the passage in Lord Wilberforce's speech on which counsel relies. The first observation which I would make is that that passage does not provide, and cannot in my view have been intended by Lord Wilberforce to provide, a universally applicable test of the existence and scope of a duty of care in the law of negligence. In this connection I would draw attention to a passage in the speech of my noble and learned friend, Lord Keith of Kinkel, in *Governors of the Peabody Donation Fund* v. *Sir Lindsay Parkinson & Co. Ltd* [1985] AC 210. After citing a passage from Lord Reid's speech in *Dorset Yacht Co. Ltd* v. *Home Office* [1970] AC 1004, 1027 and then the passage from Lord Wilberforce's speech in *Anns'* case, 751–2 now under discussion, he said, at p. 240:

'There has been a tendency in some recent cases to treat these passages as being themselves of a definitive character. This is a temptation which should be resisted.'

The second observation which I would make is that Lord Wilberforce was dealing, as is clear from what he said, with the approach to the questions of the existence and scope of a duty of care in a novel type of factual situation which was not analogous to any factual situation in which the existence of such a duty had already been held to exist. He was not, as I understand the passage, suggesting that the same approach should be adopted to the existence of a duty of care in a factual situation in which the existence of such a duty had repeatedly been held not to exist.

Simply, then, as a matter of law, the test laid down in *Anns* was confined to cases where duties of care had not been previously rejected; the *Anns* test could only be used where there was no previous litigation on the duty issue. In practice few such 'no-duty' cases can be found in the precedents, but it does seem that, as a matter of law, no duty is owed if work on one party's land affects the flow of underground water such as to harm other land and/or activities thereon. [*Langbrook Properties Ltd* v. *Surrey County Council* [1971] 1 WLR 161; *Stephens* v. *Anglian Water Authority* [1987] 1 WLR 1381.]

A larger group of cases has, meanwhile, adopted considerations of policy as being appropriate to restrict a duty of care for which there has been argument.

Duty of care – policy

GOVERNORS OF THE PEABODY DONATION FUND v. SIR LINDSAY PARKINSON & CO.

[1985] AC 210   House of Lords

Here, a large housing development was being built. It was originally planned that, due to site conditions, a system of drainage that was flexible should be used, and the local authority approved plans embodying this scheme. Subsequently, with the approval of the architects, the construction work was carried out featuring a rigid system of drainage. Two years later, this was found to be unsatisfactory, and expensive repair work had to be done. The relevant aspect of the complex ensuing litigation was a claim against the local authority for their failure to utilize their powers to insist on flexible drainage, once they had discovered that the plans were not to be followed. No duty was found to exist in such circumstances.

LORD KEITH OF KINKEL (at p. 240) referred to the keynote passages from *Home Office* and *Anns*, and went on to say:

There has been a tendency in some recent cases to treat these passages as being themselves of a definitive character. This is a temptation which should be resisted. The true question in each case is whether the particular defendant owed to the particular plaintiff a duty of care having the scope which is contended for, and whether he was in breach of that duty with consequent loss to the plaintiff. A relationship of proximity in Lord Atkin's sense must exist before any duty of care can arise, but the scope of the duty must depend on all the circumstances of the case. In *Dorset Yacht Co.* v. *Home Office* [1970] AC 1004, 1038, Lord Morris of Borth-y-Gest, after observing that at the conclusion of his speech in *Donoghue* v. *Stevenson* [1932] AC 562, Lord Atkin said that it was advantageous if the law 'is in accordance with sound common sense' and expressing the view that a special relation existed between the prison officers and the yacht company which gave rise to a duty on the former to control their charges so as to prevent them doing damage, continued, at p. 1039:

'Apart from this I would conclude that, in the situation stipulated in the present case, it would not only be fair and reasonable that a duty of care should exist but that it would be contrary to the fitness of things were it not so. I doubt whether it is necessary to say, in cases where the court is asked whether in a particular situation a duty existed, that the court is called upon to make a decision as to policy. Policy need not be invoked where reason and good sense will at once point the way. If the test as to whether in some particular situation a duty of care arises may in some cases have to be whether it is fair and reasonable that it should so arise, the court must not shrink from being the arbiter. As Lord Radcliffe said in his speech in *Davis Contractors Ltd* v. *Fareham Urban District Council* [1956] AC 696, 728, the court is "the spokesman of the fair and reasonable man".'

So in determining whether or not a duty of care of particular scope was incumbent upon a defendant it is material to take into consideration whether it is just and reasonable that it should be so.

Applying this general principle, he stated:

[The statutes] undoubtedly place certain very specific obligations upon contractors carrying out building operations, but they do nothing to detract from what is clearly the proper construction of paragraph 13(1), namely, that observance of its provisions is incumbent upon any person who puts in train a house building project. Peabody no doubt had no personal knowledge or understanding of what was going on. They relied on the advice of their architects, engineers and contractors, and in the event they were sadly let down, particularly by the architects. But it would be neither reasonable nor just, in these circumstances, to impose upon Lambeth a liability to indemnify Peabody against loss resulting from such disastrous reliance.

CURRAN v. NORTHERN IRELAND
CO-OWNERSHIP HOUSING ASSOCIATION

[1987] AC 718   House of Lords

The plaintiffs purchased a house, with financial assistance from the Northern Ireland Housing Executive, who had given the previous owner a grant for building works to improve the property. This work was found to be defective by the plaintiffs and they sought to take action against the Housing Executive, who had inspected the work on completion before authorizing payment of the grant, but they were found to owe no duty on these facts.

LORD BRIDGE (at p. 724):

> My Lords, *Anns* v. *Merton London Borough Council* [1978] AC 728 may be said to represent the high water mark of a trend in the development of the law of negligence by your Lordships' House towards the elevation of the 'neighbour-hood' principle derived from the speech of Lord Atkin in *Donoghue* v. *Stevenson* [1932] AC 562 into one of general application from which a duty of care may always be derived unless there are clear countervailing considerations to exclude it. In an article by Professors J.C. Smith and Peter Burns entitled '*Donoghue* v. *Stevenson*—The Not So Golden Anniversary' (1983) 46 MLR 147 the trend to which I have referred was cogently criticised particularly in its tendency to obscure the important distinction between misfeasance and nonfeasance.

He continued (at p. 726):

> The *Peabody* case itself and the subsequent decision of the Court of Appeal in *Investors in Industry Commercial Properties Ltd* v. *South Bedfordshire District Council* [1986] QB 1034 illustrate some of the difficulties that arise in determining the precise scope of the duty owed by local authorities exercising statutory powers of control over building operations of the kind with which the *Anns* case was concerned. It is perhaps not inappropriate to note that in *Sutherland Shire Council* v. *Heyman* (1985) 59 ALJR 564 the High Court of Australia (Gibbs CJ, Mason, Wilson, Brennan and Deane JJ) unanimously declined to follow the *Anns* case. The judgment in that case of Brennan J seems to me particularly impressive in its reasoning.
>
> I mention these considerations merely by way of introduction. I do not for a moment suggest that your Lordships' House is not bound by the *Anns* case if the ratio of that decision is applicable to the instant case. If it were thought that the law as declared by the House in the *Anns* case was so unsatisfactory as to call for reform, this might be an appropriate subject for consideration by the Law Commission. But, that said, your Lordships are, I think, entitled to be wary of effecting any extension of the principle applied in the *Anns* case

whereby, although under no statutory duty, a statutory body may be held to owe a common law duty of care to exercise its statutory powers to control the activities of third parties in such a way as to save harmless those who may be adversely affected by those activities if they are not effectively controlled.

Lord Bridge went on to consider that the statutory powers in question were created for the purpose of financial control, not to benefit the householders (see p. 233, below).

Thus, (at p. 729):

Here, in my opinion, the dictates of good sense and the consideration of what is fair and reasonable point clearly against the imposition of any duty of care owed by the Executive to the plaintiffs and it would be contrary to the fitness of things to hold them to be under any such duty. If there was no duty of care, then the allegation that the plaintiffs purchased in reliance on the improvement grant cannot assist them.

Two aspects emerge. One is the first direct, if limited, attack on *Anns*. More importantly, policy considerations are at the heart of Lord Bridge's successful sidestepping of *Anns*. Here the lack of control actually exercised by the Housing Executive is at the heart of the decision, but it may also be that the ability of the Currans to protect themselves by adequately surveying the property might also be relevant to the decision.

Subsequent cases on policy have covered many and varied situations, but common strands are becoming clear from the leading cases. One frequently mentioned argument against imposing a duty of care is where an alternative remedy is available e.g. in administrative law [e.g. *Rowling* v. *Takaro Properties* [1988] AC 473 – see below p. 223,] or through internal procedures [e.g. *Calveley* v. *Chief Constable of Merseyside* [1989] AC 1228]. The fear of potential defendants adopting unduly defensive procedures has also militated against the duty of care, as in *Hill* v. *Chief Constable of West Yorkshire* [1989] AC 453. This latter argument, so far used only in cases concerning public bodies, such as the police in *Hill* and also the decision of government ministers in *Rowling*, clearly has the potential to be raised by private defendants in a wide range of other cases and it is not easy to see how the courts could deny the argument to, say, a builder or an engineer who argues that the threat of liability results in an over-cautious approach being adopted.

Duty of care today

If there is a general test for duty of care today (see p. 27, below), it is contained within the next case, *Yuen Kun-Yeu*, where the judges attempt to be positive in creating a new order, or perhaps restoring the old order. Their discussions blend both policy and law in formulating a test for the duty of care.

YUEN KUN-YEU v.
ATTORNEY-GENERAL OF HONG KONG

[1988] AC 175   Privy Council

The plaintiffs invested in a deposit-taking company set up as part of a fraud. Their action was against a Hong Kong Government official, the Commissioner of Deposit-Taking Companies, whose powers included the suspension of dealing in shares of companies which he suspected of fraud. It was alleged that he was negligent in not using these powers, but was held to owe no duty of care to the plaintiff.

LORD KEITH OF KINKEL (at p. 191):

> Their Lordships venture to think that the two-stage test formulated by Lord Wilberforce for determining the existence of a duty of care in negligence has been elevated to a degree of importance greater than it merits, and greater perhaps than its author intended. Further, the expression of the first stage of the test carries with it a risk of misinterpretation. As Gibbs CJ pointed out in *Council of the Shire of Sutherland* v. *Heyman*, 59 ALJR 564, 570, there are two possible views of what Lord Wilberforce meant. The first view, favoured in a number of cases mentioned by Gibbs CJ, is that he meant to test the sufficiency of proximity simply by the reasonable contemplation of likely harm. The second view, favoured by Gibbs CJ himself, is that Lord Wilberforce meant the expression 'proximity or neighbourhood' to be a composite one, importing the whole concept of necessary relationship between plaintiff and defendant described by Lord Atkin in *Donoghue* v. *Stevenson* [1932] AC 562, 580. In their Lordships' opinion the second view is the correct one. As Lord Wilberforce himself observed in *McLoughlin* v. *O'Brien* [1983] 1 AC 410, 420, it is clear that foreseeability does not of itself, and automatically, lead to a duty of care. There are many other statements to the same effect. The truth is that the trilogy of cases referred to by Lord Wilberforce in *Anns* v. *Merton London Borough Council* [1978] AC 728, 751, each demonstrate particular sets of circumstances, differing in character, which were adjudged to have the effect of bringing into being a relationship apt to give rise to a duty of care. Foreseeability of harm is a necessary ingredient of such a relationship, but it is not the only one. Otherwise there would be liability in negligence on the part of one who sees another about to walk over a cliff with his head in the air, and forbears to shout a warning.
>   *Donoghue* v. *Stevenson* established that the manufacturer of a consumable product who carried on business in such a way that the product reached the

consumer in the shape in which it left the manufacturer, without any prospect of intermediate examination, owed the consumer a duty to take reasonable care that the product was free from defect likely to cause injury to health. The speech of Lord Atkin stressed not only the requirement of foreseeability of harm but also that of a close and direct relationship of proximity.

This important passage suggests that matters are rather more compli-cated than may seem to be the case from Lord Wilberforce's two-tier test. In particular, the last section makes clear that mere foreseeability of harm, though necessary, is a mere pre-condition for imposition of a duty. It is then necessary to go on to see whether the parties are in a 'close and direct' relationship of proximity. For example, I can foresee that if I release a huge cloud of noxious gas into the atmosphere, perhaps as a result of a negligently-caused explosion, during a westerly wind, I can foresee that it might spread, Chernobyl-like, over a wide area of Europe. Nevertheless, it would be difficult to see that the innumerable victims could be said to be in a 'close and direct' relationship with me. Thus, the legal test of duty of care, in restoring the original, narrow *Donoghue* test, coincides with the new policy attitude of restricting claims, especially those claims made by large numbers, against public bodies.

LORD KEITH (at p. 198):

The final matter for consideration is the argument for the Attorney-General of Hong Kong that it would be contrary to public policy to admit the plaintiffs' claim, upon grounds similar to those indicated in relation to police forces by Glidewell LJ in *Hill* v. *Chief Constable of West Yorkshire* [1988] QB 60, 75. It was maintained that if the commissioner were to be held to owe actual or potential depositors a duty of care in negligence, there would be reason to apprehend that the prospect of claims would have a seriously inhibiting effect on the work of his department. A sound judgment would be less likely to be exercised if the commissioner were to be constantly looking over his shoulder at the prospect of claims against him, and his activities would be likely to be conducted in a detrimentally defensive frame of mind. In the result, the effectiveness of his functions would be at risk of diminution. Consciousness of potential liability could lead to distortions of judgment. In addition, the principles leading to his liability would surely be equally applicable to a wide range of regulatory agencies, not only in the financial field, but also, for example, to the factory inspectorate and social workers, to name only a few. If such liability were to be desirable upon any policy grounds, it would be much better that the liability were to be introduced by the legislature, which is better suited than the judiciary to weigh up competing policy considerations.

Their Lordships are of opinion that there is much force in these arguments, but as they are satisfied that the plaintiffs' statement of claim does not disclose a cause of action against the commissioner in negligence they prefer to rest their decision upon that rather than upon the public policy argument.

However, even since *Yuen Kun-Yeu*, further dramatic change has occurred in the approach of the courts to duty of care. In order to understand this, and before we can formulate a final version of the contemporary approach to duty of care, we must at least begin to try and understand how and why negligence treats economic loss, rather than physical loss, in a different way.

Duty of care and economic loss

For negligence to be actionable, damage is needed. This is simple when a workman's leg is broken on a building site, or where a property is demolished by a passing truck. These are clear cases of physical damage. However, in some sense, purely economic loss is not really a loss at all, but a failure to make a gain. If my property, demolished as above, is an hotel, its closure during rebuilding does not cost me anything I already have, but rather merely deprives me of the chance of gain from continued trading. It is this distinction, coupled with the ever-present willingness of the tort of negligence to deny compensation because liability cannot be unlimited that makes it a consistent feature that claims for economic loss are restricted.

CATTLE v. STOCKTON WATERWORKS LTD

(1875) LR 10 QB 453   Court of Queens' Bench

The plaintiff was building a tunnel for one Knight, a landowner. The defendant company's water main leaked into the area of the works, delaying the project and reducing the plaintiff's profit. No cause of action was found to arise.

BLACKBURN J (at p. 457):

> Can Cattle sue in his own name for the loss which he has in fact sustained, in consequence of the damage, which the defendants have done to the property of Knight, causing him, Cattle, to lose money under his contract? We think he cannot.
>
> In the present case the objection is technical and against the merits, and we should be glad to avoid giving it effect. But if we did so, we should establish an authority for saying that, in such a case as that of *Fletcher* v. *Rylands* (1868)

LR 3HL 330 the defendant would be liable, not only to an action by the owner of the drowned mine, and by such of his workmen as had their tools or clothes destroyed, but also to an action by every workman and person employed in the mine, who in consequence of its stoppage made less wages than he would otherwise have done. And many similar cases to which this would apply might be suggested. It may be said that it is just that all such persons should have compensation for such a loss, and that, if the law does not give them redress, it is imperfect. Perhaps it may be so. But, as was pointed out by Coleridge, J, in *Lumley* v. *Gye* (1853) 2 E&B 216, Courts of justice should not 'allow themselves, in the pursuit of perfectly complete remedies for all wrongful acts, to transgress the bounds, within our law, in a wise consciousness as I conceive of its limited powers, has imposed on itself, of redressing only the proximate and direct consequences of wrongful acts.' In this we quite agree. No authority in favour of the plaintiff's right to sue was cited, and, as far as our knowledge goes, there was none that could have been cited.

His conclusion was thus as follows (at p. 458):

In the present case there is no pretence for saying that the defendants were malicious or had any intention to injure anyone. They were, at most, guilty of a neglect of duty, which occasioned injury to the property of Knight, but which did not injure any property of the plaintiff. The plaintiff's claim is to recover the damage which he has sustained by his contract with Knight becoming less profitable, or, it may be, a losing contract, in consequence of this injury to Knight's property. We think this does not give him any right of action.

This attitude dominated the law for the best part of a century, though it was of course formulated at a time when the tort of negligence was in a fairly primitive state of development, half a century before *Donoghue*.

Change first came in the particular area of liability for economic loss caused not, as in most cases, by a careless act but rather one caused by a negligent misstatement.

HEDLEY BYRNE & CO. LTD v. HELLER & PARTNERS

[1964] AC 465   House of Lords

The appellants were advertising agents, concerned about the financial condition of one of their clients. They asked their bank to ask the client's banker, the respondents, about the financial health of this particular client, and were assured that all was well, though 'without responsibility'. It was held that a duty of care could be owed in respect of a negligent misstatement.

LORD REID (at p. 482):

The appellants' first argument was based on *Donoghue* v. *Stevenson* [1932] AC 562. That is a very important decision, but I do not think that it has any direct bearing on this case. That decision may encourage us to develop existing lines of authority, but it cannot entitle us to disregard them. Apart altogether from authority, I would think that the law must treat negligent words differently from negligent acts. The law ought so far as possible to reflect the standards of the reasonable man, and that is what *Donoghue* v. *Stevenson* sets out to do. The most obvious difference between negligent words and negligent acts is this. Quite careful people often express definite opinions on social or informal occasions even when they see that others are likely to be influenced by them; and they often do that without taking that care which they would take if asked for their opinion professionally or in a business connection. The appellant agrees that there can be no duty of care on such occasions, and we were referred to American and South African authorities where that is recognised, although their law appears to have gone much further than ours has yet done. But it is at least unusual casually to put into circulation negligently made articles which are dangerous. A man might give a friend a negligently-prepared bottle of home-made wine and his friend's guests might drink it with dire results. But it is by no means clear that those guests would have no action against the negligent manufacturer.

Another obvious difference is that a negligently-made article will only cause one accident, and so it is not very difficult to find the necessary degree of proximity or neighbourhood between the negligent manufacturer and the person injured. But words can be broadcast with or without the consent or the foresight of the speaker or writer. It would be one thing to say that the speaker owes a duty to a limited class, but it would be going very far to say that he owes a duty to every ultimate 'consumer' who acts on those words to his detriment. It would be no use to say that a speaker or writer owes a duty but can disclaim responsibility if he wants to. He, like the manufacturer, could make it part of a contract that he is not to be liable for his negligence: but that contract would not protect him in a question with a third party, at least if the third party was unaware of it.

So it seems to me that there is good sense behind our present law that in general an innocent but negligent misrepresentation gives no cause of action. There must be something more than the mere misstatement.

Lord Reid justifies this new tortious liability (at p. 486):

A reasonable man, knowing that he was being trusted or that his skill and judgment were being relied on, would, I think, have three courses open to him. he could keep silent or decline to give the information or advice sought: or he could give an answer with a clear qualification that he accepted no responsibility for it or that it was given without that reflection or inquiry which a careful answer would require: or he could simply answer without any such qualification. If he chooses to adopt the last course he must, I think, be held to have accepted some responsibility for his answer being given carefully, or to have accepted a relationship with the inquirer which requires him to exercise such care as the circumstances require.

These passages show why it may be right to give a more restrictive liability for statements, rather than acts, and it is notable that the argument is made irrespective of reference to the economic nature of the loss incurred. The basis of the liability that comes into existence – reliance – is also clear. This means that the relationship between the parties requires a qualitative analysis – i.e. what type of relationship is it? – rather than the simpler proximity issue at the heart of a claim for physical damage.

LORD MORRIS OF BORTH-Y-GEST (at p. 494):

My Lords, it seems to me that if A assumes a responsibility to B to tender him deliberate advice, there could be a liability if the advice is negligently given. I say 'could be' because the ordinary courtesies and exchanges of life would become impossible if it were sought to attach legal obligation to every kindly and friendly act. But the principle of the matter would not appear to be in doubt. If A employs B (who might, for example, be a professional man such as an accountant or a solicitor or a doctor) for reward to give advice and if the advice is negligently given there could be a liability in B to pay damages. The fact that the advice is given in words would not, in my view, prevent liability from arising. Quite apart, however, from employment or contract there may be circumstances in which a duty to exercise care will arise if a service is voluntarily undertaken. A medical man may unexpectedly come across an unconscious man, who is a complete stranger to him, and who is in urgent need of skilled attention: if the medical man, following the fine traditions of his profession, proceeds to treat the unconscious man he must exercise reasonable skill and care in doing so.

This quote makes explicit what is implicit in Lord Reid's speech, namely that the new liability in negligent misstatement exists irrespective of whether the parties are on a contractual relationship or not (see pp. 67 et seq below).

A further point can be made in respect of *Hedley Byrne*. In spite of the endeavours of all five Law Lords to create a liability in negligent misstatement, nevertheless Hedley Byrne themselves did not benefit in any way, since the disclaimer by the respondents that their advice was given without legal responsibility was sufficient to remove reliance, or at least reasonable reliance, from the scene (see pp. 277 et seq below).

The developments in claims for economic loss remained on the sidelines of negligence liability for a considerable period. However, in the expansionist mood following *Anns*, it was inevitable that further development must occur.

JUNIOR BOOKS LTD v. VEITCHI CO. LTD

[1983] 1 AC 520   House of Lords

This action was brought by the owners of a book warehouse in Scotland. The floor of the premises turned out to be of inadequate strength and legal action in respect of the economic loss suffered was taken not against the principal contractors, but rather against their nominated sub-contractors, Veitchi, renowned experts in flooring, who were found to be under a duty of care in respect of economic loss.

LORD ROSKILL (at p. 539):

> It was strenuously argued for the appellants that for your Lordships' House now to hold that in those circumstances which I have just outlined the appellants were liable to the respondents would be to extend the duty of care owed by a manufacturer and others, to whom the principles first enunciated in *Donoghue* v. *Stevenson* have since been extended during the last half century, far beyond the limits to which the courts have hitherto extended them. The familiar 'floodgates' argument was once again brought fully into play. My Lords, although it cannot be denied that policy considerations have from time to time been allowed to play their part in the tort of negligence since it first developed as it were in its own right in the course of the last century, yet today I think its scope is best determined by considerations of principle rather than of policy. The floodgates argument is very familiar. It still may on occasion have its proper place but if principle suggests that the law should develop along a particular route and if the adoption of that particular route will accord a remedy where that remedy has hitherto been denied, I see no reason why, if it be just that the law should henceforth accord that remedy, that remedy should be denied simply because it will, in consequence of this particular development, become available to many rather than to few.

(at p. 542):

> Applying those statements of general principle as your Lordships have been enjoined to do both by Lord Reid and by Lord Wilberforce rather than to ask whether the particular situation which has arisen does or does not resemble some earlier and different situation where a duty of care has been held or has not been held to exist, I look for the reasons why, it being conceded that the appellants owed a duty of care to others not to construct the flooring so that those others were in peril of suffering loss or damage to their persons or their

property, that duty of care should not be equally owed to the respondents. The appellants, though not in direct contractual relationship with the respondents, were as nominated sub-contractors in almost as close a commercial relationship with the respondents as it is possible to envisage short of privity of contract. Why then should the appellants not be under a duty to the respondents not to expose the respondents to a possible liability to financial loss for repairing the flooring should it prove that that flooring had been negligently constructed? It is conceded that if the flooring had been so badly constructed that to avoid imminent danger the respondents had expended money upon renewing it the respondents could have recovered the cost of so doing. It seems curious that, if the appellants' work had been so bad that to avoid imminent danger expenditure had been incurred, the respondents could recover that expenditure, but that if the work was less badly done so that remedial work could be postponed they cannot do so. Yet this is seemingly the result of the appellants' contentions.

(at p. 545):

My Lords, to my mind in the instant case there is no physical damage to the flooring in the sense in which that phrase was used in *Dutton*, *Batty* and *Bowen* and some of the other cases. As my noble and learned friend, Lord Russell of Killowen, said during the argument, the question which your Lordships' House now has to decide is whether the relevant Scots and English law today extends the duty of care beyond a duty to prevent harm being done by faulty work to a duty to avoid such faults being present in the work itself. It was powerfully urged on behalf of the appellants that were your Lordships so to extend the law, a pursuer in the position of the pursuer in *Donoghue* v. *Stevenson* [1932] AC 562 could, in addition to recovering for any personal injury suffered, have also recovered for the diminished value of the offending bottle of ginger beer. Any remedy of that kind, it was argued, must lie in contract and not in delict or tort. My Lords, I seem to detect in that able argument reflections of the previous judicial approach to comparable problems before *Donoghue* v. *Stevenson* was decided. That approach usually resulted in the conclusion that in principle the proper remedy lay in contract and not outside it. But that approach and its concomitant philosophy ended in 1932 and for my part I should be reluctant to countenance its re-emergence some 50 years later in the instant case. I think today the proper control lies not in asking whether the proper remedy should lie in contract or instead in delict or tort, not in somewhat capricious judicial determination whether a particular case falls on one side of the line or the other, not in somewhat artificial distinctions between physical and economic or financial loss when the two sometimes go together and sometimes do not—it is sometimes overlooked that virtually all damage including physical damage is in one sense financial or economic for it is compensated by an award of damages—but in the first instance in establishing the relevant principles and then in deciding whether the particular case falls within or without those principles. To state this is to do no more than to restate what Lord Reid said in *Dorset Yacht Co.*

Ltd v. *Home Office* [1970] AC 1004 and Lord Wilberforce in *Anns* v. *Merton London Borough Council* [1978] AC 728. Lord Wilberforce, at p. 751, in the passage I have already quoted enunciated the two tests which have to be satisfied. The first is 'sufficient relationship of proximity', the second any considerations negativing, reducing or limiting the scope of the duty or the class of person to whom it is owed or the damages to which a breach of the duty may give rise. My Lords, it is I think in the application of those two principles that the ability to control the extent of liability in delict or in negligence lies. The history of the development of the law in the last 50 years shows that fears aroused by the floodgates argument have been unfounded. Cooke J in *Bowen* v. *Paramount Builders (Hamilton) Ltd* [1977] 1 NZLR 394, 422 described the floodgates argument as 'specious' and the argument against allowing a cause of action such as was allowed in *Dutton* v. *Bognor Regis Urban District Council* [1972] 1 QB 373, *Anns* v. *Merton London Borough Council* [1978] AC 728 and *Bowen* v. *Paramount Builders (Hamilton) Ltd* [1977] 1 NZLR 394 as 'in terrorem or doctrinaire'.

Turning back to the present appeal I therefore ask first whether there was the requisite degree of proximity so as to give rise to the relevant duty of care relied on by the respondents. I regard the following facts as of crucial importance in requiring an affirmative answer to that question.

(1)   The appellants were nominated sub-contractors.
(2)   The appellants were specialists in flooring.
(3)   The appellants knew what products were required by the respondents and their main contractors and specialised in the production of those products.
(4)   The appellants alone were responsible for the composition and construction of the flooring.
(5)   The respondents relied upon the appellants' skill and experience.
(6)   The appellants as nominated sub-contractors must have known that the respondents relied upon their skill and experience.
(7)   The relationship between the parties was as close as it could be short of actual privity of contract.
(8)   The appellants must be taken to have known that if they did the work negligently (as it must be assumed that they did) the resulting defects would at some time require remedying by the respondents expending money upon the remedial measures as a consequence of which the respondents would suffer financial or economic loss.

Lords Fraser and Russell agreed with this approach. Lord Keith, more cautiously, agreed with the decision on its particular facts, Lord Brandon dissented vigorously.

LORD BRANDON (at p. 550):

My Lords, in support of their contentions the respondents placed reliance on the broad statements relating to liability in negligence contained in the speech of Lord Wilberforce in *Anns* v. *Merton London Borough Council* [1978] AC 728, 751.

Applying that general statement of principle to the present case, it is, as I indicated earlier, common ground that the first question which Lord Wilberforce said one should ask oneself, namely, whether there is sufficient proximity between the parties to give rise to the existence of a duty of care owed by the one to the other, falls to be answered in the affirmative. Indeed, it is difficult to imagine a greater degree of proximity, in the absence of a direct contractual relationship, than that which, under the modern type of building contract, exists between a building owner and a sub-contractor nominated by him or his architect.

That first question having been answered in the affirmative, however, it is necessary, according to the views expressed by Lord Wilberforce in the passage from his opinion in *Anns* v. *Merton London Borough Council* quoted above, to ask oneself a second question, namely, whether there are any considerations which ought, inter alia, to limit the scope of the duty which exists.

To that second question I would answer that there are two important considerations which ought to limit the scope of the duty of care which it is common ground was owed by the appellants to the respondents on the assumed facts of the present case.

The first consideration is that, in *Donoghue* v. *Stevenson* itself and in all the numerous cases in which the principle of that decision has been applied to different but analogous factual situations, it has always been either stated expressly, or taken for granted, that an essential ingredient in the cause of action relied on was the existence of danger, or the threat of danger, of physical damage to persons or their property, excluding for this purpose the very piece of property from the defective condition of which such danger, or threat of danger, arises. To dispense with that essential ingredient in a cause of action of the kind concerned in the present case would, in my view, involve a radical departure from long-established authority.

The second consideration is that there is no sound policy reason for substituting the wider scope of the duty of care put forward for the respondents for the more restricted scope of such duty put forward by the appellants. The effect of accepting the respondents' contention with regard to the scope of the duty of care involved would be, in substance, to create, as between two persons who are not in any contractual relationship with each other, obligations of one of those two persons to the other which are only really appropriate as between persons who do have such a relationship between them.

In the case of a manufacturer or distributor of goods, the position would be that he warranted to the ultimate user or consumer of such goods that they were as well designed, as merchantable and as fit for their contemplated purpose as the exercise of reasonable care could make them. In the case of sub-contractors such as those concerned in the present case, the position would be that they warranted to the building owner that the flooring, when laid, would be as well designed, as free from defects of any kind and as fit for its contemplated purpose as the exercise of reasonable care could make it. In my view, the imposition of warranties of this kind on one person in favour of another, when there is no contractual relationship between them, is contrary to any sound policy requirement.

It is, I think, just worth while to consider the difficulties which would arise if the wider scope of the duty of care put forward by the respondents were accepted. In any case where complaint was made by an ultimate consumer that a product made by some persons with whom he himself had no contract was defective, by what standard or standards of quality would the question of defectiveness fall to be decided? In the case of goods bought from a retailer, it could hardly be the standard prescribed by the contract between the retailer and the wholesaler, or between the wholesaler and the distributor, or between the distributor and the manufacturer, for the terms of such contracts would not even be known to the ultimate buyer. In the case of sub-contractors such as the appellants in the present case, it could hardly be the standard prescribed by the contract between the sub-contractors and the main contractors, for, although the building owner would probably be aware of those terms, he could not, since he was not a party to such contract, rely on any standard or standards prescribed in it. It follows that the question by what standard or standards alleged defects in a product complained of by its ultimate user or consumer are to be judged remains entirely at large and cannot be given any just or satisfactory answer.

This speech has come to have greater influence than those of the majority, a point symbolized by the fact that no plaintiff has yet succeeded in an action based on *Junior Books*.

Pausing at this stage in the development of negligence claims for economic loss, it can be seen that, whether by way of a claim based on a statement or one based on an act, a clear, but restrictive, liability, based on a much more close or special relationship of proximity than that in normal negligence claims for physical damage, existed. However, the decision in *Junior Books* to deny the traditional 'floodgates' argument and to dent the long-established distinction between acts and statements, and between physical loss and economic loss, came to be regarded as a step too far.

Problems with economic loss and duty

Traditionally, if a party wishes to make a profit, they will have to enter into a contract with someone in order so to do. The law has long reflected this, with contract allowing the party to claim for future failure to earn a profit – a very intangible claim, in many ways. However, until *Hedley Byrne*, and then *Junior Books*, the law of tort had been notably reluctant so to do, as *Cattle* points out. So to remove even some of the constraints on tortious claims for economic loss was to bring the law of negligence into potential conflict with contract law. Negligence claims might be made to fill in gaps in the contractual provisions, as in *Junior Books*, but also might, if successful, actually end up contradicting the intention of the contracting parties, as for example in *Banque Financière de la Cité SA* v. *Westgate Insurance Co.* [1990] 2 All ER 947, where the effect of a tort claim, which was successful at first instance but not thereafter, would have been to impose upon an insurer a positive duty to inform their insured about dishonest conduct by an agent of the insured that would have the effect of invalidating the insurance policy in question. No normal or likely contractual provision would have imposed such a duty, and the parties would not have anticipated such a novel extension of tort doctrine. Commercially, it made no sense to impose an unwanted and unexpected tortious liability in such a case.

However, the fact that such far-reaching arguments could be plausibly made was enough to indicate to the senior members of the judiciary that the time had come to curtail the tort of negligence to a far greater extent even than had been achieved by the more limited definition of duty, from *Yuen Kun-Yeu* (p. 16, above), and the increased use of policy arguments to deny a duty (e.g. *Peabody*, p. 12, above).

The 1990 cases – problems solved?

The year 1990 may well go down into the history of the tort of negligence as a year of the same magnitude as 1932 itself, as the House of Lords battled to contain the monster that they had created, in the form of the newly vigorous tort of negligence.

CAPARO INDUSTRIES plc v. DICKMAN

[1990] 2 AC 605   House of Lords

This was a case concerning a takeover bid. The claim was made by the winner of a takeover battle against the auditors of the takeover victim. The pleasure of the plaintiff's victory faded quickly when, after acquisition, it was found that the accounts approved by the defendants were, allegedly, inaccurate and misleading. It was held that the auditors owed no duty to share purchasers.

LORD BRIDGE (at p. 617):

But since the *Anns* case a series of decisions of the Privy Council and of your Lordships' House, notably in judgments and speeches delivered by Lord Keith of Kinkel, have emphasised the inability of any single general principle to provide a practical test which can be applied to every situation to determine whether a duty of care is owed and, if so, what is its scope: see *Governors of Peabody Donation Fund* v. *Sir Lindsay Parkinson & Co. Ltd* [1985] AC 210, 239–241c; *Yuen Kun-Yeu* v. *Attorney-General of Hong Kong* [1988] AC 175, 190e–194f; *Rowling* v. *Takaro Properties Ltd* [1988] AC 473, 501d–g; *Hill* v. *Chief Constable of West Yorkshire* [1989] AC 53, 60b–d. What emerges is that, in addition to the foreseeability of damage, necessary ingredients in any situation giving rise to a duty of care are that there should exist between the party owing the duty and the party to whom it is owed a relationship characterised by the law as one of 'proximity' or 'neighbourhood' and that the situation should be one in which the court considers it fair, just and reasonable that the law should impose a duty of a given scope upon the one party for the benefit of the other. But it is implicit in the passages referred to that the concepts of proximity and fairness embodied in these additional ingredients are not susceptible of any such precise definition as would be necessary to give them utility as practical tests, but amount in effect to little more than convenient labels to attach to the features of different specific situations which, on a detailed examination of all the circumstances, the law recognises pragmatically as giving rise to a duty of care of a given scope. Whilst recognising, of course, the importance of the underlying general principles common to the whole field of negligence, I think the law has now moved in the direction of attaching greater significance to the more traditional categorisation of distinct and recognisable situations as guides to the existence, the scope and the limits of the varied duties of care which the law imposes. We must now, I think, recognise the wisdom of the words of Brennan J in the High Court of Australia in *Sutherland Shire Council* v. *Heyman* (1985) 60 ALR 1, 43–44, where he said:

'It is preferable, in my view, that the law should develop novel categories of negligence incrementally and by analogy with established categories, rather than by a massive extension of a prima facie duty of care restrained only by indefinable "considerations which ought to be negative, or to reduce or limit the scope of the duty or the class of person to whom it is owed".'

One of the most important distinctions always to be observed lies in the law's essentially different approach to the different kinds of damage which one party may have suffered in consequence of the acts or omissions of another. It is one thing to owe a duty of care to avoid causing injury to the person or property of others. It is quite another to avoid causing others to suffer purely economic loss.

(at p. 623):

These considerations amply justify the conclusion that auditors of a public company's accounts owe no duty of care to members of the public at large who rely upon the accounts in deciding to buy shares in the company. If a duty of care were owed so widely, it is difficult to see any reason why it should not equally extend to all who rely on the accounts in relation to other dealings with a company as lenders or merchants extending credit to the company. A claim that such a duty was owed by auditors to a bank lending to a company was emphatically and convincingly rejected by Millett J in *Al Saudi Banque* v. *Clarke Pixley* [1990] Ch 313. The only support for an unlimited duty of care owed by auditors for the accuracy of their accounts to all who may foreseeably rely upon them is to be found in some jurisdictions in the United States of America where there are striking differences in the law in different states. In this jurisdiction I have no doubt that the creation of such an unlimited duty would be a legislative step which it would be for Parliament, not the courts, to take.

LORD ROSKILL (at p. 628):

I agree with your Lordships that it has now to be accepted that there is no simple formula or touchstone to which recourse can be had in order to provide in every case a ready answer to the questions whether, given certain facts, the law will or will not impose liability for negligence or in cases where such liability can be shown to exist, determine the extent of that liability. Phrases such as 'foreseeability', 'proximity', 'neighbourhood', 'just and reasonable', 'fairness', 'voluntary acceptance of risk', or 'voluntary assumption of responsibility' will be found used from time to time in the different cases. But, as your Lordships have said, such phrases are not precise definitions. At best they are but labels or phrases descriptive of the very different factual situations which can exist in particular cases and which must be carefully examined in each case before it can be pragmatically determined whether a duty of care exists and, if so, what is the scope and extent of that duty. If this conclusion involves a return to the traditional categorisation of cases as pointing to the existence and scope of any duty of care, as my noble and learned friend Lord Bridge of Harwich, suggests, I think this is infinitely preferable to recourse to somewhat wide generalisations which leave their practical application matters of difficulty and uncertainty.

LORD OLIVER (at p. 633):

I think that it has to be recognised that to search for any single formula which will serve as a general test of liability is to pursue a will-o'-the wisp. The fact is that once one discards, as it is now clear that one must, the concept of foreseeability of harm as the single exclusive test – even a prima facie test – of the existence of the duty of care, the attempt to state some general principle which will determine liability in an infinite variety of circumstances serves not to clarify the law but merely to bedevil its development in a way which corresponds with practicality and common sense.

After citing *Sutherland* v. *Heyman* (1985) 59 ALJR 564 and *Hedley Byrne* v. *Heller*, he went on (at p. 635):

Perhaps, therefore, the most that can be attempted is a broad categorisation of the decided cases according to the type of situation in which liability has been established in the past in order to found an argument by analogy. Thus, for instance, cases can be classified according to whether what is complained of is the failure to prevent the infliction of damage by the act of the third party (such as *Dorset Yacht Co. Ltd* v. *Home Office* [1970] AC 1004, *P. Perl (Exporters) Ltd* v. *Camden London Borough Council* [1984] QB 342, *Smith* v. *Littlewoods Organisation Ltd* [1987] AC 241 and, indeed, *Anns* v. *Merton London Borough Council* [1978] AC 728 itself), in failure to perform properly a statutory duty claimed to have been imposed for the protection of the plaintiff either as a member of a class or as a member of the public (such as the *Anns* case, *Ministry of Housing and Local Government* v. *Sharp* [1970] 2 QB 223, *Yuen Kun-Yeu* v. *Attorney-General of Hong Kong* [1988] AC 175) or in the making by the defendant of some statement or advice which has been communicated, directly or indirectly, to the plaintiff and upon which he has relied. Such categories are not, of course, exhaustive. Sometimes they overlap as in the *Anns* case, and there are cases which do not readily fit into easily definable categories (such as *Ross* v. *Caunters* [1980] Ch 297). Nevertheless, it is, I think, permissible to regard negligent statements or advice as a separate category displaying common features from which it is possible to find at least guidelines by which a test for the existence of the relationship which is essential to ground liability can be deduced.

The full effect of these vital passages is slow to become clear. In particular, Lord Oliver, especially, appears at one stage to deny that any general principles of negligence exist, thus hurling us back to the days before *Donoghue* v. *Stevenson*. But this cannot be.

Lord Bridge invites us to proceed incrementally from case to case in creating novel claims for negligence. So if we are faced in the future with a case where a surveyor has fully inspected a building and then placed a value on it, then assuming there were no relevant precedents, how could we proceed? It may be that Lord Bridge is suggesting that we look at *Caparo*, and identify that auditors and surveyors are both professionals who value things, and so extend the *Caparo* decision in that direction. Surely, however, the way in which the common law has always proceeded, and should always proceed, is by way of identifying the principles that underlie the decision – proximity between the parties, reliance by one on the other's skill, etc. – and developing the law thus. Lord Oliver's denial of the role of such principles is not merely regressive, but rather simply misleading.

The case as a whole, however is not so restrictive in its approach. This becomes clear at the stage when all the judgments are placed together. Lord Bridge recognizes the concepts of foreseeability and proximity that lead towards the imposition of a duty as being proper ones to use, albeit excessively vague. Lord Oliver, in turn, insists that there are examples where, indeed, the general principles do not apply, but on closer inspection they turn out to be exceptions which are, and have been for long, recognized in any event as cases where different and more restrictive duty concepts apply. This is done simply by varying the level of proximity needed, at least in relation to economic loss cases where much higher levels of proximity are needed. So *Caparo* creates much heat and noise, but after the excitement fades, it is clear that it is a case which does not fundamentally change duty, but simply, as will be seen (at p. 250, below) reduces somewhat the scope of the already restrictive liability for negligent misrepresentation.

MURPHY v. BRENTWOOD DISTRICT COUNCIL

[1991] 1 AC 398   House of Lords

This was a building case, with a claim by the first purchaser of a dwelling-house made in respect of the defective foundations of the property against the local authority. The claim was unsuccessful.

LORD MACKAY (at p. 486):

> We are asked to depart from the judgment of this House in *Anns* v. *Merton London Borough Council* [1978] AC 728 under the practice statement of 1966 (*Practice Statement (Judicial Precedent)* [1966] 1 WLR 1234). That decision was taken after very full consideration by a committee consisting of most eminent members of this House. In those circumstances I would be very slow to accede to the suggestion that we should now depart from it. However, the decision was taken as a preliminary issue of law and accordingly the facts had not at that stage been examined in detail and the House proceeded upon the basis of the facts stated in the pleadings supplemented by such further facts and documents as had been agreed between the parties. Under the head 'Nature of the damages recoverable and arising of the cause of action' Lord Wilberforce said, at p. 759:

> > 'There are many questions here which do not directly arise at this stage and which may never arise if the actions are tried. But some conclusions are necessary if we are to deal with the issue as to limitation.'

> When one attempts to apply the proposition established by the decision to detailed factual situations difficulties arise and this was clearly anticipated by Lord Wilberforce when he said, at p. 760:

> > 'We are not concerned at this stage with any issue relating to remedial action nor are we called upon to decide upon what the measure of the damages should be; such questions, possibly very difficult in some cases, will be for the court to decide. It is sufficient to say that a cause of action arises at the point I have indicated.'

That point was when damage to the house had occurred resulting in there being a present or imminent danger to the health or safety of persons occupying it.

As I read the speech of Lord Wilberforce the cause of action which he holds could arise in the circumstances of that case can only do so when damage occurs to the house in question as a result of the weakness of the foundations and therefore no cause of action arises before that damage has occurred even if as a result of information obtained about the foundations it may become apparent to an owner that such damage is likely.

The person to whom the duty is owed is an owner or occupier of the house who is such when the damage occurs. And therefore an owner or occupier who becomes aware of the possibility of damage arising from a defective foundation would not be within the class of persons upon whom the right of action is conferred.

As has been demonstrated in the speeches of my noble and learned friends, the result of applying these qualifications to different factual circumstances is to require distinctions to be made which have no justification on any reasonable principle and can only be described as capricious. It cannot be right for this House to leave the law in that state.

Two options call for consideration. The first is to remove altogether the qualifications on the cause of action which *Anns* held to exist. This would be in itself a departure from *Anns* since these qualifications are inherent in the decision. The other option is to go back to the law as it was before *Anns* was decided and this would involve also overruling *Dutton* v. *Bognor Regis Urban District Council* [1972] 1 QB 373.

Faced with the choice I am of the opinion that it is relevant to take into account that Parliament has made provisions in the Defective Premises Act 1972 imposing on builders and others undertaking work in the provision of dwellings obligations relating to the quality of their work and the fitness for habitation of the dwelling. For this House in its judicial capacity to create a large new area of responsibility on local authorities in respect of defective buildings would in my opinion not be a proper exercise of judicial power. I am confirmed in this view by the consideration that it is not suggested, and does not appear to have been suggested in *Anns*, that the Public Health Act 1936, in particular Part II, manifests any intention to create statutory rights in favour of owners or occupiers of premises against the local authority charged with responsibility under the Act. The basis of the decision in *Anns* is that the common law will impose a duty in the interests of the safety and health of owners and occupiers of buildings since that was the purpose for which the Act of 1936 was enacted. While of course I accept that duties at common law may arise in respect of the exercise of statutory powers or the discharge of statutory duties I find difficulty in reconciling a common law duty to take reasonable care that plans should conform with byelaws or regulations with the statute which has imposed on the local authority the duty not to pass plans unless they comply with the byelaws or regulations and to pass them if they do.

In these circumstances I have reached the clear conclusion that the proper exercise of the judicial function requires this House now to depart from *Anns* in so far as it affirmed a private law duty of care to avoid damage to property which causes present or imminent danger to the health and safety of owners, or occupiers, resting upon local authorities in relation to their function of supervising compliance with building byelaws or regulations, that *Dutton* v. *Bognor Regis Urban District Council* should be overruled and that all decisions subsequent to *Anns* which purported to follow it should be overruled. I accordingly reach the same conclusion as do my noble and learned friends.

The formal overruling of *Anns* was agreed by a full House of seven law lords in this important case. We shall return and discuss its direct implications for builders in Chapter 8, and for local authority liability in Chapter 10. For now, though, although it might seem that the landmark decision, upon which so much of the modern law of negligence has been built, has been demolished, it is worth noting that the terms of Lord Mackay's disavowal are confined to the narrow effect of the case on local authority liability. It is equally clear that the *Anns* two-stage test for duty remains, in an attenuated form, as part of the basis of the current three-stage duty test in *Yuen Kun-Yeu*, which in turn appears to have survived the attacks on the tort of negligence in both *Caparo* and *Murphy*. In the mainstream case of physical injury, foresight and proximity must be positively satisfied and policy must be examined, with special, more restrictive, rules applying in economic loss cases, and in the other exceptional cases identified in *Caparo*.

Economic loss redefined

A further restriction placed on negligence claims recently has been to redefine what is regarded as economic loss. By categorizing more incidents as being losses that are economic in character, more cases fall within the highly restricted liability that exists in negligence for such losses.

D & F ESTATES v. CHURCH COMMISSIONERS FOR ENGLAND

[1989] AC 177   House of Lords

The plaintiffs leased a flat from the defendants. During routine repairs, it was discovered that some of the plaster work was loose, as a result of negligent work by sub-contractors. No claim was allowed in respect of renewal of the plaster work.

LORD BRIDGE (at p. 206):

> If the defect is discovered before any damage is done, the loss sustained by the owner of the structure, who has to repair or demolish it to avoid a potential

source of danger to third parties, would seem to be purely economic. Thus, if I acquire a property with a dangerously defective garden wall which is attributable to the bad workmanship of the original builder, it is difficult to see any basis in principle on which I can sustain an action in tort against the builder for the cost of either repairing or demolishing the wall. No physical damage has been caused. All that has happened is that the defect in the wall has been discovered in time to prevent damage occurring.

Fuller citation of this case on this and other issues is in Chapter 7, below.

So, the nature of the loss is defined not by examination of the loss, but rather by investigating its cause. A crack in the wall of a house caused by my lorry colliding with it is regarded as physical damage, because it is caused by an outside agency. However, the same crack, caused by defective workmanship in the construction process, will be seen as economic loss at least in the absence of any danger. This seemingly bizarre result has been confirmed by the House of Lords recently.

MURPHY v. BRENTWOOD DISTRICT COUNCIL (see p. 31, above)

LORD BRIDGE (at p. 475):

> If a manufacturer negligently puts into circulation a chattel containing a latent defect which renders it dangerous to persons or property, the manufacturer, on the well known principles established by *Donoghue* v. *Stevenson* [1932] AC 562, will be liable in tort for injury to persons or damage to property which the chattel causes. But if a manufacturer produces and sells a chattel which is merely defective in quality, even to the extent that it is valueless for the purpose for which it is intended, the manufacturer's liability at common law arises only under and by reference to the terms of any contract to which he is a party in relation to the chattel; the common law does not impose on him any liability in tort to persons to whom he owes no duty in contract but who, having acquired the chattel, suffer economic loss because the chattel is defective in quality. If a dangerous defect in a chattel is discovered before it causes any personal injury or damage to property, because the danger is now known and the chattel cannot safely be used unless the defect is repaired, the defect becomes merely a defect in quality. The chattel is either capable of repair at economic cost or it is worthless and must be scrapped. In either case the loss sustained by the owner or hirer of the chattel is purely economic. It is recoverable against any party who owes the loser a relevant contractual duty. But it is not recoverable in tort in the absence of a special relationship of proximity imposing on the tortfeasor a duty of care to safeguard the plaintiff from economic loss. There is no such special relationship between the manufacturer of a chattel and a remote owner or hirer.
>
> I believe that these principles are equally applicable to buildings. If a builder erects a structure containing a latent defect which renders it dangerous to

persons or property, he will be liable in tort for injury to persons or damage to property resulting from that dangerous defect. But if the defect becomes apparent before any injury or damage has been caused, the loss sustained by the building owner is purely economic. If the defect can be repaired at economic cost, that is the measure of the loss. If the building cannot be repaired, it may have to be abandoned as unfit for occupation and therefore valueless. These economic losses are recoverable if they flow from breach of a relevant contractual duty, but, here again, in the absence of a special relationship of proximity they are not recoverable in tort.

Duty of care – summary

In considering whether a duty of care exists in a novel case today, we must still start with basic concepts of 'neighbourhood' – are the parties in sufficient proximity? If economic loss, are they highly proximate? Is harm a foreseeable result of negligence? It is then necessary to further enquire whether any past precedents exist to hinder the imposition of a duty, and to look at whether policy ideas preclude the presence of a duty. In doing this, we have not, in most respects, moved very far from the starting-point of the tort of negligence in 1932.

Chapter 2

Breach of Duty

For all the legal and intellectual energy expended on the issue of duty of care in negligence, it is an issue that does not arise in many mainstream cases; no-one doubts that a solicitor or surveyor owes a duty to his client or a doctor to his patient, and the cases where duty is in issue are often more peripheral. However, the next stage of the negligence action, the issue of breach of duty, is of central importance in almost every case, because the question at the breach of duty stage is whether the conduct of the defendant was in fact negligent, or in other words whether the defendant was at fault in the conduct of his affairs, and therefore liable in tort.

The basic yardstick by which the defendant's conduct is measured in all negligence cases is with reference to what may be reasonably expected from the 'reasonable man'. The defendant who cannot attain the standard of the reasonable man will be held to be negligent; no liability attaches to the defendant who succeeds in reaching the reasonable man's standard. Crucially, the level of the reasonable man's conduct is not at some impossibly high level; rather, he is regarded as being the epitome of ordinariness. He is the 'man in the street' or 'Mr Average'; he is not endowed with any special skill or foresight. He is 'free from both over-apprehension and from over-confidence', according to Lord Macmillan in *Glasgow Corporation* v. *Muir* [1943] AC 448 at 457, and his standard therefore should be a relatively easy one to attain. He will weigh up risks and having assessed them, will take reasonable precautions against those risks that are substantial enough to merit such precautions.

However, it would clearly be inappropriate to use this test without some adaptation in cases of professional liability in construction cases. It is self-evident that building owners expect their buildings to be designed and built with something more than the skill of the ordinary man, and so it is necessary to adapt the test accordingly; the basic approach is to compare the acts of the professional or skilled defendant with the approach of the reasonable practitioner of the particular profession or skill in question.

Standard of care of professional defendants

BOLAM v. FRIERN HOSPITAL MANAGEMENT COMMITTEE

[1957] 1 WLR 582 High Court

Bolam was a patient in a mental hospital. He was given electro-convulsive therapy without any restraint or sedation. As a result he was injured. At this stage in medical development, such treatment was normal but opinion was divided as to whether restraint or sedation should be used. McNair J directed a jury in the following terms.

MC NAIR J (at p. 586):

I must tell you what in law we mean by 'negligence'. In the ordinary case which does not involve any special skill, negligence in law means a failure to do some act which a reasonable man in the circumstances would do, or the doing of some act which a reasonable man in the circumstances would not do; and if that failure or the doing of that act results in injury, then there is a cause of action. How do you test whether this act or failure is negligent? In an ordinary case it is generally said you judge it by the action of the man in the street. He is the ordinary man. In one case it has been said you judge it by the conduct of the man on the top of a Clapham omnibus. He is the ordinary man. But where you get a situation which involves the use of some special skill or competence then the test as to whether there has been negligence or not is not the test of the man on the top of a Clapham omnibus, because he has not got this special skill. The test is the standard of the ordinary skilled man exercising and professing to have that special skill. A man need not possess the highest expert skill; it is well established law that it is sufficient if he exercises the ordinary skill of an ordinary competent man exercising that particular art. I do not think that I quarrel much with any of the submissions in law which have been put before you by counsel. [Counsel for Bolam] put it in this way, that in the case of a medical man, negligence means failure to act in accordance with the standards of reasonably competent medical men at the time. That is a perfectly accurate statement, as long as it is remembered that there may be one or more perfectly proper standards; and if he conforms with one of those proper standards, then he is not negligent. [Counsel for Bolam] also was quite right, in my judgment, in saying that a mere personal belief that a particular technique is best is no defence unless that belief is based on reasonable grounds. That again is unexceptionable. But the emphasis which is laid by the defence is on this aspect of negligence, that the real question you have to make up your minds about on each of the three major topics is whether the defendants, in acting in the way they did, were acting in accordance with a practice of competent respected professional opinion. [Counsel for Friern Hospital Management Committee] submitted that if you are satisfied that they were acting in accordance with a practice of a competent body of professional opinion, then it would be wrong for you to hold that negligence was established. In a recent Scottish case, *Hunter* v. *Hanley* [1955] SLT 213, 217,Lord President Clyde said:

> 'In the realm of diagnosis and treatment there is ample scope for genuine difference of opinion and one man clearly is not negligent merely because his conclusion differs from that of other professional men, nor because he has displayed less skill or knowledge than others would have shown. The true test for establishing negligence in diagnosis or treatment on the part of a doctor is whether he has been proved to be guilty of such failure as no doctor of ordinary skill would be guilty of, if acting with ordinary care.'

If that statement of the true test is qualified by the words 'in all the circumstances', [counsel for Bolam] would not seek to say that that expression of opinion does not accord with the English law. It is just a question of expression. I myself would prefer to put it this way, that he is not guilty of negligence if he has acted in accordance with a practice accepted as proper by a responsible body of medical men skilled in that particular art. I do not think there is much difference in sense. It is just a different way of expressing the same thought. Putting it the other way round, a man is not negligent, if he is acting in accordance with such a practice, merely because there is a body of opinion who would take a contrary view. At the same time, that does not mean that a medical man can obstinately and pig-headedly carry on with some old technique if it has been proved to be contrary to what is really substantially the whole of informed medical opinion. Otherwise you might get men today saying: 'I do not believe in anaesthetics. I do not believe in antiseptics. I am going to continue to do my surgery in the way it was done in the eighteenth century'. That clearly would be wrong.

Before I get to the details of the case, it is right to say this, that it is not essential for you to decide which of two practices is the better practice, as long as you accept that what the defendants did was in accordance with a practice accepted by responsible persons; if the result of the evidence is that you are satisfied that this practice is better than the practice spoken of on the other side, then it is really a stronger case.

This passage has become enshrined in the law as a classic exposition of breach of duty in a professional negligence case, and its central role has been affirmed recently (in *Gold* v. *Haringey Area Health Authority* [1988] QB 481). The key point to note is that the court, perhaps reflecting the lack of their specialist knowledge of medicine or any other profession, is not engaged in a quest to discover whether the defendant was objectively correct in his *modus operandi*. The sole function of the court is to decide whether there is *a* responsible body of professional opinion which supports what the defendant did, and this is irrespective of the size of that body of opinion, the presence of other bodies of opinion or the objective correctness of their view. The court's sole function is to establish, qualitatively, the 'responsibility' (or otherwise) of that body of opinion that is being quoted to support the actions of the defendant.

This, however, could in theory (and in the past in practice too) lead to rather sterile courtroom disputes, where each side to the dispute brings along their own flock of expert witnesses to, respectively, lend support to or fuel the attack on the conduct of the defendant. The House of Lords has now acted to stop such an unsatisfactory exercise.

MAYNARD v. WEST MIDLANDS REGIONAL HEALTH AUTHORITY

[1984] 1 WLR 634   House of Lords

The plaintiff was in hospital probably suffering from tuberculosis but possibly suffering from more severe cancer-related illnesses. An operation was carried out to investigate this latter possibility which carried with it the risk of nerve injury. Such injury occurred, and the operation revealed that there was no risk of the cancer-related illnesses. Expert evidence was provided on both sides as to the appropriateness of the course of treatment undergone by the plaintiff, and the trial judge, Comyn J found for the plaintiff, stating that he preferred the evidence of the expert witnesses called in her support. The Court of appeal reversed this decision, and the House of Lords agreed.

LORD SCARMAN (at p. 638):

> The present case may be classified as one of clinical judgment. Two distinguished consultants, a physician and a surgeon experienced in the treatment of chest diseases, formed a judgment as to what was, in their opinion, in the best interests of their patient. They recognised that tuberculosis was the most likely diagnosis. But, in their opinion, there was an unusual factor, viz. swollen glands in the mediastinum unaccompanied by any evidence of lesion in the lungs. Hodgkin's disease, carcinoma and sarcoidosis were, therefore, possibilities. The danger they thought was Hodgkin's disease; though unlikely, it was, if present, a killer (as treatment was understood in 1970) unless remedial steps were taken in its early stage. They, therefore, decided on mediastinoscopy, an operative procedure which would provide them with a biopsy from the swollen glands which could be subjected to immediate microscopic examination. It is said that the evidence of tuberculosis was so strong that it was unreasonable and wrong to defer diagnosis and to put their patient to the risks of the operation. The case against them is not mistake or carelessness in performing the operation, which it is admitted was properly carried out, but an error of judgment in requiring the operation to be undertaken.
>
> A case which is based on an allegation that a fully considered decision of two consultants in the field of their special skill was negligent clearly presents

certain difficulties of proof. It is not enough to show that there is a body of competent professional opinion which considers that theirs was a wrong decision, if there also exists a body of professional opinion, equally competent, which supports the decision as reasonable in the circumstances. It is not enough to show that subsequent events show that the operation need never have been performed, if at the time the decision to operate was taken it was reasonable in the sense that a responsible body of medical opinion would have accepted it as proper. I do not think that the words of Lord President Clyde in *Hunter* v. *Hanley*, 1955 SLT 213, 217 can be bettered:

> 'In the realm of diagnosis and treatment there is ample scope for genuine difference of opinion and one man clearly is not negligent merely because his conclusion differs from that of other professional men . . . The true test for establishing negligence in diagnosis or treatment on the part of a doctor is whether he has been proved to be guilty of such failure as no doctor of ordinary skill would be guilty of if acting with ordinary care . . .'

> I would only add that a doctor who professes to exercise a special skill must exercise the ordinary skill of his speciality. Differences of opinion and practice exist, and will always exist, in the medical as in other professions. There is seldom any one answer exclusive of all others to problems of professional judgment. A court may prefer one body of opinion to the other: but that is no basis for a conclusion of negligence.

He went on (at p. 639):

> I have to say that a judge's 'preference' for one body of distinguished professional opinion to another also professionally distinguished is not sufficient to establish negligence in a practitioner whose actions have received the seal of approval of those whose opinions, truthfully expressed, honestly held, were not preferred. If this was the real reason for the judge's finding, he erred in law even though elsewhere in his judgment he stated the law correctly. For in the realm of diagnosis and treatment negligence is not established by preferring one respectable body of professional opinion to another. Failure to exercise the ordinary skill of a doctor (in the appropriate speciality, if he be a specialist) is necessary.

Lord Scarman went on to emphasize (at p. 640) that the doctors in the case had the support of 'a strong body of evidence given by distinguished medical men supporting and approving of what he did in the case . . .'. This was enough to satisfy the *Bolam* test, and to find the defendants not liable in negligence.

This decision is of profound importance for professional defendants. It confirms explicitly what was implicit in *Bolam*, namely that it is simply enough for the defendant to point to some responsible support for his view, without the court being allowed to concern itself with the correctness of that support, or the presence of alternative, even better, views. *Maynard* also, in passing, provides authority for the sensible proposition that the precise standard of care applies not just to the general profession or trade being followed, but rather to the particular specialism therein.

Both the *Bolam* test and its interpretation in *Maynard* tilt the balance in an action towards the professional defendant. And the trend of other decisions in this area is generally the same. In particular, the courts will not confer the benefit of hindsight on the activities of the professional or skilled man, and also will allow him to make mistakes.

ROE v. MINISTER OF HEALTH

[1954] 2 QB 66 Court of Appeal

The plaintiff was undergoing a routine minor operation but became paralysed after being given an anaesthetic. This, it was discovered, had been contaminated by phenol, in which it had been stored, the contamination occurring invisibly, through molecular flaws. This hazard, well-known by the time of the trial, was an unknown risk at the time of the accident, six years previously. The plaintiff's claim failed.

MORRIS LJ (at p. 92):

It is now known that there could be cracks not ordinarily detectable. But care has to be exercised to ensure that conduct in 1947 is only judged in the light of knowledge which then was or ought reasonably to have been possessed. In this connexion the then-existing state of medical literature must be had in mind. The question arises whether Dr Graham [the anaesthetist] was negligent in not adopting some different technique. I cannot think that he was. I think that a consideration of the evidence in the case negatives the view that Dr Graham was negligent and I see no reason to differ from the conclusions which were reached on this part of the case by the judge.

WHITEHOUSE v. JORDAN

[1981] 1 WLR 246 House of Lords

The infant plaintiff was born with severe brain damage. This was caused by the use of force during a forceps delivery by the defendant. There had been a clear medical need to induce the birth of the baby and the defendant had been faced with a difficult choice between two risky courses of action, viz. forceps delivery or birth by Caesarean section. The plaintiff's claim failed.

LORD EDMUND-DAVIES (at p. 257):

> The principal questions calling for decision are: (a) in what manner did Mr Jordan use the forceps, and (b) was that manner consistent with the degree of skill which a member of his profession is required by law to exercise? Surprising though it is at this late stage in the development of the law of negligence, counsel for Mr Jordan persisted in submitting that his client should be completely exculpated were the answer to question (b) 'Well, at worst he was guilty of an error of clinical judgment'. My Lords, it is high time that the unacceptability of such an answer be finally exposed. To say that a surgeon committed an error of clinical judgment is wholly ambiguous, for, while some such errors may be completely consistent with the due exercise of professional skill, other acts or omissions in the course of exercising 'clinical judgment' may be so glaringly below proper standards as to make a finding of negligence inevitable.

LORD FRASER (at p. 263):

> Referring to medical men, Lord Denning said [1980] 1 All ER 650, 658:
>
> > 'If they are found to be liable [sc. for negligence] whenever they do not effect a cure, or whenever anything untoward happens, it would do a great disservice to the profession itself.'
>
> That is undoubtedly correct, but he went on to say this: 'We must say, and say firmly, that, in a professional man, an error of judgment is not negligent.' Having regard to the context, I think that the learned Master of the Rolls must have meant to say that an error of judgment 'is not *necessarily* negligent.' But in my respectful opinion, the statement as it stands is not an accurate statement of law. Merely to describe something as an error of judgment tells us nothing about whether it is negligent or not. The true position is that an error of judgment may, or may not, be negligent; it depends on the nature of the error. If it is one that would not have been made by a reasonably competent professional man professing to have the standard and type of skill that the defendant held himself out as having, and acting with ordinary care, then it is negligent. If, on the other hand, it is an error that such a man, acting with ordinary care, might have made, then it is not negligent.

Pragmatically, this must be correct. Regrettable as it may be, all professionals are honest enough to admit that they are not infallible; the tort of negligence only arises if the defendant falls below the usual standard of fallibility.

Different professions – same standard?

This series of cases indicates a very generous attitude to professional defendants. They can make mistakes, take advantage of divided professional opinion, be in a small minority, and yet still not be held liable in negligence. However, it may be significant that all the cases are ones involving medical negligence. In principle, one would assume that cases on the standard of care of one group of professionals should apply equally to other professionals, unless meaningful distinctions can be drawn between the different groups.

However, there remains a suspicion that the courts are more deferential to medical opinions than those expressed by members of other professions. This is a proposition that is at present impossible to prove, but which is frequently asserted by practitioners in this field of litigation.

There is no doubt that the courts claim to use the same tests over the full range of professions. As Lord Denning stated in *Greaves* v. *Baynham Meikle* [1975] 1 WLR 1095, 'in the ordinary employment of a professional man, whether it is a medical man, a lawyer, or an accountant, an architect or an engineer, his duty is to use reasonable care and skill'. Likewise, in *Governors of the Hospitals for Sick Children* v. *McLaughlin and Harvey* [1987] CILL 371, Judge Newey noted that the *Bolam* test, though founded in a medical case, had been regularly used in *Greaves* and many other construction cases.

Nevertheless, in the application of these tests, the possibility exists for variation between different areas of professional activity, and the suspicion lurks that the courts may be more willing to impose their own view and less willing to accept the profession's view in construction cases than in, for example, medical ones.

Varying the standard of care

So far it has been seen that there is a simple general test for breach of duty, albeit one whose application will vary in accordance with the infinite variety of fact situations which might arise. However, there are certain exceptional cases where the courts depart from the usual standard of care, viz. compliance with normal standards of behaviour of a comparable reasonable man.

(a) Raising the standard

Exceptionally, compliance with normal practice is not enough, in rare cases where a court decides that to follow the normal practice is itself negligent.

CAVANAGH v. ULSTER WEAVING CO.

[1960] AC 145 House of Lords

The plaintiff was injured when he fell from a wet rooftop ladder with no handrail while wearing rubber boots with wet and slippery soles and carrying a bucket of cement. Evidence was adduced that this was in accordance with normal practice. The plaintiff succeeded in his claim.

LORD KEITH OF AVONHOLM (at p. 165):

> The ruling principle is that an employer is bound to take reasonable care for the safety of his workmen, and all other rules or formulas must be taken subject to this principle. All that Lord Dunedin meant (in *Morton* v. *William Dixon Ltd* [1909] SC 89), in my opinion, was that if a plaintiff was complaining of a particular act of omission as constituting negligence, there were two ways in which he might endeavour to show this: by proving that the omission complained of was a precaution commonly undertaken by others in like circumstances, or was so obvious 'that it would be folly in anyone to neglect to provide it'. It is on the second alternative that discussion has generally turned. But I see no particular difficulty. There is no magic in the word 'folly'. It gives the formula the characteristic that was described by Lord Normand in *Paris* v. *Stepney Borough Council* [1951] AC 367 as 'trenchant'. But the language could be phrased otherwise without any loss of meaning. Lord Dunedin might equally have said: 'It would be stupid not to provide it', or 'that no sensible man would fail to provide it', or 'that common sense would dictate that it should be provided'. Lord Cooper himself, who was particularly averse from watering down Lord Dunedin's language, on three separate occasions in his judgment in *Gallagher* v. *Balfour Beatty & Co.* [1951] SC 712 used 'inexcusable' as the equivalent of 'folly'. With this may be read the passage from Lord Normand's judgment in *Paris* v. *Stepney Borough Council* that the formula 'does not detract from the test of the conduct and judgment of the reasonable and prudent man'. In *General Cleaning Contractors Ltd* v. *Christmas* [1953] AC 180 my noble and learned friend Lord Tucker said: 'It is true that in some cases there may be precautions which are so obvious that no evidence is required on the subject'. In *Morris* v. *West Hartlepool Steam Navigation Co. Ltd* [1956] AC 552 Lord Cohen said that he agreed with Parker LJ that 'folly' was not to be read as 'ridiculous' and did not think that that was the sense in which Lord Dunedin used it. I refrain from quoting observations in a similar sense from others of their Lordships in the cases cited. Lord Dunedin cannot, in my opinion, have intended to depart from or modify the fundamental principle that an employer is bound to take reasonable care for the safety of his workmen, and in every case the question is whether the circumstances are such as to entitle judge or jury to say that there has or has not been a failure to exercise such reasonable care. It is immaterial, in my opinion, whether the alleged failure in duty is in respect of an act of omission or an act of commission. But where it is an act of omission that is alleged, I

think it will be found, in the absence of evidence of practice, that the circumstances will rarely, if ever, lead judge or jury to hold that there was negligence unless the precaution which it is suggested should have been taken is one of a relatively simple nature which can readily be understood and commends itself to common intelligence as something to be required.

This passage emphasizes the rare nature of a finding that the general practice of a trade or profession is itself at fault. It is right that such a finding should be rare, given the difficulty inherent in judges substituting their non-expert judgment for that of those skilled in a particular trade or profession.

(b)   Standard of care of novices and trainees

This important area has been uncertain for several years, but a clear answer has now been given, that the newly-qualified and thus inexperienced professional must reach the standard of care of the reasonable practitioner of that profession.

WILSHER v. ESSEX AREA HEALTH AUTHORITY

[1987] QB 730   Court of Appeal

(The case went on to the House of Lords on another issue: see pp. 52 below.)

A baby born prematurely was under specialist treatment at a hospital. A junior doctor, new to that type of work, was alleged to be at fault in his treatment of the baby. The question arose as to what standard of care should be expected from the doctor.

MUSTILL LJ (at p. 750):

The second proposition, advanced on behalf of the defendants, directs attention to the personal position of the individual member of the staff about whom the complaint is made. What is expected of him is as much as, but no more than, can reasonably be required of a person having his formal qualifications and practical experience. If correct, this proposition entails that the standard of care which the patient is entitled to demand will vary according to the chance of recruitment and rostering. The patient's right to complain of faulty treatment will be more limited if he has been entrusted to the care of a doctor who is a complete novice in the particular field (unless perhaps he can point to some fault of supervision in a person further up the hierarchy) than if he has been in the hands of a doctor who has already spent months on the same ward: and his prospects of holding the health authority vicariously liable for the consequences of any mistreatment will be correspondingly reduced.

To my mind, this notion of a duty tailored to the actor, rather than to the act which he elects to perform, has no place in the law of tort. Indeed, the defendants did not contend that it could be justified by any reported authority on the general law of tort. Instead, it was suggested that the medical profession is a special case. Public hospital medicine has always been organised so that young doctors and nurses learn on the job. If the hospitals abstained from using inexperienced people, they could not staff their wards and theatres, and the junior staff could never learn. The longer-term interests of patients as a whole are best served by maintaining the present system, even if this may diminish the legal rights of the individual patient: for, after all, medicine is about curing, not litigation.

I acknowledge the appeal of this argument, and recognise that a young hospital doctor, who must get on to the wards in order to qualify without necessarily being able to decide what kind of patient he is going to meet, is not in the same position as another professional man who has a real choice whether or not to practise in a particular field. Nevertheless, I cannot accept that there should be a special rule for doctors in public hospitals – I emphasise *public*, since presumably those employed in private hospitals would be in a different category. Doctors are not the only people who gain their experience, not only from lectures or from watching others perform, but from tackling live clients or customers, and no case was cited to us which suggested that any such variable duty of care was imposed on others in a similar position. To my mind, it would be a false step to subordinate the legitimate expectation of the patient that he will receive from each person concerned with his care a degree of skill appropriate to the task which he undertakes, to an understandable wish to minimise the psychological and financial pressures on hard-pressed young doctors.

For my part, I prefer the third of the propositions which have been canvassed. This relates the duty of care not to the individual, but to the post which he occupies. I would differentiate 'post' from 'rank' or 'status'. In a case such as the present, the standard is not just that of the averagely competent and well-informed junior houseman (or whatever the position of the doctor) but of such a person who fills a post in a unit offering a highly specialised service.

The Court of Appeal also thought that a direct liability should be placed on the employer who allows such an inexperienced person to be placed in a key post.

Sir Nicolas Browne-Wilkinson VC (at p. 777):

In my judgment, if the standard of care required of such a doctor is that he should have the skill required of the post he occupies, the young houseman or the doctor seeking to obtain specialist skill in a special unit would be held liable for shortcomings in the treatment without any personal fault on his part at all. Of course, such a doctor would be negligent if he undertook treatment for which he knows he lacks the necessary experience and skill. But one of the chief hazards of inexperience is that one does not always know the risks which exist. In my judgment, so long as the English law rests liability on personal fault, a doctor who has properly accepted a post in a hospital in order to gain necessary experience should only be held liable for acts or omissions which a careful doctor with his qualifications and experience would not have done or omitted. It follows that, in my view, the health authority could not be held vicariously liable (and I stress the word *vicariously*) for the acts of such a learner who has come up to those standards, notwithstanding that the post he held required greater experience than he in fact possessed.

The only argument to the contrary (and it is a formidable one) is that such a standard of care would mean that the rights of a patient entering hospital will depend on the experience of the doctor who treats him. This, I agree, would be wholly unsatisfactory. But, in my judgment, it is not the law. I agree with the comments of Mustill LJ as to the confusion which has been caused in this case both by the pleading and the argument below which blurred the distinction between the vicarious liability of the health authority for the negligence of his doctors and the direct liability of the health authority for negligently failing to provide skilled treatment of the kind that it was offering to the public. In my judgment, a health authority which so conducts its hospital that it fails to provide doctors of sufficient skill and experience to give the treatment offered at the hospital may be directly liable in negligence to the patient. Although we were told in argument that no case has ever been decided on this ground and that it is not the practice to formulate claims in this way, I can see no reason why, in principle, the health authority should not be so liable if its organisation is at fault.

Wilsher is undoubtedly correct, not least because a subjective test, tailoring each defendant's standard of care to his particular level of ability and experience, would be very difficult and time-consuming to use in litigation. However, it should be noted that *Wilsher* goes very much against the trend of the other cases in this section, which tend to favour the professional defendant for the most part. Even the unqualified, if they hold themselves out as having qualifications, have to reach the appropriate professional standard of care.

FREEMAN v. MARSHALL & CO.

(1966) 200 EG 777 High Court

The defendant surveyor, who was an unqualified 'estate agent, valuer and surveyor' failed to appreciate evidence of rising damp and rot in a basement he was inspecting. In his defence, he contended that he only had a basic understanding of building technology and structures involved in routine estate agency work.

LAWTON J (at p. 777):

> [He] . . . 'had no organised course of training as a surveyor and had never passed any professional examination in surveying. He was a member of the Valuers Institution through election, not examination . . . In fairness to him, he claimed only to have a working knowledge of structures from the point of view of buying and selling, but if he held himself out in practice as a surveyor he must be deemed to have the skills of a surveyor and be adjudged upon them.'

Proof of breach

Practice, as well as the rules of law, tends not to help the plaintiff in an action against a construction or other professional. What does such a plaintiff have to do in order to succeed? A plaintiff in any negligence claim has the burden of proof placed on them, i.e. it is for the plaintiff to prove, on the balance of probabilities, that the defendant is under a duty, is in breach and has caused the relevant damage. As far as breach of duty by a professional defendant is concerned, the plaintiff has in effect to establish three things:

(i)   what in fact happened;
(ii) what the reasonable man in the defendant's position should have done; and
(iii) then establish by deduction that the defendant fell below that standard.

These issues can often cause difficulty. Proving what has actually happened is often far from easy. The plaintiff may be impaired by injuries suffered in the tort, and the reasons for, say, the collapse of a building may be concealed both by the fact of the destruction of the building and also by the likelihood of something complex and/or highly technical being the cause. Similar problems of technicality plague the plaintiff at the second stage too, with an (often) non-expert plaintiff having to fight through potentially conflicting expert testimony to discover what the proper professional standard should have been.

In one way, however, the courts redress the balance in favour of the plaintiff to some extent by use of the maxim *res ipsa loquitur*. The effect of this is simply to lessen the burden of proof placed on the plaintiff in a case where negligence seems so likely to be the explanation for what has occurred that negligence, as a cause, can be said to speak for itself. Such an approach, first clearly adopted in *Scott* v. *London and St Katherine Docks* (1865) 3 H & C 596 has now evolved, and the principle will apply in circumstances when: (a) negligence is the most likely cause of the incident; and (b) that negligence is by the defendant.

Both these make the maxim's use problematic in many building negligence cases. Buildings can often be damaged without negligence, for example by changing soil and weather conditions while, even if negligence appears to be a likely cause, the wide range of contractors, sub-contractors, architects, surveyors, engineers, etc., involved will make it difficult in many cases to isolate which particular party is the one whose negligence is so self-evident.

A further problem with using *res ipsa loquitur* is that even if a plaintiff is able to apply the maxim, its effect is relatively limited. According to the recent Privy Council decision in *Ng Chun Pin* v. *Lee Chuen Tat* [1988] RTR 298, *res ipsa loquitur* does not actually reverse the burden of proof, but merely creates a presumption of negligence which, if the defendant is able to rebut it by means of a reasonable suggestion of a non-negligent explanation, results in the burden of proof then returning to the plaintiff. So plaintiffs in construction cases will not often be able to make use of this 'shortcut' to proof of breach and will also find it, even if utilizable, of only very limited effect.

In summary, a plaintiff fighting an action against a construction professional has an uphill task in establishing a breach of duty by the defendant. The onus lies on the plaintiff to prove what actually happened and what should have happened, against a background of legal rules which have the effect of giving ample opportunity to the defendant to escape liability, even though he may be wrong, or in a minority of professional opinion.

Chapter 3

Damage

The third element that makes up the tort of negligence is damage. Simply, there must be some; the tort of negligence does not apply until harm has resulted. This has significant impact for limitation periods (see pp. 79 below) and also causes the difficulties with economic loss (see pp. 18 above). Complications arise, however, when it is realized that it is essential for the plaintiff to establish not just the fact of damage, but also that the damage is as a consequence of the breach of duty. This takes two forms. Firstly, it must be clear that the breach caused the damage, and then it must also be shown that the end result is not too remote a consequence from the viewpoint of the breach of duty.

Causation

What should be a simple question of fact has, like much else in negligence, become more complicated in recent years. The basic idea of causation is that it is necessary to show that but for the breach of duty, the damage that is the subject of the claim would not have occurred. However, application of this simple approach is hampered by the range of different causes that might be applicable. For instance, if an ordinary dwelling-house starts to develop cracks, a whole range of different causes may be possible. Obviously, poor workmanship by the builder will be one possible explanation, but equally the builder may have been loyally following a poor design. Other possible explanations will of course be quite independent of negligence – weather or soil conditions, creeping tree roots, etc. So separating out the true cause from all the other possibilities can be a complex task.

The basic test for causation is a simple factual one. The connection between breach and damage must be proven. Sometimes this cannot be done.

BARNETT v. CHELSEA AND KENSINGTON
HOSPITAL MANAGEMENT COMMITTEE

[1969] 1 QB 428   Court of Appeal

The plaintiff nightwatchman was attacked by robbers and treated at a hospital. Subsequently his condition deteriorated, he returned to the hospital but was refused treatment. This was regarded as a negligent failure to treat the plaintiff. However the plaintiff's claim failed on causation grounds.

NIELD J (at p. 438):

It remains to consider whether it is shown that the deceased's death was caused by that negligence or whether, as the defendants have said, the deceased must have died in any event. In his concluding submission [counsel for Barnett] submitted that the casualty officer should have examined the deceased and had he done so he would have caused tests to be made which would have indicated the treatment required and that, since the defendants were at fault in these respects, therefore the onus of proof passed to the defendants to show that the appropriate treatment would have failed, and authorities were cited to me. I find myself unable to accept that argument, and I am of the view that the onus of proof remains upon the plaintiff, and I have in mind (without quoting it) the decision cited by [counsel for the defendants] in *Bonnington Castings Ltd* v. *Wardlaw* [1956] 1 AC 613. However, were it otherwise and the onus did pass to the defendants, then I would find that they have discharged it, as I would proceed to show.

There has been put before me a timetable which I think is of much importance. The deceased attended at the casualty department at five or ten minutes past eight in the morning. If the casualty officer had got up and dressed and come to see the three men and examined them and decided to admit them, the deceased (and Dr Lockett [an expert witness] agreed with this) could not have been in bed in a ward before 11 a.m. I accept Dr Goulding's evidence [an expert witness] that an intravenous drip would not have been set up before 12 noon, and if potassium loss was suspected it could not have been discovered until 12.30 p.m. Dr Lockett, dealing with this, said: 'If this man had not been treated until after 12 noon the chances of survival were not good.'

Without going into detail into the considerable volume of technical evidence which has been put before me, it seems to me to be the case that when death results from arsenical poisoning it is brought about by two conditions; on the one hand dehydration and on the other disturbance of the enzyme processes. If the principal condition is one of enzyme disturbance – as I am of the view it was here – then the only method of treatment which is likely to succeed is the use of the specific antidote which is commonly called B.A.L. Dr Goulding said in the course of his evidence:

'The only way to deal with this is to use the specific B.A.L. I see no reasonable prospect of the deceased being given B.A.L. before the time at which he died' – and at a later point in his evidence – 'I feel that even if fluid loss had been discovered death would have been caused by the enzyme disturbance. Death might have occurred later.'

I regard that evidence as very moderate, and it might be a true assessment of the situation to say that there was no chance of B.A.L. being administered before the death of the deceased.

For those reasons, I find that the plaintiff has failed to establish, on the balance of probabilities, that the defendants' negligence caused the death of the deceased.

This shows the operation of the traditional 'but for' test. The hapless plaintiff did not die as a result of the negligent non-treatment; it could not be said that 'but for' the negligent non-treatment, he would have remained alive. As a result, the plaintiff could not establish the base-line of causation. The case is also clear in insisting that the plaintiff has the onus of proving that there has been congestion. This has been an unsettled point at times, but has recently been reasserted by the House of Lords.

WILSHER v. ESSEX AREA HEALTH AUTHORITY

[1988] AC 1074   House of Lords

This is the next stage of the litigation noted above at pp. 45. The sole issue for the House of Lords to consider was whether the appropriate test had been used in deciding whether the alleged negligence of the defendants had caused the blindness of the prematurely born plaintiff, given that there were several other possible explanations for the blindness that did not involve negligence. The lower courts had used the test that the causation principle was satisfied if there was any increase in the risk of injury as a result of the negligence, stemming from *McGhee* v. *National Coal Board* [1973] 1 WLR 1.

LORD BRIDGE OF HARWICH (at p. 1090):

The conclusion I draw . . . is that *McGhee* v. *National Coal Board* [1973] 1 WLR 1 laid down no new principle of law whatever. On the contrary, it affirmed the principle that the onus of proving causation lies on the pursuer or plaintiff. Adopting a robust and pragmatic approach to the undisputed primary facts of the case, the majority concluded that it was a legitimate inference of fact that the defenders' negligence had materially contributed to the pursuer's injury. The decision, in my opinion, is of no greater significance than that and to attempt to extract from it some esoteric principle which in some way modifies, as a matter of law, the nature of the burden of proof of causation which a plaintiff or pursuer must discharge once he has established a relevant breach of duty is a fruitless one.

In the Court of Appeal in the instant case Sir Nicolas Browne-Wilkinson VC, being in a minority, expressed his view on causation with understandable caution. But I am quite unable to find any fault with the following passage in his dissenting judgment [1987] QB 730, 779:

'To apply the principle in *McGhee* v. *National Coal Board* to the present case would constitute an extension of that principle. In the *McGhee* case there was no doubt that the pursuer's dermatitis was physically caused by brick dust: the only question was whether the continued presence of such brick dust on the pursuer's skin after the time when he should have been provided with a shower caused or materially contributed to the dermatitis which he contracted. There was only one possible agent which could have caused the dermatitis, viz., brick dust, and there was no doubt that the dermatitis from which he suffered was caused by that brick dust.

'In the present case the question is different. There are a number of different agents which could have caused the RLF. Excess oxygen was one of them. The defendants failed to take reasonable precautions to prevent one of the possible causative agents (e.g. excess oxygen) from causing RLF. But no one can tell in this case whether excess oxygen did or did not cause or contribute to the RLF suffered by the plaintiff. The plaintiff's RLF may have been caused by some completely different agent or agents, e.g. hypercarbia, intraventricular haemorrhage, apnoea or patent ductus arteriosus. In addition to oxygen, each of those conditions has been implicated as a possible cause of RLF. This baby suffered from each of those conditions at various times in the first two months of his life. There is no satisfactory evidence that excess oxygen is more likely than any of those other four candidates to have caused RLF in this baby. To my mind, the occurrence of RLF following a failure to take a necessary precaution to prevent excess oxygen causing RLF provides no evidence and raises no presumption that it was excess oxygen rather than one or more of the four other possible agents which caused or contributed to RLF in this case.

'The position, to my mind, is wholly different from that in the *McGhee* case where there was only one candidate (brick dust) which could have caused the dermatitis, and the failure to take a precaution against brick dust causing dermatitis was followed by dermatitis caused by brick dust. In such a case, I can see the common sense, if not the logic, of holding that, in the absence of any other evidence, the failure to take the precaution caused or contributed to the dermatitis. To the extent that certain members of the House of Lords decided the question on inferences from evidence or presumptions, I do not consider that the present case falls within their reasoning. A failure to take preventative measures against one out of five possible causes is no evidence as to which of those five caused the injury.'

The case was then ordered to be re-heard, using the reasserted approach to causation. It is clear that the case increased the burden on plaintiffs in construction cases, who must establish, and not merely assert, a material link between breach and damage. The risk of damage is not enough; material proof of its cause must be established.

Remoteness of damage

The simple 'but-for' test of causation has severe limitations. Great long chains of events can be traced back to a distant cause, in terms of pure logic, which make little sense of where the issue of real responsibility falls to be allocated. So the tort of negligence has had to devise methods of limiting liability. For a long period, liability was limited by the decision in *The Polemis* [1921] 3 KB 560 to damage that was a direct consequence of the breach, but this could give quirky results. These were corrected by the Privy Council and the current formula is to restrict remote claims with reference to the concept of foreseeability of damage.

WAGON MOUND (No. 1)

[1961] AC 388   Privy Council

An oil spill during the refuelling of a ship spread to a nearby repair wharf. In spite of advice that no danger would arise from the continuation of welding operations, a spark ignited the oil, and the resulting fire caused considerable damage. It was held that, on the basis of its unforeseeability, the damage did not give rise to liability.

VISCOUNT SIMONDS (at p. 422):

> Enough has been said to show that the authority of *Polemis* [1921] 3 KB 560 has been severely shaken though lip-service has from time to time been paid to it. In their Lordships' opinion it should no longer be regarded as good law. It is not probable that many cases will for that reason have a different result, though it is hoped that the law will be thereby simplified, and that in some cases, at least, palpable injustice will be avoided. For it does not seem consonant with current ideas of justice or morality that for an act of negligence, however slight or venial, which results in some trivial foreseeable damage the actor should be liable for all consequences however unforeseeable and however grave, so long as they can be said to be 'direct'. It is a principle of civil liability, subject only to qualifications which have no present relevance, that a man must be considered to be responsible for the probable consequences of his act. To demand more of him is too harsh a rule, to demand less is to ignore that civilised order requires the observance of a minimum standard of behaviour.

This concept applied to the slowly developing law of negligence has led to a great variety of expressions which can, as it appears to their Lordships, be harmonised with little difficulty with the single exception of the so-called rule in *Polemis*. For, if it is asked why a man should be responsible for the natural or necessary or probable consequences of his act (or any other similar description of them) the answer is that it is not because they are natural or necessary or probable, but because, since they have this quality, it is judged by the standard of the reasonable man that he ought to have foreseen them. Thus it is that over and over again it has happened that in different judgments in the same case, and sometimes in a single judgment, liability for a consequence has been imposed on the ground that it was reasonably foreseeable or, alternatively, on the ground that it was natural or necessary or probable. The two grounds have been treated as coterminous, and so they largely are. But, where they are not, the question arises to which the wrong answer was given in *Polemis*. For, if some limitation must be imposed upon the consequences for which the negligent actor is to be held responsible – and all are agreed that some limitation there must be – why should that test (reasonable foreseeability) be rejected which, since he is judged by what the reasonable man ought to foresee, corresponds with the common conscience of mankind, and a test (the 'direct' consequence) be substituted which leads to nowhere but the never-ending and insoluble problems of causation.

(at p. 425):

It is, no doubt, proper when considering tortious liability for negligence to analyse its elements and to say that the plaintiff must prove a duty owed to him by the defendant, a breach of that duty by the defendant, and consequent damage. But there can be no liability until the damage has been done. It is not the act but the consequences on which tortious liability is founded. Just as (as it has been said) there is no such thing as negligence in the air, so there is no such thing as liability in the air. Suppose an action brought by A for damage caused by the carelessness (a neutral word) of B, for example, a fire caused by the careless spillage of oil. It may, of course, become relevant to know what duty B owed to A, but the only liability that is in question is the liability for damage by fire. It is vain to isolate the liability from its context and to say that B is or is not liable, and then to ask for what damage he is liable. For his liability is in respect of that damage and no other. If, as admittedly it is, B's liability (culpability) depends on the reasonable foreseeability of the consequent damage, how is that to be determined except by the foreseeability of the damage which in fact happened – the damage in suit? And, if that damage is unforeseeable so as to displace liability at large, how can the liability be restored so as to make compensation payable?

(at p. 426):

Their Lordships conclude this part of the case with some general observations. They have been concerned primarily to displace the proposition that unforeseeability is irrelevant if damage is 'direct'. In doing so they have inevitably insisted that the essential factor in determining liability is whether the damage is of such a kind as the reasonable man should have foreseen. This accords with the general view thus stated by Lord Atkin in *Donoghue* v. *Stevenson* [1932] AC 562: 'The liability for negligence, whether you style it such or treat it as in other systems as a species of "culpa", is no doubt based upon a general public sentiment of moral wrongdoing for which the offender must pay.' It is a departure from this sovereign principle if liability is made to depend solely on the damage being the 'direct' or 'natural' consequence of the precedent act. Who knows or can be assumed to know all the processes of nature? But if it would be wrong that a man should be held liable for damage unpredictable by a reasonable man because it was 'direct' or 'natural', equally it would be wrong that he should escape liability, however 'indirect' the damage, if he foresaw or could reasonably foresee the intervening events which led to its being done: cf. *Woods* v. *Duncan* [1946] AC 401. Thus foreseeability becomes the effective test.

This decision represents an important and largely successful attempt to harmonize the test for remoteness of damage with the tests of foreseeability employed elsewhere in negligence. However, its subsequent acceptance into the mainstream of English law has seen it come to represent very little by way of restriction on plaintiff's claims. Indeed, the courts were very quick to emphasize how flexible the new approach should be.

HUGHES v. THE LORD ADVOCATE

[1963] AC 837   House of Lords

A manhole was left open and unguarded by Post Office workers. Two small boys entered the manhole aided by the gas lamps left by the workman. They safely regained the surface but then the lamp fell back into the hole, an explosion occurred, and the appellant then also fell into the hole, and was injured. His claim succeeded.

LORD PEARCE (at p. 857):

The defenders are therefore liable for all the foreseeable consequences of their neglect. When an accident is of a different type and kind from anything that a defender could have foreseen he is not liable for it (see *The Wagon Mound* [1961] AC 388). But to demand too great precision in the test of foreseeability would be unfair to the pursuer since the facets of misadventure are innumerable (see *Miller* v. *South of Scotland Electricity Board* [1958] SC (HL) 20; *Harvey* v. *Singer Manufacturing Co. Ltd* [1960] SC 155). In the case of an allurement to children it is particularly hard to foresee with precision the exact shape of the disaster that will arise. The allurement in this case was the combination of a red paraffin lamp, a ladder, a partially closed tent, and a cavernous hole within it, a setting well fitted to inspire some juvenile adventure that might end in calamity. The obvious risks were burning and conflagration and a fall. All these in fact occurred, but unexpectedly the mishandled lamp instead of causing an ordinary conflagration produced a violent explosion. Did the explosion create an accident and damage of a different type from the misadventure and damage that could be foreseen? In my judgment it did not. The accident was but a variant of the foreseeable. It was, to quote the words of Denning LJ in *Roe* v. *Minister of Health* [1954] 2 QB 66 'within the risk created by the negligence'. No unforeseeable, extraneous, initial occurrence fired the train. The children's entry into the tent with the ladder, the descent into the hole, the mishandling of the lamp, were all foreseeable. The greater part of the path to injury had thus been trodden, and the mishandled lamp was quite likely at that stage to spill and cause a conflagration. Instead, by some curious chance of combustion, it exploded and no conflagration occurred, it would seem, until after the explosion. There was thus an unexpected manifestation of the apprehended physical dangers. But it would be, I think, too narrow a view to hold that those who created the risk of fire are excused from the liability for the damage by fire because it came by way of explosive combustion. The resulting damage, though severe, was not greater than or different in kind from that which might have been produced had the lamp spilled and produced a more normal conflagration in the hole.

Neither the precise method whereby the harm is caused, or its extent, have to be foreseen, so long as the general type of damage is foreseeable.

BRADFORD v. ROBINSON RENTALS

[1967] 1 WLR 337 High Court

The plaintiff was ordered to drive his employer's unheated van from Exeter to Bedford and back in exceptionally severe winter weather. He suffered frostbite, and claimed successfully.

REES J (at p. 344):

So far as the principles of law applicable to this case are concerned, they may be shortly stated. The defendants, as the plaintiff's employers, were under a duty at common law to take reasonable steps to avoid exposing the plaintiff to a reasonably foreseeable risk of injury.

It was strongly argued on behalf of the defendants that injury to his health suffered by the plaintiff in this case by 'frost-bite' or cold injury was not reasonably foreseeable. There was no evidence that before the plaintiff started the journey either the plaintiff himself or the defendants' servants, the Exeter branch manager or the senior engineer, actually contemplated that the plaintiff might suffer from 'frost-bite' if he were required to carry out the journey. I am, however, satisfied that any reasonable employer in possession of all the facts known to the branch manager and senior engineer on 8 January would have realised – and they must have realised – that if the plaintiff was required to carry out the journey he would certainly be subjected to a real risk of some injury to his health arising from prolonged exposure to an exceptional degree of cold.

No doubt the kinds of injury to health due to prolonged exposure to an exceptional degree of cold are commonly thought to include, for example, that the victim might suffer from a common cold or in a severe case from pneumonia, or that he might suffer from chilblains on his hands and feet. The question I have to consider is whether the plaintiff has established that the injury to his health by 'frost-bite' (and I use the lay term for convenience), which is admittedly unusual in this country, is nevertheless of the type and kind of injury which was reasonably foreseeable.

The law does not require that the precise nature of the injury must be reasonably foreseeable before liability for its consequences is attributed. The point is thus dealt with in a convenient way in Salmond on Torts, 14th ed. (1965) at p. 719:

> '(i) *Type of damage must be foreseen.* It has been made plain that the precise details of the accident, or the exact concatenation of circumstances, need not be foreseen. It is sufficient if the type, kind, degree or order of harm could have been foreseen in a general way. The question is, was the accident a variant of the perils originally brought about by the defendant's negligence? The law of negligence has not been fragmented into a number of distinct torts.'

That extract conveniently states the principle to be followed in the present case since *The Wagon Mound* [1961] AC 388.

Valuable statements of the same principle are to be found in *Hughes* v. *Lord Advocate* [1963] AC 837. That case related not to the duty owed by an employer to his employee but to the duty owed by the Post Office to children allured to an open manhole in a carriageway protected by a canvas shelter containing some lighted lamps. Two children played with the lamps, of which one fell into the manhole and caused an explosion, thus severely burning one of the children. It was contended that although some injury by burning might have been foreseen, yet the explosion causing the massive burning could not have been reasonably foreseen. Lord Jenkins said:

> 'It is true that the duty of care expected in cases of this sort is confined to reasonably foreseeable dangers, but it does not necessarily follow that liability is escaped because the danger actually materialising is not identical with the danger reasonably foreseen and guarded against. Each case much depends on its own particular facts.'

And Lord Guest said:

> 'In order to establish a coherent chain of causation it is not necessary that

the precise details leading up to the accident should have been reasonably foreseeable: it is sufficient if the accident which occurred is of a type which should have been foreseeable by a reasonably careful person . . .'

Now what were the facts known to the defendants through the manager and the senior engineer which would reasonably lead them to foresee injury to the health of the plaintiff by exposure to cold? My findings of fact may be summarised thus. They knew that the weather during 9 and 10 January was likely to be severe with temperatures at or about freezing point with ice and snow on the roads; that the round journey of about 450 to 500 miles in two days would be likely to involve the plaintiff in at least 20 hours of driving, when the old Austin van was unheated and the new Austin van was also likely to be unheated, so that certainly the senior engineer should know that the plaintiff would be required to drive with a window open in both directions; that the defective radiator and the lack of anti-freeze liquid in the old Austin meant that the plaintiff, on the outward journey, would be required to stop and add water at intervals which might well be frequent; that the plaintiff was being sent on a task wholly outside his normal daily duties which involved frequent stops at customers' houses. From all these facts it is plain, in my judgment, that the defendants knew that the plaintiff was being called upon to carry out an unusual task which would be likely to expose him for prolonged periods to extreme cold and considerable fatigue. They also knew that the plaintiff, then aged about 57 years and whom they rightly esteemed as a sensible and conscientious man, took and vehemently expressed the view that the journey was hazardous and ought not to be undertaken by him.

I have taken into account the evidence of the expert witness called by the defendants who said that, up to the year 1963, only about 10 per cent of the commercial vehicles were fitted with heaters, although about 40 to 50 is the present percentage, and that during the snow-bound period which included January 1963, commercial vehicles did continue to use the roads.

In all these circumstances I hold that the defendants did, by sending the plaintiff out on this journey, expose him to a reasonably foreseeable risk of injury arising from exposure to severe cold and fatigue. That breach of duty caused the plaintiff to suffer from 'frost-bite' or cold injury with serious consequences. Even if there had been – and there is not – evidence that the plaintiff was abnormally susceptible to 'frost-bite' as opposed to the more common sequels of prolonged exposure to severe cold and fatigue, he would be entitled to succeed on the footing that a tortfeasor must take his victim as he finds him (see *Smith* v. *Leech Brain & Co. Ltd* [1962] 2 QB 405).

In most cases of building negligence, the issue of foreseeability of damage is not likely to arise in such a clear form. The more likely situation where it might be at issue is in relation to the financial costs arising as a consequence of the negligent building work. A plaintiff whose house is unsafe may seek to be compensated for moving into a nearby luxury hotel; a business whose factory has to be shut may incur huge expenses on alternative premises or urgent imports.

The normal principle in assessing the measure of damages in tort also fails to address the problem. The classic statement here is that of Lord Wright in *Liesbosch Dredger* v. *SS Edison (Owners)* [1933] AC 449 at 459 (see also below) where he said that recovery was allowed of 'such a sum as will replace (him) so far as can be done by compensation in money, in the same position as if the loss had not been inflicted'. However, here too the 'but-for' test does not give a sensible answer since but for the tort, the losses incurred would not have been suffered.

Instead, the courts have come to recognize that the contractual principle of mitigation of damage applies equally in tort cases, and excessive claims made by the plaintiff are regarded as being caused by his own acts, and not by the defendants, as shown by *Darbishire* v. *Warren* [1963] 1 WLR 1067 where it was held unreasonable to spend more on repairing a car than the car itself was worth.

Particular problems occur when the plaintiff's losses are worsened as a result of his own financial problems which render him unable to respond adequately to the damage, and thus worsen it. The traditional view has been a harsh one.

LIESBOSCH DREDGER v. SS EDISON (OWNERS)

[1933] AC 449   House of Lords

The plaintiff's dredger was sunk by the defendant's cargo ship. At the time it was carrying out contract works. In order to avoid heavy penalties if the contract was delayed, and because of their inability to purchase another dredger, they had to hire another dredger, at great expense.

LORD WRIGHT (at p. 460):

> I think it desirable to examine the claim made by the appellants, which found favour with the Registrar and Langton J, and which in effect is that all their circumstances, in particular their want of means, must be taken into account and hence the damages must be based on their actual loss, provided only that, as the Registrar and the judge have found, they acted reasonably in the unfortunate predicament in which they were placed, even though but for their financial embarrassment they could have replaced the *Liesbosch* at a moderate price and with comparatively short delay. In my judgment the appellants are not entitled to recover damages on this basis. The respondents' tortious act involved the physical loss of the dredger; that loss must somehow be reduced to terms of money. But the appellants' actual loss in so far as it was due to their impecuniosity arose from that impecuniosity as a separate and concurrent cause, extraneous to and distinct in character from the tort; the impecuniosity was not traceable to the respondents' acts, and in my opinion was outside the

legal purview of the consequences of these acts. The law cannot take account of everything that follows a wrongful act; it regards some subsequent matters as outside the scope of its selection, because 'it were infinite for the law to judge the cause of causes', or consequences of consequences. Thus the loss of a ship by collision due to the other vessel's sole fault, may force the shipowner into bankruptcy and that again may involve his family in suffering, loss of education or opportunities in life, but no such loss could be recovered from the wrongdoer. In the varied web of affairs, the law must abstract some consequences as relevant, not perhaps on grounds of pure logic but simply for practical reasons. In the present case if the appellants' financial embarrassment is to be regarded as a consequence of the respondents' tort, I think it is too remote, but I prefer to regard it as an independent cause, though its operative effect was conditioned by the loss of the dredger.

However, this case fell to be re-examined in a useful construction case late in 1979.

DODD PROPERTIES LTD v. CANTERBURY CITY COUNCIL

[1980] 1 WLR 433 Court of Appeal

The plaintiffs' garage was damaged by pile-driving carried out on the defendants' nearby site. Repairs in 1970 would have cost a total of £15,783, but no repairs had been done by the time of the trial in 1978, when the cost of repair had escalated to £42,278. The plaintiffs' lack of resources to carry out this work was a major factor in this delay. At first instance, Cantley J awarded the lower figure, but the Court of Appeal increased the award to the higher figure.

MEGAW LJ (at p. 450):

The general principle, referred to in many authorities, has recently been recognised by Lord Wilberforce, in *Miliangos* v. *George Frank (Textiles) Ltd* [1976] AC 443, 468, namely, that '. . . as a general rule in English law damages for tort or for breach of contract are assessed as at the date of the breach . . .'. But in the very passage in which this 'general rule' is there stated, it is stressed that it is not a universal rule. That it is subject to many exceptions and qualifications is clear. Cantley J in the present case rightly recognised that that was so, in the passage from his judgment which I have recently read.

Indeed, where, as in the present case, there is serious structural damage to a building, it would be patently absurd, and contrary to the general principle on which damages fall to be assessed, that a plaintiff, in a time of rising prices, should be limited to recovery on the basis of the prices of repair at the time of wrongdoing, on the facts here being two years, at least, before the time when, acting with all reasonable speed, he could first have been able to put the repairs in hand. Once that is accepted, as it must be, little of practical reality remains in postulating that, in a tort such as this, the 'general rule' is applicable. The damages are not required by English law to be assessed as at the date of breach.

The true rule is that, where there is a material difference between the cost of repair at the date of the wrongful act and the cost of repair when the repairs can, having regard to all relevant circumstances, first reasonably be undertaken, it is the latter time by reference to which the cost of repair is to be taken in assessing damages. That rule conforms with the broad and fundamental principle as to damages, as stated in Lord Blackburn's speech in *Livingstone* v. *Rawyards Coal Co.* (1880) 5 App Cas 25, 39, where he said that the measure of damages is

'... that sum of money which will put the party who has been injured, or who has suffered, in the same position as he would have been in if he had not sustained the wrong for which he is now getting his compensation or reparation'.

In any case of doubt, it is desirable that the judge, having decided provisionally as to the amount of damages, should, before finally deciding, consider whether the amount conforms with the requirement of Lord Blackburn's fundamental principle. If it appears not to conform, the judge should examine the question again to see whether the particular case falls within one of the exceptions of which Lord Blackburn gave examples, or whether he is obliged by some binding authority to arrive at a result which is inconsistent with the fundamental principle. I propose to carry out that exercise later in this judgment.

Cantley J has held, in a passage which I have already read, that as a commercial decision, judged exclusively from the plaintiffs' point of view, it was reasonable to postpone incurring expense of the repairs up to – for so I understand what Cantley J says – the time when the action had been heard and liability decided, resulting in a judgment which, when complied with, would have put the plaintiffs in funds. The reasons why that deferment of repairs was reasonable from the plaintiffs' point of view included the fact, not that they were 'impecunious', meaning poverty-stricken or unable to raise the necessary money, but that the provision of the money for repairs would have involved for them a measure of 'financial stringency'.

(and at p. 453):

It was not 'financial stringency', let alone 'impecuniousness' as in *The Liesbosch*, which on any fair view, on Cantley J's findings, was *the* cause, or even, I think, an effective cause, of the decision to postpone repairs. The 'financial stringency' which would have been created by carrying out the repairs was merely one factor among a number of factors which together produced the result that commercial good sense pointed towards deferment of the repairs. The second reason which I would mention is that, once it is accepted that the plaintiff was not in any breach of any duty owed by him to the defendant in failing to carry out repairs earlier than the time when it was reasonable for the repairs to be put in hand, this becomes, for all practical purposes, if not in theory, equated with a plaintiff's ordinary duty to mitigate his damages.

This extract confirms both a general and a narrow point. The general point is that the normal, though not invariable rule at the time of the breach of duty in premises litigation is that damages are awarded on the basis of the costs of repair work necessary to put the plaintiff back in the position he would have been in had the tort not occurred. However, this principle, especially in relation to the time of the calculation, can be varied by reference to reasonably foreseeable circumstances. Here, it made sound commercial sense to allow the plaintiffs to wait until the time of trial, given the defendants' obdurate refusal to accept any liability and the plaintiffs' 'financial stringency'. This latter phrase is likely to become a neat way of defusing the *Liesbosch* principle, which will be confined to cases of the most extreme and unforeseeable 'impecuniosity'.

Negligence – summary

The tort of negligence, as can be seen from this brief overview has recently undergone considerable major changes. However, the duty stage, for all the recent upheaval, still remains broadly favourable to plaintiffs in straightforward cases of direct physical harm, though equally plaintiffs suffering harm indirectly, especially if the losses fall within the category of those now defined as economic in nature, will face a much harder struggle to show that there is a duty of care. However, the next stage, the key question of breach of duty, is broadly much more favourable to defendants, especially those professing special skills. This by the way questions the necessity for all the recent curtailment of duty of care, with the aim of helping defendants as a class be free from the burden of liability, since the breach test would often have the same effect in most cases.

Similarly, the elements that go to make up 'damage' have different effects. The tightening up of the causation test tends to make the plaintiff's task more difficult; the very loose foreseeability test tends to make it easier, though in the majority of cases the fact of damage and its cause will be readily apparent.

As can be seen from keynote cases such as *Anns* and *Murphy*, the law of negligence today has itself been shaped by construction disputes. Equally, the broad ambit of the tort means that cases from every quarter of human activity have come to shape the general principles of negligence that permeate the construction indusry in so many different ways.

PART B
General Issues in the Law of Tort

The purpose of this part of the book is to point to several areas of law which affect the application of the tort of negligence to all aspects of construction litigation. If they have a common feature it is that they tend to raise the question of the way in which negligence liability is affected by the presence of contractual arrangements. Many potential parties to negligence litigation are in a contractual relationship; many others such as sub-contractors, though not privy to the same contract, are nevertheless connected within a web of contracts. A builds a property for B and B then sells it to C; this is a classic situation where A and C are not in a contractual relationship, so C must sue in tort, but where the presence of the respective contractual arrangements may still affect the conduct of the litigation.

Chapter 4

Interaction of Tort and Contract

Can parties to a contract sue in tort?

The logical first question is whether the mere presence of an A–B contract prevents an A–B tort action. For years, the law on what may be thought to be an essential point has fluctuated and today no clear answer may be given.

BAGOT v. STEVENS SCANLAN

[1966] 1 QB 197   High Court

The defendants were a firm of architects supervising the laying of a drainage system on the land of the plaintiff. The parties were in a contractual relationship, and this was held to preclude the operation of the law of tort.

DIPLOCK LJ (at p. 204):

> It seems to me that, in this case, the relationship which created the duty of exercising reasonable skill and care by the architects to their clients arose out of the contract and not otherwise. The complaint that is made against them is of a failure to do the very thing which they contracted to do. That was the relationship which gave rise to the duty which was broken. It was a contractual relationship, a contractual duty, and any action brought for failure to comply with that duty is, in my view, an action founded on contract. It is also, in my view, an action founded upon contract alone.

An unequivocal message! But one which, though based on much clear authority such as *Groom* v. *Crocker* [1939] 1 KB 194, was already beginning to appear out of date. For in *Hedley Byrne* v. *Heller* [1964] AC 465 (see pp. 19, above), Lord Morris of Borth-y-Gest had stated that the new liability for negligent misstatements was one that arose 'quite irrespective of contract' (at p. 502), and this more flexible view of the interrelationship of the two forms of obligation came to take greater prominence, culminating in a careful analysis by Oliver J (as he then was).

MIDLAND BANK TRUST CO. v. HETT, STUBBS & KEMP

[1979] Ch 384   High Court

A father agreed to grant to his son an option on land (i.e. a right to acquire it in the future). They saw their solicitors, the defendants, who drafted the relevant documents but, allegedly negligently, failed to register it, so that when father and son had a disagreement, the former was able to cancel the option. It was held that an action could be both in tort and in contract on these facts.

OLIVER J (at p. 411):

> Now if there was, in the House of Lords' decision in *Hedley Byrne* [1964] AC 465, anything that was inconsistent with and destroyed the reasoning upon which *Groom* v. *Crocker* [1939] 1 KB 194 was based, I should not, I apprehend, be any longer bound by it nor bound to follow those subsequent cases at first instance which nevertheless applied it – and that would include even a case carrying the high authority of Diplock LJ, who, in *Bagot* v. *Stevens Scanlan & Co. Ltd* [1966] 1 QB 197, was sitting as an additional judge of the Queen's Bench Division. I have to confess that, unassisted by prior authority, I think that I should have felt myself compelled to the conclusion that the ratio of *Groom* v. *Crocker* [1939] 1 KB 94 – which, of course, was decided before the doctrine of tortious negligence arising from special relationships enunciated in the *Hedley Byrne* case had been fully developed – could not stand alongside that overriding decision of the House of Lords.
>
> The principle was stated by Lord Morris of Borth-y-Gest as a perfectly general one and it is difficult to see why it should be excluded by the fact that the relationship of dependence and reliance between the parties is a contractual one rather than one gratuitously assumed, in the absence, of course, of contractual terms excluding or restricting the general duties which the law implies. Logically, as it seems to me, this could be so only if there is read into every contract not only an implied term to employ reasonable care and skill in the performance of the contract, but a further term to the effect that the contract shall be the conclusive and exclusive source of all duties owed by one party to the other to the exclusion of any further or more extensive duties which the general law would otherwise impose. Lord Morris of Borth-y-Gest expressed the principle when he said, at pp. 502–3:
>
> > 'My Lords, I consider that it follows and that it should now be regarded as

settled that if someone possessed of a special skill undertakes, quite irrespective of contract, to apply that skill for the assistance of another person who relies upon such skill, a duty of care will arise. The fact that the service is to be given by means of or by the instrumentality of words can make no difference. Furthermore, if in a sphere in which a person is so placed that others could reasonably rely upon his judgment or his skill or upon his ability to make careful inquiry, a person takes it upon himself to give information or advice to, or allows his information or advice to be passed on to, another person who, as he knows or should know, will place reliance upon it, then a duty of care will arise.'

Now this was a perfectly general statement of principle and even if it be treated as qualified to the extent indicated in the majority view of the Board in *Mutual Life and Citizens' Assurance Co. Ltd* v. *Evatt* [1971] AC 793 there seems to me to be no ground for confining its operation to non-contractual relationships. Indeed, the exposition of the principle in the majority judgment of Lord Diplock in that case at pp. 801–3, suggests just the contrary, for Lord Diplock there ascribes the origin of the principle to the duty imposed by the law by reason merely of the carrying on of particular professions or trades and summarised in the maxim 'spondet peritiam artis et imperitia culpae adnumeratur'.

That there is no such restriction is, I think, clear not only from the speech of Lord Morris of Borth-y-Gest but from other speeches as well.

(and at p. 416):

Nor, if I may respectfully say so, do I follow the argument advanced by counsel in *Bagot* v. *Stevens Scanlan & Co. Ltd* [1966] 1 QB 197, 200, that the cause of action there was necessarily in contract alone because the architects in that case had 'failed to do the very thing which they contracted to do'. Well, so they had, but the form of the breach cannot affect the nature of the duty, nor does an obligation imposed by law become an obligation different in quality simply because the obligee agrees to accept money for its performance.

At p. 419, Oliver J went on to dismiss arguments that there is a general implied term in every solicitor's contract or a general principle of law that operates to exclude tort as between parties to a contract.

Contractual relationships are a matter of agreement in individual cases and, apart from statute, I know of no authority for importing into contracts some universal term which applies whether the parties could be contemplated as intending it or not . . . There is not and never has been any rule of law that a person having alternative claims must frame his action in one or the other. If I have a contract with my dentist to extract a tooth, I am not thereby precluded from suing him in tort if he negligently shatters my jaw: *Edwards* v. *Mallan* [1908] 1 KB 1002; nor does the contractual duty assumed by a local authority to answer inquiries prevent its being sued both in tort and in contract if it does so carelessly.

These are simply the crucial points of thirty pages of careful discussion of this issue, and most commentators thought that the issue was settled. A simple view might be that much energy had been wasted on drafting clauses in contracts excluding the operation of the tort of negligence (see pp. 92, below) if the mere fact of the contract itself precluded any role for the tort.

However, this area of the law was rocked to its core by subsequent comments of Lord Scarman, at a time when the expanded role of negligence was increasingly being questioned.

TAI HING COTTON MILL LTD v. LIU CHONG HING BANK LTD

[1986] AC 80   Privy Council

A company's tax accounting procedures enabled its (hitherto) trusted accounts clerk to embezzle $HK 5.5 million. The company sued its bankers, who counterclaimed that the company was in breach both of contract and of its tortious duty of care. It was held that the company was under no contractual duty to its bank and, therefore, under no duty in tort either.

LORD SCARMAN (at p. 107):

> Their Lordships do not believe that there is anything to the advantage of the law's development in searching for a liability in tort where the parties are in a contractual relationship. This is particularly so in a commercial relationship. Though it is possible as a matter of legal semantics to conduct an analysis of the rights and duties inherent in some contractual relationships including that of banker and customer either as a matter of contract law when the question will be what, if any, terms are to be implied or as a matter of tort law when the task will be to identify a duty arising from the proximity and character of the relationship between the parties, their Lordships believe it to be correct in principle and necessary for the avoidance of confusion in the law to adhere to the contractual analysis: on principle because it is a relationship in which the parties have, subject to a few exceptions, the right to determine their obligations to each other, and for the avoidance of confusion because different consequences do follow according to whether liability arises from contract or tort, e.g. in the limitation of action.

This robust assertion was made without reference to the recent cases to the contrary, and left the courts in a quandary as to which route to follow.

Gradually courts have begun to follow the lead of the *Tai Hing* decision, but only cautiously to date.

NATIONAL BANK OF GREECE SA v. PINIOS SHIPPING (No 1)

[1990] 1 AC 637   Court of Appeal

The plaintiff bank had problems with a shipowner defaulting in making payments on the vessel 'Maira'. It was agreed that agents should manage the vessel, but they failed to adequately insure it. Pinios alleged that the bank was careless in failing to ensure that proper insurance was obtained. No such contractual term was found to exist, and no liability in tort was allowed either.

LLOYD LJ (at p. 650):

> Turning from contract to tort, [counsel for Pinios Shipping] argues strenuously that even if he fails in contract he is entitled to succeed in tort. He relies on *Dorset Yacht Co. Ltd* v. *Home Office* [1970] AC 1004, and the much discussed and increasingly precarious dictum of Lord Wilberforce in *Anns* v. *Merton London Borough Council* [1978] AC 728, 751. But those were cases where there was no contract between the parties. So it was tort, or nothing. Here there is a contract, and a most elaborate contract at that.
>
> Now I accept that in a large class of cases it always was, and maybe still is, possible for the plaintiff to sue either in contract or tort. The obvious example would be actions against innkeepers and the like, and those exercising a common calling. In *Boorman* v. *Brown* (1842) 3 QB 511, 525–6, Tindall CJ, delivering the judgment of the Court of Exchequer Chamber, said:
>
> > 'That there is a large class of cases in which the foundation of the action springs out of privity of contract between the parties, but in which, nevertheless, the remedy for the breach, or non-performance, is indifferently either assumpsit or case upon tort, is not disputed. Such are actions against attorneys, surgeons and other professional men, for want of competent skill or proper care in the service they undertake to render . . . The principle in all these cases would seem to be that the contract creates a duty, and the neglect to perform that duty, or the non-feasance is a ground of action upon a tort.'

In the House of Lords, Lord Campbell said (1844) 11 Cl. & Fin. 1, 44:

> 'wherever there is a contract, and something is to be done in the course of the employment which is the subject of that contract, if there is a breach of duty in the course of that employment, the [party injured] may either recover in tort or in contract.'

See also *Esso Petroleum Co. Ltd* v. *Mardon* [1976] QB 801, 819, *per* Lord Denning MR.

But so far as I know it has never been the law that a plaintiff who has the choice of suing in contract or tort can fail in contract yet nevertheless succeed in tort; and, if it ever was the law, it has ceased to be the law since *Tai Hing Cotton Mill Ltd* v. *Liu Chong Hing Bank Ltd* [1986] AC 80. In that case the bank advanced very much the same argument as had been advanced by [Counsel for Pinios Shipping]. But the argument was rejected.

(and at p. 651):

Nothing in the subsequent cases at first instance on which [counsel for Pinios Shipping] relied throws any doubt on the appropriateness of Lord Scarman's observations to the present case. I would hold without hesitation that if, in a case such as the present, the plaintiff fails in contract, he must necessarily fail in tort.

The position would be different if the contract and the tort lay in different fields. Thus, if, to take a simple example, I give my employee a lift home, and injure him by my careless driving, then obviously he will not be prevented from recovering from me in tort, because of the existence between us of a contract of employment. But that is not this case.

(The case went to the House of Lords on another issue).

So this endorsement of *Tai Hing* is couched in terms of this particular case, where it is important to note that no contractual obligation arose and therefore to impose a tortious obligation would be to extend the liability and to go against the apparent intention of the parties. This use of *Tai Hing* may be less appropriate in the context of a construction contract where obligations do already arise between contracting parties. However, the courts have tended to closely examine the precise ambit of the terms of those contracts, and have tended here too to limit the role of tort.

SURREY HEATH BOROUGH COUNCIL v. LOVELL CONSTRUCTION LTD

(1988) 42 BLR 25   High Court (Official Referee's Business)

The plaintiff engaged the defendant to construct a new office block. It was alleged that careless welding work by the defendant caused a fire which delayed completion. It was held that the terms of the contract did not permit the Council to claim for certain of the losses suffered, and also that a tort claim in respect of these losses would not succeed.

JUDGE FOX-ANDREWS QC (at p. 45):

> [Counsel for the plaintiff] concedes that if a contract expressly deals with the subject matter of a claim there is no room for a parallel duty in tort, but he submits if it does not there is room for a duty in tort.
>
> I am satisfied that the contract makes full provision for all the claims that Surrey Heath seek to advance.
>
> However, even if I am wrong in that finding, I would nevertheless hold that in the circumstances of this particular contract the obligations should be determined solely in a contractual context.

Although the judge's reasoning is not totally clear, it is apparent that he is seeking to avoid filling gaps in the contract with liability in tort. It is understandable to prevent tort in this way running contrary to the express intention of the parties in a case between two businesses, though equally it could be argued that Lovell could have inserted an exclusion clause to positively deal with possible tort liability.

UNIVERSITY COURT OF THE UNIVERSITY OF GLASGOW v. WHITFIELD AND LAING

(1988) 42 BLR 66   High Court (Official Referee's Business)

A new University art gallery began to suffer damage from water ingress. Whitfield had designed the building which was constructed by Laing. The question arose whether the contractors owed an obligation to warn of the design defects which they should have noticed as construction proceeded. No such duty was owed.

JUDGE BOWSHER QC:

> The third party submits further that the third party owed no duty to the plaintiffs in tort relevant to the matters under consideration. [Counsel for Laing] submits that as the third party has entered into a most detailed contract with the plaintiffs it would be wrong to search for a liability in tort, particularly where the attempt to find a liability in tort is made in order to go outside the statutory provisions which are not retrospective. [She] refers to the detailed provisions of the JCT Standard Form of Contract 1963 Edition as revised in 1971 with a Scottish Appendix and incorporating voluminous bills of quantities containing in particular a description of the works making it clear that the third party had no design responsibility under the contract.
>
> [She] relies on *Tai Hing Cotton Mill Limited* v. *Liu Chong Hing Bank Limited* [1986] AC 80 and in particular the speech of Lord Scarman beginning at page 96H and especially at pages 107B–H. [She] relies in particular on the passage:

'Their Lordships do not, however accept that the parties' mutual obligations in tort can be any greater than those to be found expressly or by necessary implication in their contract.'

I accept the submission [by counsel for University Court of the University of Glasgow] in reply to the effect that there can be and very often is concurrent liability both in contract and in tort: *Batty* v. *Metropolitan Property* [1978] QB 554; (1977) 7 BLR 1 and *Midland Bank* v. *Hett Stubbs* [1979] Ch 384. But it does not necessarily follow that the concurrent liabilities in tort and contract are identical and indeed it is unlikely to be so. Here there is a contractual duty governing the relationship between the parties regarding damage arising out of a defective building. As I understand [the] submission [by counsel for Laing], she suggests that while there may be a concurrent duty in tort it is limited to a duty of care to avoid acts or omissions which are liable to cause damage to persons or to some property other than the defective building being created. I think that submission is well founded in the present state of the development of the law. I refer to the discussion of the most recent authorities in the judgment of Bingham LJ in *Simaan General Contracting Company* v. *Pilkington Glass Co.* [1988] 2 WLR 761; 41 BLR 48. I shall mention that decision again at a later point in this judgment.

It follows, since no danger to persons or to other property is alleged, that in support of the claim to contribution, the defendant is unable to rely on any liability in tort owed by the third party to the plaintiff, whether concerning alleged bad workmanship or an alleged duty to warn the plaintiff of bad design.

Even if that broad conclusion were wrong, it follows from *Tai Hing Cotton Mill* v. *Liu Chong Hing Bank*[1986] AC 80 that where there is a detailed contract of the nature found here, there is no room for the implication of a duty to warn about possible defects in design. I should add that I formed that conclusion before recalling a decision much discussed in years past, *Lynch* v. *Thorne* [1956] 1 WLR 303. Since that case does not alter my decision, I have not thought it necessary to invite the parties to present further argument upon it. In that case where there was an express contract as to the way in which a house was to be constructed and the builder had complied with those express terms, it was held that there was no room for any term that the walls would be waterproof. Lord Evershed MR stated at page 306:

'Where there is a written contract expressly setting forth the bargain between the parties it is, as a general rule, also well established that you only imply terms under the necessity of some compulsion.'

If, as I take to be the position in the circumstances of this particular case, there was no room for the implication in the contract of an implied duty to warn the building owner of defects in the architect's design, it follows from the *Tai Hing* decision that there should be no wider duty in tort. If that alternative way of looking at the case were the only way, there would of course remain for consideration a duty in tort regarding bad workmanship, but for the wider reasons given above, I do not regard that way of putting the case as being open to the defendant.

Again the assumption is made that the parties to a construction contract have carefully decided on what liabilities are to arise (which is probably true) and have consciously decided not to create other possible liabilities (which is more controversial). Full consideration of this issue has been given now by the Court of Appeal.

GREATER NOTTINGHAM CO-OPERATIVE SOCIETY LTD v. CEMENTATION PILINGS & FOUNDATIONS LTD

[1989] QB 71   Court of Appeal

Work on the plaintiff's land caused damage to an adjacent restaurant. Having paid compensation, the plaintiff sought to seek an indemnity from their sub-contractors, and also damages in respect of consequent delays. The sub-contractors had provided warranties in a collateral agreement which did not extend to cover these particular losses. No tort claim for these economic losses was permitted.

PURCHAS LJ (at p. 95):

With one vital exception the position in the present case marches step by step with the 'checklist' to be found in the speech of Lord Roskill in *Junior Books Ltd* v. *Veitchi Co. Ltd* [1983] 1 AC 520, 546, namely

(1)   the appellants were nominated sub-contractors;
(2)   the appellants were specialists;
(3)   the appellants were fully appraised of the requirements and responsibilities of the main contract;
(4)   the appellants were solely responsible for the provision of piles;
(5)   the employers relied upon the appellants' skill and experience, saving only the intervention of expert engineers and architects;
(6)   the appellants must have known that the employers relied upon their skill and experience but as regards item;
(7)   in the present case there was actual privity of contract, which was not present in *Junior Books Ltd* v. *Veitchi Co. Ltd*

The last mentioned crucial distinction was relied upon by each party as having a diametrically opposite impact upon liability. [Counsel for Greater Nottingham Co-operative Society Ltd] submitted that the existence of the direct contract made the present case a fortiori to *Junior Books Ltd* v. *Veitchi Co. Ltd*; whereas [counsel for Cementation Pilings & Foundations Ltd] submitted that the limited relationship and liability provided for in the direct contract restricted liability in tort so that the latter was coterminous with the former.

(and at p. 99):

The terms of the direct contractual relationship between the sub-contractors and the employers involve the warranties already set out in this judgment and no other obligations imposed upon the sub-contractors by way of a direct duty towards the employers. In line with the approach of Robert Goff LJ, in *Muirhead* v. *Industrial Tank Specialities Ltd* [1986] QB 507 and that of Bingham LJ in *Simaan General Contracting Co.* v. *Pilkington Glass Ltd (No. 2)* [1988] QB 758 in considering whether there should be a concurrent but more extensive liability in tort as between the two parties arising out of the execution of the contract, it is relevant to bear in mind – (a) the parties had an actual opportunity to define their relationship by means of contract and took it; and (b) that the general contractual structure as between the employers, the main contractors and the sub-contractors as well as the professional advisers provided a channel of claim which was open to the employers such as Bingham LJ mentioned in *Simaan General Contracting Co.* v. *Pilkington Glass Ltd (No. 2)* as being available in that case to the Sheikh. Although this is new ground, doing the best I can to distil from the mass of authorities which have already been considered in detail in the two judgments of Robert Goff LJ and Bingham LJ, I do not believe that it would be in accordance with present policy to extend *Junior Books Ltd* v. *Veitchi Co. Ltd* [1983] 1 AC 520 rather than to restrict it. This does give rise to an apparent inconsistency, namely the effect of enhancing the close relationship upon which Lord Roskill based his duty in tort in *Junior Books Ltd.* v. *Veitchi Co. Ltd* by adding a direct contractual relationship does not confirm a duty to avoid economic loss but negatives that liability. But in this compartment of consideration it is not only the proximity of the relationship giving rise to reliance which is critical but also the policy of the law as to whether or not in these circumstances damages for pecuniary loss ought to be recoverable.

The argument on the other side is, of course, that the collateral contract between the employers and the sub-contractors was restricted to the specific topics therein set out. These, as will appear from the extract already cited in this judgment, placed upon the sub-contractors a contractual duty not to expose the employers to claims under the main contract for extensions of time under clause 23 and reciprocal undertakings by the employers to ensure that the sub-contractors were not prejudiced by any act on their part under the terms of the main contract in clauses 25, 27 and 30. If the parties intended that there should be a contractual restriction of the duty in tort owed by the one to the other it would have been expected that specific provision would be made in the terms of the contract. Bearing in mind the authorities which establish quite clearly that the duty in tort is capable of running coincidentally with liability in contract there is nothing in the terms of the collateral contract to destroy a liability which otherwise would have arisen in tort.

Thus it is said, reverting to Lord Roskill's checklist, the presence of the contract, far from restricting the ambit of the duty in tort to pay damages for pecuniary loss which otherwise would have arisen from the close proximity of the parties and the reliance of the one upon the other, the existence of the contract both establishes that special relationship and was a justification for awarding pecuniary damages where these flowed according to the ordinary criteria of damages in tort from the established negligence of [the negligent builder]. This appears to be a part of the path towards Pandora's Box hitherto untrod.

The arguments are finely balanced; and, not without some hesitation, I have come to the conclusion that [the] submissions [by Counsel for Cementation Pilings and Foundations Ltd] are the more persuasive. In order to establish what might be called the *Hedley Byrne* type of liability, it must be possible to cull from the close relationship of the parties the assumption by the tortfeasor of a duty not to cause pecuniary loss to the victim. In *Hedley Byrne & Co. Ltd v. Heller and Partners Ltd* [1964] AC 465 the relationship was not affected by a direct contractual relationship and this was also the position in *Junior Books Ltd v. Veitchi Co. Ltd* [1983] 1 AC 520, and there was, therefore, no contractual influence on the relationship. In the present case the tortfeasor had contracted to be liable for failure to use reasonable skill and care in the design of the pile driving operation and in the selection of materials and goods (clause A(1)(a) and (b)); but the contract was significantly silent as to liability for the manner in which the work was executed. Once it is established that there is no general liability in tort for pecuniary loss dissociated from physical damage (see *per* Robert Goff LJ in *Muirhead* v. *Industrial Tank Specialities Ltd* [1986] QB 507 and Bingham LJ in *Simaan General Contracting Co.* v. *Pilkington Glass Ltd (No. 2)* [1988] QB 758) it would be difficult to construct a special obligation of this nature in tort to which liabilities created by a collateral contract did not extend.

MANN LJ (at p. 109):

I ask myself whether it is just and reasonable to impose a duty in tort where the parties are united by a contract which is notably silent upon the liability which it is sought to enforce by tort. In my judgment it is not. I would adopt the words used by Cumming-Bruce LJ in delivering the judgment of the court in *William Hill Organisation Ltd* v. *Bernard Sunley & Sons Ltd* (1982) 22 BLR 1, 30: 'The plaintiffs are not entitled to claim a remedy in tort which is wider than the obligations assumed by the defendants under their contract.'

I recognise that in *Junior Books Ltd* v. *Veitchi Co. Ltd* [1983] 1 AC 520, 546, Lord Roskill regarded one of the factors which were of crucial importance in giving rise to the duty of care as 'The relationship between the parties was as close as it could be short of actual privity of contract.' However where there is a privity, then in my view the rights and obligations of the parties in regard to economic loss should be solely dependent upon the terms of the privity. I recognise that to breach a contract may also be a delictual act. That is a proposition quite different from asserting that there can be a duty in tort giving rise to a liability to compensate for economic loss in circumstances where the contract between the parties is silent. Contractual silence in my view is adverse to the establishment of a close relationship for the purposes of the law of tort in regard to economic loss.

This case, as Purchas LJ is at pains to point out, is a finely balanced one. Again the contractual setting is seen as precluding the operation of tort, but great emphasis is laid on the fact that the tort claim would require the especially high levels of proximity and reliance necessary for a *Junior Books*-type claim for economic loss. So, here too, there is not an unequivocal adoption of Lord Scarman's dicta in *Tai Hing* but rather an acceptance of the need to carefully examine the contractual terms and prevent the use of negligence to contradict what the parties have agreed.

However, it must be arguable that there are other cases where to impose a tortious duty would be to fulfil or further the intention of the parties to a contract. Cases such as *Tai Hing* are the exception, where contract provides for a low standard of care. Normally use is made of contract to achieve a high standard and/or a definite result (e.g. 'the building will be finished by 31 December or 1991'). In such a case, the (lesser) tort duty, to take reasonable care, complements rather than contradicts the intentions of the contracting parties and helps the achievement of the aims of the contract.

Therefore in the type of case under discussion, the parties should equally be taken as intending to retain their tortious obligations, unless expressly excluded, and accordingly the *Tai Hing* should not be adopted here. This seems compatible with some recent cases such as *London Congregational Union Inc.* v. *Harriss* [1988] 1 All ER 15, but remains a question finally to be decided by the courts. The Court of Appeal, in *Johnstone* v. *Bloomsbury Health Authority* [1991] 2 WLR 1362 asserts that tort is not available to parties to a contract, but again this is where the use of tort would go beyond the express terms of the agreement.

Why tort not contract? – limitation and damages

The foregoing discussion has coyly failed to face directly the question of why the parties to a contract might seek to rely instead on the often less onerous obligations imposed by the law of tort. The answer, as hinted in some of the above cases, lies in the differences that exist between tort and contract and, in particular, in the more favourable rules that exist in relation to time limits and damages.

Limitation of actions

Limitation Act 1980 s.2

An action founded on tort shall not be brought after the expiration of six years from the date on which the cause of action accrued.

Limitation Act 1980 s.5

An action founded on simple contract shall not be brought after the expiration of six years from the date on which the cause of action accrued.

Expressed thus, this looks more like a similarity than a difference. The crucial distinction comes in assessing the vital date *when* the cause of action accrues; for tort this is when the cause of action is complete and, as we have seen, damage is an essential component in an action founded in negligence. However, an action in contract is complete when the breach occurs, and there is no need to wait for damage. (*Bagot* v. *Stevens Scanlan & Co Ltd* [1966] 1 QB 197, pp 67 above, is clearly correct on this point.)

Obviously in many cases of defective building work breach and damage will be simultaneous. However, this is not so in cases of latent damage. If defective foundations are laid in 1991, that is a breach of contract and time begins to run forthwith. However, damage may not actually occur until, for instance, soil movements put pressure on the foundations in 1997, and the tort claim will only run from then, the very year in which the period for a contract claim expires. The facts of *Bagot*, *Midland Bank* and even *Anns* v. *Merton* itself all demonstrate this point.

However, there is a further problem for the plaintiff. Even if, in the example, damage occurs to the foundations in 1997, a further period may elapse before the damage is discovered, perhaps through investigation of cracks in the walls. On the one hand, it is harsh to the plaintiff to allow his cause of action to expire without (in an extreme case) any knowledge that he has one; equally, there are dangers to the defendant if he may be liable for in the future when a long-standing latent defect is discovered.

For a period, the law allowed the plaintiff's time to run from the date when the damage became discoverable, but this could lead to extremely disadvantageous burdens to be placed on the defendant. The extreme example is provided by *Dennis* v. *Charnwood District Council* [1983] QB 409 where the defendant local authority was held liable for work carried out by its miniscule predecessor, the Barrow-upon-Soar Rural District Council, in approving plans for a bungalow in 1955, the defects in which did not become fully apparent until 1976. The practical difficulty of defending such a stale claim, with memories faded and records perhaps missing, suggest that sympathy for the defendant is appropriate.

The pendulum swung the other way, however, in *Pirelli General Cable Works Ltd* v. *Oscar Faber & Co.* [1983] 2 AC 1. Cracks in the plaintiff's tall chimney occurred in 1969/70 but were not discovered until 1977. The House of Lords held that the date of actual damage was the relevant date, and therefore Pirelli's claim was time-barred, though an exception to this was said to possibly exist when a building was 'doomed from the start', a gnomic phrase that led to much unsuccessful litigation. A compromise clearly needed to be reached between these two opposing views and the ruminations of the Law Reform Committee led to legislation that, for recent claims (accruing after 18 September 1986 when the legislation came into force), replaces the *Pirelli* ruling.

Latent Damage Act 1986 s.1

This created a new section in the Limitations Act 1980, s. 14A

14A (1) This section applied to any action for damages for negligence, other than one to which section 11 of this Act applies, where the starting date for reckoning the period of limitation under subsection (4)(b) below falls after the date on which the cause of action accrued.

(2) Section 2 of this Act shall not apply to an action to which this section applies.

(3) An action to which this section applies shall not be brought after the expiration of the period applicable in accordance with subsection (4) below.

(4) That period is either—
(a) six years from the date on which the cause of action accrued; or
(b) three years from the starting date as defined by subsection (5) below, if that period expires later than the period mentioned in paragraph (a) above.

(5) For the purposes of this section, the starting date for reckoning the period of limitation under subsection (4)(b) above is the earliest date on which the plaintiff or any person in whom the cause of action was vested before him first had both the knowledge required for bringing an action for damages in respect of the relevant damage and a right to bring such an action.

(6) In subsection (5) above 'the knowledge required for bringing an action for damages in respect of the relevant damage' means knowledge both—

        (a) of the material facts about the damage in respect of which damages are claimed; and

        (b) of the other facts relevant to the current action mentioned in subsection (8) below.

(7)   For the purposes of subsection (6)(a) above, the material facts about the damage are such facts about the damage as would lead a reasonable person who had suffered such damage to consider it sufficiently serious to justify his instituting proceedings for damages against a defendant who did not dispute liability and was able to satisfy a judgment.

(8)   The other facts referred to in subsection (6)(b) above are—

        (a) that the damage was attributable in whole or in part to the act or omission which is alleged to constitute negligence; and

        (b) the identity of the defendant; and

        (c) if it is alleged that the act or omission was that of a person other than the defendant, the identity of that person and the additional facts supporting the bringing of an action against the defendant.

(9)   Knowing that any acts or omissions did or did not, as a matter of law, involve negligence is irrelevant for the purposes of subsection (5) above.

(10)   For the purposes of this section a person's knowledge includes knowledge which he might reasonably have been expected to acquire—

        (a) from facts observable or ascertainable by him; or

        (b) from facts ascertainable by him with the help of appropriate expert advice which it is reasonable for him to seek;

but a person shall not be taken by virtue of this subsection to have knowledge of a fact ascertainable only with the help of expert advice so long as he has taken all reasonable steps to obtain (and, where appropriate, to act on) that advice.

14B (1)   An action for damages for negligence, other than one to which section 11 of this Act applies, shall not be brought after the expiration of fifteen years from the date (or, if more than one, from the last of the dates) on which there occurred any act or omission—

        (a) which is alleged to constitute negligence; and

        (b) to which the damage in respect of which damages are claimed is alleged to be attributable (in whole or in part).

(2)   This section bars the right of action in a case to which subsection (1) above applies notwithstanding that—

        (a) the cause of action has not yet accrued; or

        (b) where section 14A of this Act applies to the action, the date which is for the purposes of that section the starting date for

reckoning the period mentioned in subsection (4)(b) of that
section has not yet occurred;

before the end of the period of limitation prescribed by this
section.

This miracle of legislative circumlocution has the following effect:

(i)   the basic claim period is still six years from date of damage (ss. 2,
      14A(4)(a));
(ii) in cases of latent damage, a further period of three years will run
      from the discoverability of damage beyond the end of the basic claim
      period (s.14A(4)(b) and (5));
(iii) however this latter claim cannot be brought more than fifteen years
      from the date of breach (s.14B) – in effect a 'long stop' provision to
      provide long-term protection for defendants.

An example may clarify matters. Building work is carried out badly in
1991. Actual damage occurs in 1997 but is not discoverable until 2004.
The contract claim expires in 1997 and the basic tort claim in 2003.
However, the new three-year period from discoverability then arises in
2004, but is peremptorily cut off by the 'long-stop' provision in 2006,
finally enabling the worried defendant to destroy his records.

Even the simplest case, where the issue is when the damage occurs, can
give rise to problems. Buildings tend to become damaged slowly and
progressively, as little cracks gradually become bigger. The courts seem to
understand this, and adopt a pragmatic approach.

LONDON BOROUGH OF BROMLEY v. RUSH AND TOMPKINS LTD

(1985) 34 BLR 94   High Court (Official Referee's Business)

The plaintiffs occupied an office block built by the defendants. The
preliminary issue arose as to whether the claim was time-barred. The
block was completed in 1967; exterior cracks began to occur in 1975 due
to corrosion of the steel reinforcement in the concrete. The writ was issued
in 1980. It was held that the claim was time-barred, since the hidden
corrosion was sufficient to count as damage.

JUDGE STABB QC (at p. 104):

If the nature and extent of the damage necessary to give rise to a cause of
action is a matter of degree, I should apply, as a test, that it should be relevant

and significant. For example, any damage to the fabric of the building, which necessitates remedial work, must plainly be relevant and significant. In contrast, not every defect will lead to damage that is relevant and significant. As an example, it has been said that in some buildings where corrosion to .pipes is bound to occur, what is called a corrosion allowance is made in the thickness of the pipes installed, so that any corrosion does not affect the purpose for which the pipe is to be used during its expected life. Accordingly, when the corrosion takes place, the pipe can be said to be damaged, but the damage would not be relevant and significant. In *Pirelli* v. *Oscar Faber* the crack in the chimney in 1970 showed that damage then existed which must have been relevant and significant, in the sense that the crack was an indication of the unsuitability of the chimney.

It is the second question, however, which gives rise to the difficulty in this case. It has been suggested that there are two types of case to consider. First, the type of case where the defect in a building may not give rise to damage, unless and until it is put to the test of compliance with conditions which the designer or builder was under a duty to foresee. Examples of this type of case are those in which defective foundations carry the required load until exceptional but foreseeable weather conditions affect the subsoil and cause the building to settle and so become damaged; or where a building is negligently not built to its design-imposed loading and only fails when subjected to such imposed loading some years after the completion of its construction. In each instance, the cause of action only arises when damage consequent on the defect is caused by some factor, which may or may not arise. Such was the case in *Pirelli*, where damage resulting from the defective design was caused only by the user of the chimney. An illustration of the same principle is to be found in the case of *Dove* v. *Banham Locks* [1983] 1 WLR 1436, where it was only the subsequent burglary which gave rise to any damage resulting from the defective lock previously supplied.

Secondly, there is the type of case in which the initial defect needs no 'trigger' or other factor to produce its consequent damage, that is a defect which in the course of time will inevitably result in damage. The present case is an example of this. All the experts were agreed that with the building as constructed, the steel reinforcement, with the passage of time, would inevitably begin to corrode, that such corrosion was irreversible without corrective remedial measures being taken, and that the inevitable consequences of such corrosion was damage to the building. It was contended on behalf of the defendants that in such a type of case the defect itself constitutes the damage, and that the building can then be said to fall into the category of doomed from the start. It is clear that Lord Fraser in the *Pirelli* case regarded the 'doomed from the start' category as being very much the exception, and I respectfully agree with the observation of Judge Smout in the case of *Kensington and Chelsea Area Health Authority* v. *Wettern Composites Ltd and Others* (1984) 31 BLR 57; [1985] 1 All ER 346 when he said (at 31 BLR 73) that to be doomed from the start suggested not merely an inevitability of damage but an impending inevitability. I do not accept that the latent defect in this case can properly be equated with the damage, and I think that [counsel for London Borough of Bromley] is right when he says that the distinction sought to be drawn between the two types of case that I have mentioned is really a distinction without a difference. In each case, some factor has to happen to make the latent damage patent. In the one type of case it is some specific outside factor; in the other, it is simply the passage of time.

At what stage, then, is it to be said in this case that relevant and significant damage to the building first occurred. It seems to me that, on the evidence and on the submissions of counsel, I have to decide upon one of the following stages:

The first:	depassivation and commencement of corrosion of the steel reinforcement;
Secondly:	the appearance of hairline cracks in the exterior of the building consequent upon the corrosion of the steel reinforcement;
Thirdly:	the appearance of enlarged cracks in the exterior of the building consequent upon such corrosion, which indicated a necessity to repair to avoid possible danger to the public; and
Fourthly:	the spalling of concrete resulting from such corrosion.

I think that I can reject numbers (1) and (4) at the outset. So far as (1) is concerned, at that stage damage to the steel reinforcement would have begun, and admittedly was to continue to progress, but at that stage, the corrosion did not affect the ability of the steel structurally to reinforce the concrete, nor was there any adverse effect upon the concrete itself. It was the stage of the onset of corrosion. Accordingly, in my view, the damage then was not significant. As to (4) I have already stated that, in my view, spalling of concrete did not occur and would not have occurred until the end of 1976 or early 1977. But before that, in 1975 and early 1976 serious cracks in the concrete had already appeared, and plainly, therefore, damage to the building existed before concrete was expected to spall.

This leaves the two stages; the appearance of hairline cracks and the appearance of the enlarged cracks. I had the advantage of hearing the evidence of three expert consulting engineers of standing and distinction, and it seems to me that, in the end, there was very little difference in their respective assessments of the timing of the progress of the corrosion. All were agreed that corrosion in the building would have been rapid and that depassivation would have occurred by 1969/1970. The experts called on behalf of the defendants both agreed that hairline cracks would then have appeared by 1972. The expert called on behalf of the first defendant thought that spalling would have occurred four years later. If one goes back four years from the time when it was later estimated that spalling would have occurred, namely late 1976 or early 1977, that places his estimated time for the appearance of the hairline cracks at early 1972 or early 1973. The expert called on behalf of the second defendant considered that serious cracking would have occurred three to four years after the hairline cracking, that is to say in 1975/1976, which is precisely when they did in fact appear. The expert called on behalf of the plaintiffs thought that one should allow three to four years from hairline cracks to spalling. This would place the time of hairline cracks between, at the earliest, the end of 1972 or, at the latest, early 1974.

On this evidence, I am persuaded that on the balance of probabilities, hairline cracks in the concrete, caused by the progress of corrosion of the steel reinforcement, came into existence during the period 1972/1973, and certainly before 5 March 1974. I have, moreover, come to the conclusion that the hairline cracking of the concrete, even though not discovered and not reasonably discoverable by the plaintiffs, was the first manifestation of the existence of relevant and significant damage in this building, and that accordingly the plaintiffs' cause of action accrued more than six years before the issue of the writ in March 1980.

Different claims create different problems. The period when claims were based on a threat to health and safety saw the sensible ruling that time ran from the date of that threat becoming imminent (see e.g. *Jones* v. *Stroud District Council* [1986] 1 WLR 1141). Claims for economic loss also need to have their own rules and the claim runs from the date when the loss of value is incurred, which is usually the date that the premises in question are acquired, at what later turns out to be an over-valuation.

DEPARTMENT OF THE ENVIRONMENT v. ESSEX GOODMAN AND SUGGITT

(1985) 276 EG 308   High Court (Official Referee's Business)

The plaintiffs leased an office building in July 1975. The defendants included the architects and engineers, and the surveyors. The building turned out to be defective and a writ issued in January 1982. It was held that the plaintiffs' claim was time-barred.

JUDGE HAWSER QC (at p. 316):

> If the damage had not occurred at the date of the report, the third defendants would not be liable at all. If it was then in existence and reasonably discoverable, they would have been liable immediately the plaintiffs committed themselves to the lease.
>
> *Pirelli* dealt with engineers and by analogy with the architect or the builders. It laid down the principle that the cause of action in such cases arose when the damage which gave rise to it, and which actually completed the cause of action, occurred, even though that damage was not reasonably discoverable until a later date. Thus, the test was not, as had hitherto been thought in building cases, discoverability but the actual occurrence of the damage.
>
> The present relationship between the plaintiffs and the defendants is, however, in my view, quite different. The surveyors were employed to find out whether any damage existed and if they failed to use reasonable care and skill

to do so and to report it to the plaintiffs they would have been liable as at the date when the plaintiffs acted upon the report. If the defects in design, which the third defendants ought to have discovered, existed at all, it seems to me that they must have existed at the time of the survey for liability to arise. If the damage occurred subsequently, that is after July of 1975, or if it could not have been discovered by the exercise of reasonable care and skill, the third defendants would not be liable, since they would have complied with their duty.

I think that this governs the position and that accordingly the time for the statute of limitations to run is from the date when the report was acted upon by the plaintiffs, that is the end of July 1975.

Two further provisions may also appear to extend the usual period.

Limitation Act 1980 s.32

32 (1) Subject to [subsections (3) and (4A)] below, where in the case of any action for which a period of limitation is prescribed by this Act, either—

(a) the action is based upon the fraud of the defendant; or

(b) any fact relevant to the plaintiff's right of action has been deliberately concealed from him by the defendant; or

(c) the action is for relief from the consequences of a mistake;

the period of limitation shall not begin to run until the plaintiff has discovered the fraud, concealment or mistake (as the case may be) or could with reasonable diligence have discovered it.

References in this subsection to the defendant include references to the defendant's agent and to any person through whom the defendant claims and his agent.

(2) For the purposes of subsection (1) above, deliberate commission of a breach of duty in circumstances in which it is unlikely to be discovered for some time amounts to deliberate concealment of the facts involved in that breach of duty.

(3) Nothing in this section shall enable any action—

(a) to recover, or recover the value of, any property; or

(b) to enforce any charge against, or set aside any transaction affecting, any property;

to be brought against the purchaser of the property or any person claiming through him in any case where the property has been purchased for valuable consideration by an innocent third party since the fraud or concealment or (as the case may be) the transaction in which the mistake was made took place.

(4) A purchaser is an innocent third party for the purposes of this section—

(a) in the case of fraud or concealment of any fact relevant to the plaintiff's right of action, if he was not a party to the fraud or (as the case may be) to the concealment of that fact and did not at the time of the purchase know or have reason to believe that the fraud or concealment had taken place; and

(b) in the case of mistake, if he did not at the time of the purchase know or have reason to believe that the mistake had been made.

[(4A) Subsection (1) above shall not apply in relation to the time limit prescribed by section 11A(3) of this Act or in relation to that time limit as applied by virtue of section 12(1) of this Act.]

[(5) Sections 14A and 14B of this Act shall not apply to any action to which subsection (1)(b) above applies (and accordingly the period of limitation referred to in that subsection, in any case to which either of those sections would otherwise apply, is the period applicable under section 2 of this Act).]

This section has been interpreted broadly.

APPLEGATE v. MOSS

[1971] 1 QB 406   Court of Appeal

A developer was being sued (in contract) for defective foundations, which were not discovered until after the normal time limit had expired. It was held that although the developer had done nothing deliberate to harm the plaintiff's interests, his actions had had the effect of concealing the poor work and thus the claim was not time-barred.

LORD DENNING MR (at p. 413):

It has long been settled that 'fraud' in this context does not necessarily involve any moral turpitude: see *Beaman* v. *ARTS Ltd* [1949] 1 KB 550. It is sufficient if what was done was unconscionable: see *Kitchen* v. *Royal Air Force Association* [1958] 1 WLR 563: a test which was applied in the case of a building contract in *Clark* v. *Woor* [1965] 1 WLR 650. Those cases show that 'fraud' is not used in the common law sense. It is used in the equitable sense to denote conduct by the defendant or his agent such that it would be 'against conscience' for him to avail himself of the lapse of time. The section applies whenever the conduct of the defendant or his agent has been such as to hide from the plaintiff the existence of his right of action, in such circumstances that it would be inequitable to allow the defendant to rely on the lapse of time as a bar to the claim. Applied to a building contract, it means that if a builder does his work badly, so that it is likely to give rise to trouble thereafter, and then covers up his bad work so that it is not discovered for some years, then he cannot rely on the statute as a bar to the claim. The right of action is concealed by 'fraud' in the sense in which 'fraud' is used in this section.

It is plain that the right of action here was concealed by the fraud of someone. The builder put in rubbishy foundations and then covered them up.

Compare this case with the more lenient decision in:

KALISZEWSKA v. JOHN CLAGUE AND PARTNERS

(1984) 5 Con LR 62   High Court (Official Referee's Business)

The plaintiff's bungalow, designed by the defendants, was damaged by subsidence. This was a result of the defendant's negligence in not taking into account the implications of there being trees on the site. This was not regarded as 'concealed fraud'.

JUDGE WHITE (at p. 86):

> It is argued that here was an architect who deliberately rejected current wisdom as idealistic and took a risk which he could not rationally justify because of his client's pocket without informing his client of these facts. If this was a true reflection of the situation at the time I would accept that this was deliberate concealment. Those passages have to be considered, however, in the context of the defendant's evidence as a whole. Taking his evidence as a whole I have no doubt that the defendant's design was defective, not because he knowingly took the risk of providing inadequate foundations, but simply because he was on this particular occasion incompetent. The truth, as I have already found, is that he did not appreciate the importance of the trees on this site. He simply thought that they were not large enough to create a problem. If the trees had been large poplars the danger of their presence on the clay would have registered with him, and he would have taken precautions – made boreholes or at the least a close examination of the foundation trenches. Unfortunately, he was not sufficiently conscious of the published material to realise that trees of any kind in the numbers that there were on this site would inevitably lead to the problems which have occurred. I bear in mind that he did provide deeper foundations than was his usual practice in the area and reinforced them. I am sure he thought that this was sufficient for the site. At one stage he accepted that his state of knowledge was behind the published material, commenting that so was that of every average practitioner. Unfortunately, on this occasion with the particular problems on this site, the knowledge which he should have had was lacking.
>
> This is a straightforward case of an honest blunder, a defendant unaware that he was committing a wrong. I am not satisfied that he knew there was a risk of his foundations not being adequate. I am sure he thought they were, with his local knowledge, judging as best he could what extra should be

provided for the site. His state of knowledge simply was not in the standards of the time adequate enough to alert him to the full extent of the problem that faced him.

So it seems that the defendant's knowledge of, and silence concerning, the defect will be enough to bring him within the ambit of s. 32. On the other hand, blissful ignorance, as in *Kaliszewska*, will not trigger the application of the section.

Latent Damage Act 1986 s. 3

(1) Subject to the following provisions of this section, where—
  (a) a cause of action ('the original cause of action') has accrued to any person in respect of any negligence to which damage to any property in which he has an interest is attributable (in whole or in part); and
  (b) another person acquires an interest in that property after the date on which the original cause of action accrued but before the material facts about the damage have become known to any person who, at the time when he first has knowledge of those facts, has any interest in the property;

  a fresh cause of action in respect of that negligence shall accrue to that other person on the date on which he acquires his interest in the property.

(2) A cause of action accruing to any person by virtue of subsection (1) above—
  (a) shall be treated as if based on breach of a duty of care at common law owed to the person to whom it accrues; and
  (b) shall be treated for the purposes of section 14A of the 1980 Act (special time limit for negligence actions where facts relevant to cause of action are not known at date of accrual) as having accrued on the date on which the original cause of action accrued.

This is to simply assert that a new claim and new time-limit may arise for the subsequent owners of a property from their acquisition of it, though this will still be subject to the overall time limits imposed elsewhere by the 1986 Act.

To summarize, there remain advantages to plaintiffs in choosing to sue tortious remedies against defendants in latent damage claims, with both claims based on 'damage' and those based on 'discoverability' potentially subsisting after the end of the contractual time-limit. It would be highly disadvantageous to plaintiffs in construction cases involving latent damage if the widespread adoption of the *Tai Hing* were to deny them the use of tortious limitation periods, and thus rob them of any remedy.

It can be argued that the Latent Damage Act 1986 will quickly wane in importance. Claims in tort for latent damage by the first purchaser could fall foul of the *Tai Hing* if that principle continues to be more widely adopted. More significantly, it is clear that cases of latent damage will, after *Murphy* v. *Brentwood*, fall to be treated as economic loss cases and the difficulties of claiming under that head of damages are well known. So the generosity of parliament in preserving the rights of property owners faced with latent defects therein has been outflanked by the judges in redefining the nature of the loss so that claims will be few and far between.

Damages

The principal difference here may be simply expressed. The tort of negligence, as we have seen (pp. 54 above), has a very wide test of foreseeability of damage, but the leading case in contract law imposes a more restrictive view, and justifies this divergence cogently.

THE HERON II

[1969] 1 AC 350   House of Lords

A cargo of sugar en route from Constanza to Basrah was delayed. The market price of sugar fell between the scheduled date and the actual date. This loss was held to be claimable.

LORD REID (at p. 382):

> So the question for decision is whether a plaintiff can recover as damages for breach of contract a loss of a kind which the defendant, when he made the contract, ought to have realised was not unlikely to result from a breach of contract causing delay in delivery. I use the words 'not unlikely' as denoting a degree of probability considerably less than an even chance but nevertheless not very unusual and easily foreseeable.

(and at p. 385):

> The modern rule of tort is quite different and it imposes a much wider liability. The defendant will be liable for any type of damage which is reasonably foreseeable as liable to happen even in the most unusual case, unless the risk is so small that a reasonable man would in the whole circumstances feel justified in neglecting it. And there is good reason for the difference. In contract, if one party wishes to protect himself against a risk which to the other party would appear unusual, he can direct the other party's attention to it before the contract is made, and I need not stop to consider in what

circumstances the other party will then be held to have accepted responsibility in that event. But in tort there is no opportunity for the injured party to protect himself in that way, and the tortfeasor cannot reasonably complain if he has to pay for some very unusual but nevertheless foreseeable damage which results from his wrongdoing.

So this area gives a further justification for preferring a tort remedy to one framed in contract. An alternative approach was mooted by Lord Denning MR in *Parsons (Livestock) Ltd* v. *Uttley Ingham* [1978] QB 791. He thought that the two tests of foreseeability should be split between physical loss (broad test) and economic loss (narrow test), irrespective of whether the claim was framed in tort to contract. This idea has not been developed subsequently.

Chapter 5

Contractual Terms and Negligence Liability

Contract terms, in particular clauses excluding or limiting liability, can affect tortious liability in two ways, directly, by showing that the parties have agreed to reduce or remove their tortious obligations or, more subtly, as a background factor curtailing the impact of the duty of care.

Many of the traditional problems surrounding exclusion clauses do not cause concern in construction cases. The potential unfairness to often ignorant consumers of wide-ranging standard-form non-negotiable exclusion clauses is not likely to occur in building cases. Exclusion clauses are not widely used by professional men, and the typical construction contract is carefully negotiated by well-advised parties. However, some of the restrictions placed on exclusion clauses, in the interests of consumer protection, could potentially be relevant in the construction field and are now noted.

Common law controls

The courts have been at pains to ensure that exclusion clauses are clearly incorporated into contracts (e.g. *Thornton* v. *Shoe Lane Parking Ltd* [1971] 2 QB 163) and are also keen to ensure that their terms clearly cover the form of liability that is being utilized.

WHITE v. JOHN WARWICK & CO.

[1953] 1 WLR 1285   Court of Appeal

The plaintiff newsvendor hired a cycle from the defendants and was badly injured when the saddle suddenly tilted forward. An exclusion clause provided that 'Nothing in this agreement shall render the owners liable for any personal injury . . .' The clause was held to be ineffective to prevent an action for personal injury framed in negligence.

DENNING LJ (at p. 1293):

> In my opinion, the claim for negligence in this case is founded in tort and not on contract. This can be seen by considering what would be the position if, instead of the newsvendor himself, it was his servant who had been riding the cycle and had been injured. If the servant could show that the defendants had negligently sent out a defective machine for immediate use, he would have had a cause of action in negligence on the principle stated in *Donoghue* v. *Stevenson* [1932] AC 562, and, as against the servant, the exemption clause would be no defence. That shows that the defendants owed a duty of care to the servant. A fortiori they owed a like duty to the newsvendor himself. In either case a breach of that duty is a tort, which can be established without relying on any contract at all. It is true that the newsvendor in the present case could also rely on a contract, if he had wished, but he is not bound to do so; and if he can avoid the exemption clause by framing his claim in tort he is, in my judgment, entitled to do so.

(and at p. 1295):

> The present case turns on the construction of the exemption clause. In my judgment, it exempts the defendants from liability in contract, but not from liability in tort. If the plaintiff can make out his cause of action in negligence, he is, in my opinion, entitled to do so, although the same facts also give a cause of action in contract from which the defendants are exempt.

Statutory controls

The main forum for control of exclusion clauses is now in the form of the Unfair Contract Terms Act 1977. This controls attempts to exclude, limit or otherwise restrict liabilities incurred in the course of business (ss.13(1); 1(3)), and its controls on exclusion of liability in negligence are found in s.2.

Unfair Contract Terms Act 1977 s.2

2.(1) A person cannot by reference to any contract term or to a notice given to persons generally or to particular persons exclude or restrict his liability for death or personal injury resulting from negligence.

(2) In the case of other loss or damage, a person cannot so exclude or restrict his liability for negligence except in so far as the term or notice satisfies the requirement of reasonableness.

(3) Where a contract term or notice purports to exclude or restrict liability

for negligence a person's agreement to or awareness of it is not of itself to be taken as indicating his voluntary acceptance of any risk.

This is clear as regards damage to life and limb, but begs several questions in relation to property damage, which is likely to be a major part of any claim.

Unfair Contract Terms Act 1977 s. 11

11.(1) In relation to a contract term, the requirement of reasonableness for the purposes of this Part of this Act, section 3 of the Misrepresentation Act 1967 and section 3 of the Misrepresentation Act (Northern Ireland) 1967 is that the term shall have been a fair and reasonable one to be included having regard to the circumstances which were, or ought reasonably to have been, known to or in the contemplation of the parties when the contract was made.

[S. 11(2) is omitted]

(3) In relation to a notice (not being a notice having contractual effect), the requirement of reasonableness under this Act is that it should be fair and reasonable to allow reliance on it, having regard to all the circumstances obtaining when the liability arose or (but for the notice) would have arisen.

(4) Where by reference to a contract term or notice a person seeks to restrict liability to a specified sum of money, and the question arises (under this or any other Act) whether the term or notice satisfies the requirement of reasonableness, regard shall be had in particular (but without prejudice to subsection (2) above in the case of contract terms) to—
(a) the resources which he could expect to be available to him for the purpose of meeting the liability should it arise; and
(b) how far it was open to him to cover himself by insurance.

(5) It is for those claiming that a contract term or notice satisfies the requirement of reasonableness to show that it does.

'Reasonableness' is bound to be vague; its interpretation will always be enmeshed in the facts of the cases. However, the factors outlined in Schedule 2 of the Act for use in sale of goods cases, but which may be used, as relevant, in other cases, show that the courts will be keen to look at the genuineness or otherwise of the bargain, and the reality of the consent given. Factors there listed include:

(a) the relative strength of the parties' bargaining positions;

(b) whether an inducement, such as reduced price, was given to the customer accepting the exclusion clause;
(c) the customer's knowledge of the term and its extent; and
(d) whether it was a 'one-off' order.

This suggests that exclusion clauses in a typical commercial construction contract, freely bargained between broadly equal parties will be seen as reasonable and hence will remain effective.

Indirect impact of contractual terms

The main area where a negligence action has its uses is in the absence of a contract between the parties. That is not to say, however, that such cases are not free from contractual influences; many will be set against the background of other contracts. One view is that this background can preclude the operation of tort altogether.

ERNST & WHINNEY v.
WILLARD ENGINEERING (DAGENHAM) LTD

(1987) 40 BLR 67   High Court (Official Referee's Business)

The plaintiff firm took over the lease of an office block and found that the air conditioning was defective and in need of repair. Action was taken against those responsible for the installation of the plant.

JUDGE DAVIES QC (at p. 73):

> The issues before me raise important questions regarding liability in tort, in what is predominantly a contractual setting. This is a perennial problem which assumes an accentuated importance in cases of this kind. This is because they so frequently involve contractual relationships on more than one plane. I refer to the chain of contracts, sub-contracts and sub-sub-contracts which go into the construction of the building in the first place, and then the other contractual chain whereby the building owner's interest in the building is transferred along the line to lessees and purchasers. This is a case in point, where the action is brought by the assignees of the original lessees of the building owner, against the mechanical consultant employed by the latter to design and supervise the installation of the building's heating system, and against the sub-contractor who was employed by the building owner's main contractor, to instal it; in addition the sub-contractor entered into a warranty agreement with the building owners.

and at p. 80, after citing Lord Scarman's remarks in *Tai Hing*, pp. 70, above:

> In my view, his words are equally pertinent to a situation like this one where one has a fasciculus of contracts. This was a contractual situation from beginning to end. The damage alleged is wholly comprehended within the defendants' contracts to construct and the contractual obligations of the plaintiffs and the various lessees and the transfers concerned. In my view there is no basis or justification for any action for negligence in tort.

This surely goes too far. It is doubtful whether the mere presence of a contract precludes the operation of tort. Here it is argued that the background of a 'fasciculus' (network) of contracts precludes any role for tort. However, *Tai Hing* itself only seeks to restrict tort as between contracting parties, and, simply, here the parties were not in such a relationship and therefore *Tai Hing* has no bearing on the matter. It is the classic fallacy, exposed in *Donoghue* v. *Stevenson* itself, to allow the lack of a contract to interfere with tortious liability. Equally, other factors would probably preclude the use of tort in the case now, in particular the purely qualitative character of the defects, in effect economic loss, coupled with the absence of any close relationship of proximity or reliance between the parties. However, other cases have more carefully examined the role of the contracts in such cases where a tortious duty is being invoked.

SOUTHERN WATER AUTHORITY v. CAREY

[1985] 2 All ER 1077   High Court (Official Referee's Business)

The plaintiffs contracted for a new sewage works. By this contract, neither main contractors nor their sub-contractors, servants or agents were to be liable for a wide range of defects. The plaintiffs sought to sue, *inter alia*, sub-contractors who were responsible for equipment which failed. It was held that the parties to the action were not within a contractual relationship and also that the plaintiffs could not sue in negligence.

JUDGE SMOUT QC (at p. 1085):

> No one would doubt that in an ordinary building case as between the sub-contractors and the building owner who has suffered damage there is a sufficient relationship of proximity that in the reasonable contemplation of the sub-contractor carelessness on his part may be likely to cause damage to

the building owner. Thus a prima facie duty of care lies on the sub-contractor. So also in this case. But one has to go on to consider whether there are any considerations which ought to negative or to reduce or limit the scope of that duty. And merely to ask the question in the context of this case seems to me to foretell the answer. Did not the plaintiffs' predecessor as building owner, as it were, itself stipulate that the sub-contractors should have a measure of protection following on the issue of appropriate taking-over certificates? We must look to see the nature of such limitation clause to consider whether or not it it is relevant in defining the scope of the duty in tort. The contractual setting may not necessarily be overriding, but it is relevant in the consideration of the scope of the duty in tort for it indicates the extent of the liability which the plaintiffs' predecessor wished to impose. To put it more crudely, and I hope I do no injustice to counsel for the third defendants to say that that is how he emphasised the matter, the contractual setting defines the area of risk which the plaintiffs' predecessor chose to accept and for which it may or may not have sought commercial insurance.

In *Junior Books Ltd* v. *Veitchi Co Ltd* [1982] 3 All ER 201 at 210–11 [1983] 1 AC 520 at 541–2, Lord Roskill cited the passage from Lord Wilberforce's speech in *Anns* v. *Merton London Borough* [1978] AC 728 to which I have referred and with the agreement of Lord Fraser and Lord Russell expressed the majority view of the House of Lords in his approval of Lord Wilberforce's proposition. Significantly, Lord Roskill added this ([1982] 3 All ER 201 at 214, [1983] 1 AC 520 at 546):

'During the argument it was asked what the position would be in a case where there was a relevant exclusion clause in the main contract. My Lords, that question does not arise for decision in the instant appeal, but in principle I would venture the view that such a clause according to the manner in which it was worded might in some circumstances limit the duty of care just as in the *Hedley Byrne* case [1963] 2 All ER 575, [1964] AC 465 the plaintiffs were ultimately defeated by the defendants' disclaimer of responsibility.'

The case has now arisen where there is such a limitation that is directly in point. While the terms of cl 30(vi) may, if literally interpreted, exceed the bounds of common sense the intent is clear, namely that the sub-contractor whose works have been so completed as to be the subject of a valid taking-over certificate should be protected in respect of those works from any liability in tort to the plaintiffs. As the plaintiffs' predecessor did so choose to limit the scope of the sub-contractors' liability, I see no reason why such limitation should not be honoured.

So the fact of the adjacent contracts, and the expectations engendered thereby (and presumably the associated decisions as to insurance coverage necessary) were enough to count as policy factors reducing what would otherwise (then) have been a clear duty of care. This approach has now also been adopted by the Court of Appeal.

NORWICH CITY COUNCIL v. HARVEY

[1989] 1 WLR 828   Court of Appeal

The plaintiff employed a main contractor to extend their swimming pool. The plaintiff was stated by the contract to accept the risk of fire. Roofing work was delegated to a sub-contractor on similar terms. The sub-contractor caused a fire by his carelessness, but the plaintiff was held to be unable to sue him in negligence.

MAY LJ (at p. 833):

> I trust I do no injustice to the plaintiff's argument in this appeal if I put it shortly in this way. There is no dispute between the employer and the main contractor that the former accepted the risk of fire damage: see *James Archdale & Co. Ltd* v. *Comservices Ltd* [1954] 1 WLR 459 and *Scottish Special Housing Association* v. *Wimpey Construction UK Ltd* [1986] 1 WLR 995. However clause 20[C] does not give rise to any obligation on the employer to indemnify the sub-contractor. That clause is primarily concerned to see that the works were completed. It was intended to operate only for the mutual benefit of the employer and the main contractor. If the judge and the sub-contractor are right, the latter obtains protection which the rules of privity do not provide. Undoubtedly the sub-contractor owed duties of care in respect of damage by fire to other persons and in respect of other property (for instance the lawful visitor, employees of the employer, or other buildings outside the site); in those circumstances it is impracticable juridically to draw a sensible line between the plaintiff on the one hand and others on the other to whom a duty of care was owed. The employer had no effective control over the terms upon which the relevant sub-contract was let and no direct contractual control over either the sub-contractor or any employee of its.
>
> In addition, the plaintiff pointed to the position of the first defendant, the sub-contractor's employee. Ex hypothesi he was careless and even if his employer be held to have owed no duty to the building employer, on what grounds can it be said that the employee himself owed no such duty? In my opinion, however, this particular point does not take the matter very much further. If in principle the sub-contractor owed no specific duty to the building owner in respect of damage by fire, then neither in my opinion can any of its employees have done so.

(and at p. 836):

> In my opinion the present state of the law on the question whether or not a duty of care exists is that, save where there is already good authority that in the circumstances there is such a duty, it will only exist in novel situations where not only is there foreseeability of harm, but also such a close and direct relation between the parties concerned, not confined to mere physical proximity, to the extent contemplated by Lord Atkin in his speech in

Donoghue v. *Stevenson* [1932] AC 562. Further, a court should also have regard to what it considers just and reasonable in all the circumstances and facts of the case.

In the instant case it is clear that as between the plaintiff and the main contractor the former accepted the risk of damage by fire to its premises arising out of and in the course of the building works. Further, although there was no privity between the plaintiff and the sub-contractor, it is equally clear from the documents passing between the main contractor and the sub-contractor to which I have already referred that the sub-contractor contracted on a like basis. In *Scottish Special Housing Association* v. *Wimpey Construction UK Ltd* [1986] 1 WLR 995 the House of Lords had to consider whether, as between the employer and main contractor under a contract in precisely the same terms as those of the instant case, it was in truth intended that the employer should bear the whole risk of damage by fire, even fire caused by the contractor's negligence. The position of sub-contractors was not strictly in issue in the *Scottish Housing* case, which I cannot think the House did not appreciate, but having considered the terms of clauses 18, 19 and 20[C] of the same standard form as was used in the instant case Lord Keith of Kinkel, in a speech with which the remainder of their Lordships agreed, said, at p. 999:

'I have found it impossible to resist the conclusion that it is intended that the employer shall bear the whole risk of damage by fire, including fire caused by the negligence of the contractor or that of sub-contractors.'

As Lord Keith went on to point out, a similar conclusion was arrived at by the Court of Appeal in England in *James Archdale & Co. Ltd* v. *Comservices Ltd* [1954] 1 WLR 459 upon the construction of similarly but not identically worded corresponding clauses in a predecessor of the standard form used in *Scottish Special Housing Association* v. *Wimpey Construction UK Ltd* . . . and the instant case. Again the issue only arose in the earlier case as between employer and main contractor, but approaching the question on the basis of what is just and reasonable I do not think that the mere fact that there is no strict privity between the employer and the sub-contractor should prevent the latter from relying upon the clear basis upon which all the parties contracted in relation to damage to the employer's building caused by fire, even when due to the negligence of the contractors or sub-contractors.

PACIFIC ASSOCIATES INC. v. BAXTER

[1990] 1 QB 993   Court of Appeal

The plaintiffs contracted with the Ruler of Dubai to carry out major dredging works in Dubai. The defendants were by this contract appointed to supervise the work as consultant engineers. This contract stated that the defendants were not to be held liable to the plaintiffs. Disputes arose and eventually the plaintiffs sought to sue in negligence for their losses on the project.

PURCHAS LJ (at p. 1020):

It is now necessary to consider the position of the contractor against the circumstances prevailing at the time when its relationship with the engineer was established, namely, at tender stage, and to pose the question, 'Did the contractor rely on any assumption of liability in tort appearing to be accepted on the part of the engineer which would afford to the contractor remedies beyond those which it acquired under the terms of the contract in respect of which it was to tender?' One must start with the proposition that if the contractor had required an indemnity or extra-contractual protection in respect of defaults by the engineer or insolvency on the part of the employer then it was open to the contractor to have stipulated for such protection. On the contrary, by accepting the invitation to tender on the terms disclosed in the document 'Instructions to Tenderers' and the contractual documents submitted therewith the contractor must be taken to accept the role to be played by the engineer as defined in the contract.

The terms of the contract provided a three-stage process under which the contractor obtained payment for his work, the third stage of which (G.C. 67) included a reference to one or more independent arbitrators who are given the power 'to open up review and revise any decision, opinion, direction, certificate or valuation of the engineer'. In the case of withholding by the engineer of any certificate the resort to arbitration is not postponed until after completion of the works. However in this case, as a matter of history, the claim under G.C. 67 was not made until after second certificate of completion. The opening words of G.C. 67 are extremely wide: 'If any dispute or difference of any kind whatsoever shall arise between the employer or the engineer and the contractor . . .' No function of the engineer under or in connection with the contract was mentioned to the court during the course of argument which would escape this clause. Clause P.C. 84 refers to P.C. 83 (deduction of sums certified due from the contractor to the employer) and other matters in which the engineer's decision is made final by the contract but the purpose of the clause is to ensure that the engineer 'will act independently of and entirely unfettered by the employer'. It was not argued that this clause affected in any way the operation of the arbitration provisions in G.C. 67.

P.C. 85 gives very wide discretionary powers to the engineer in the discharge of his function under the contract which are quite incompatible with the role of an arbitrator. Wide as they may be, however, they are subject to the provisions of G.C. 67, and it is under this clause that arbitration is introduced. In my judgment, as I have already indicated, there is no question of arbitral immunity being enjoyed under the terms of this contract by the engineer.

It remains only to consider shortly the significance of P.C. 86. This is admittedly an important part of the contractual structure against which the contractor accepted the engineer in his role under the contract. The clause provides specifically that 'neither the engineer nor any of his staff shall be in any way personally liable for the acts or obligations under the contract'. This can only refer to his own acts in performing the obligations imposed on the engineer under the contract. The question is whether the protection of this clause extends to the negligent performance of those functions. In this context it is not necessary to establish that the negligence on the part of the person relying on the exclusion clause is the only negligence to which reference can be made as is the case with such a clause in ordinary contracts. The presence of the reservation is given its normal role to play in the overall consideration of what responsibility was accepted by the proposed tortfeasor. This is dealt with in the speech of Lord Reid in *Hedley Byrne & Co. Ltd* v. *Heller and Partners Ltd* [1964] AC 465, 492–3:

'It appears to me that the only possible distinction in the present case is that here there was no adequate disclaimer of responsibility. But here the appellants' bank, who were their agents in making the inquiry, began by saying that "they wanted to know in confidence and without responsibility on our part", that is, on the part of the respondents. So I cannot see how the appellants can now be entitled to disregard that and maintain that the respondents did incur a responsibility to them.

The appellants founded on a number of cases in contract where very clear words were required to exclude the duty of care which would otherwise have flowed from the contract. To that argument there are, I think, two answers. In the case of a contract it is necessary to exclude liability for negligence, but in this case the question is whether an undertaking to assume a duty to take care can be inferred: and that is a very different matter. And, secondly, even in cases of contract general words may be sufficient if there was no other kind of liability to be excluded except liability for negligence: the general rule is that a party is not exempted from liability for negligence "unless adequate words are used" – *per* Scrutton LJ in *Rutter* v. *Palmer* [1922] 2 KB 87, 92, CA. It being admitted that there was here a duty to give an honest reply, I do not see what further liability there could be to exclude except liability for negligence: there being no contract that was no question of warranty.'

In accepting the invitation to tender with the complete contractual framework including the disclaimed in P.C. 86, it would, in my judgment, be impossible either to support the contention that the engineer was holding himself out to accept a duty of care with the consequential liability for pecuniary loss outside the provisions afforded to the contractor under the contract, or to support the contention that the contractor relied in any way on such an assumption of responsibility on the part of the engineer in any way to bolster or extend its rights. I think that I should say that even if P.C. 86 were not included in the contract in this case, the provisions of G.C. 67 would, in my view, be effective to exclude the creation of any direct duty on the engineer towards the contractor.

(and at p. 1023):

I have come to the conclusion, for the reasons already stated, that no liability can be established in tort under which the engineer owed a direct duty to the contractor in the circumstances disclosed in this case. I emphasise, however, that in coming to this conclusion it does depend on the particular circumstances of the case not the least of which were the contractual provisions in the contract which afforded an avenue enabling the contractor to recover from the employer. I see no justification for superimposing on this contractual structure an additional liability in tort as between the engineer and the contractor. In coming to this conclusion I have taken into account:

(i)   that although there was a degree of proximity established in the obvious sense that the contractor under the terms of the contract relied on the engineer performing their duties in supervising the execution of the works, this is not the same quality of proximity required to establish a duty of care in the *Hedley Byrne* sense;

(ii)   that the duty on the engineer to perform in accordance with the contract arose out of some contractual relationship, unspecified, existing between the employer and the engineer, which gave rise to that duty;

(iii)   that there was no direct contractual relationship between the engineer and the contractor;

(iv)   that under the contract, the contractor could challenge in the fullest sense the performance of his duties by the engineer by claiming against the employer for sums due to the contractor including extra expenses and interest on outstanding sums due;

(v)   that the contractor, when tendering for the contract, was content to offer for the works on the terms set out in the invitation to tender which incorporated the full terms of the proposed contract.

In these circumstances the following propositions appear to me to be established.

(i)   The engineer remains under contractual obligations to the employer, which give rise to a duty to exercise skill and care and in appropriate circumstances to act fairly between the employer and the contractor. If the engineer is in breach of this duty they are liable to the employer for economic loss directly flowing from the breach. Whether this action lies in contract or tort or both will only be a relevant question in very exceptional circumstances. Their Lordships did not specifically consider this question in *Sutcliffe* v. *Thackrah* [1974] AC 727.

(ii)   There is no reason to infer that the contractor was relying on any right to recover damages in the form of economic loss arising from any breach of duty in (i) above, other than by pursuing its remedies against the employer under the contract.

(iii)   There is no reason to infer that the engineer ever assumed or appeared to assume a direct responsibility to the contractor for any economic loss that might be occasioned to the contractor as a result of any breach of their duty as in (i) above.

(iv)   There is, therefore, no basis on which a duty of care to prevent economic loss can be imposed on the engineer in favour of the contractor which would be for all practical purposes coterminous with the rights to be enjoyed by the contractor under the contract.

These two cases show that the ideas first formulated by Judge Smout in *Southern Water* have now borne fruit to a great extent. The knowledge of the parties as to how the various responsibilities are to be allocated in all the contracts will now be a factor limiting the duty of care, though clearly that knowledge is vital. It is a spectacular circumvention of the doctrine of privity of contract that relies on contract terms not being effective as against third parties as contract terms, but rather is evidence of all the surrounding circumstances which go to the reasonableness or otherwise of imposing a tortious duty of care. This approach seems to make some good sense in the tortious context and certainly confirms with the commercial reality that knowledge of risk of liability and provision of insurance go hand-in-hand.

However, the new approach runs completely contrary to the tenor of the leading House of Lords decision in *Midland Silicones Ltd* v. *Scruttons Ltd* [1962] AC 446, that third parties could not take the benefit of provisions in contracts to which they are not privy, even when not attempting to sue in contract. Equally it does accord closely with the vigorous and cogent dissent of Lord Denning in the same case. It is evident that the higher courts will soon have to consider whether to endorse the approach in the cases extracted above, and thus to finally remove the so-called 'privity of contract fallacy', viz. that privity issues can affect a party who is not seeking to frame a claim based on the binding nature of a contract to which he is not a party but is rather seeking to use it as evidence, no more, in a claim based on breach of tortious duty.

The other difficulty is that in some respects these cases do come close to resurrecting the old 'privity of contract fallacy'. The parties in each case have carefully set up their various contracts, and could easily be assumed to have left gaps to be filled in by the law of tort, the presence of which in construction law is, after all, now well-known and which can be excluded easily in commercial agreements. The successful defendants in these cases are running the same argument that failed for Stevenson against Donoghue i.e. that the absence of direct contractual provision should lead to the denial of a tort remedy; the rejection of this approach in *Donoghue* was clear and its revival surprising.

Perhaps the key distinction that emerges centres on the type of case. *Donoghue* was a consumer case and would not, in all likelihood, be at all aware of the presence or absence of detailed contractual provisions. These commercial contract cases are very different with the parties, it may be reasonable to imagine, positively aware of the issues of risk and liability that they will face. It may be reasonable to use such actual knowledge to circumvent what would otherwise normally have been a duty situation, in a relatively narrow range of commercial cases. However, in so deciding, it is clear that a court will soon consciously need to decide whether this approach is contrary to the Lord's decision in *Midland Silicones* or alternatively theirs in *Donoghue*, and how to proceed in the light of such findings. It seems likely that this will be an area where careful examination of both the knowledge and the intentions of the parties will have to be undertaken.

Chapter 6

Contribution Issues

Multi-party cases, as are so many instances of premises litigation, throw up obvious problems of assessing the respective share of blame to be allocated between the parties. These issues are primarily the province of the law of tort, but can also invoke contractual points at times.

Joint tortfeasors

Civil Liability (Contribution) Act 1978 ss.1, 2

1.(1) Subject to the following provisions of this section, any person liable in respect of any damage suffered by another person may recover contribution from any other person liable in respect of the same damage (whether jointly with him or otherwise).

(2) A person shall be entitled to recover contribution by virtue of subsection (1) above notwithstanding that he has ceased to be liable in respect of the damage in question since the time when the damage occurred, provided that he was so liable immediately before he made or was ordered or agreed to make the payment in respect of which the contribution is sought.

(3) A person shall be liable to make contribution by virtue of subsection (1) above notwithstanding that he has ceased to be liable in r spect of the damage in question since the time when the damage occurred, unless he ceased to be liable by virtue of the expiry of a period of limitation or prescription which extinguished the right on which the claim against him in respect of the damage was based.

(4) A person who has made or agreed to make any payment in bona fide settlement or compromise of any claim made against him in respect of any damage (including a payment into court which has been accepted) shall be entitled to recover contribution in accordance with this section without regard to whether or not he himself is or ever was liable in respect of the damage, provided, however, that he would have been

liable assuming that the factual basis of the claim against him could be established.

(5) A judgment given in any action brought in any part of the United Kingdom by or on behalf of the person who suffered the damage in question against any person from whom contribution is sought under this section shall be conclusive in the proceedings for contribution as to any issue determined by that judgment in favour of the person from whom the contribution is sought.

(6) References in this section to a person's liability in respect of any damage are references to any such liability which has been or could be established in an action brought against him in England and Wales by or on behalf of the person who suffered the damage; but it is immaterial whether any issue arising in any such action was or would be determined (in accordance with the rules of private international law) by reference to the law of a country outside England and Wales.

2.(1) Subject to subsection (3) below, in any proceedings for contribution under section 1 above the amount of the contribution recoverable from any person shall be such as may be found by the court to be just and equitable having regard to the extent of that person's responsibility for the damage in question.

(2) Subject to subsection (3) below, the court shall have power in any such proceedings to exempt any person from liability to make contribution, or to direct that the contribution to be recovered from any person shall amount to a complete indemnity.

(3) Where the amount of the damages which have or might have been awarded in respect of the damage in question in any action brought in England and Wales by or on behalf of the person who suffered it against the person from whom the contribution is sought was or would have been subject to—
   (a) any limit imposed by or under any enactment or by any agreement made before the damage occurred;
   (b) any reduction by virtue of section 1 of the Law Reform (Contributory Negligence) Act 1945 or section 5 of the Fatal Accidents Act 1976; or
   (c) any corresponding limit or reduction under the law of a country outside England and Wales;
   the person from whom the contribution is sought shall not by virtue of any contribution awarded under section 1 above be required to pay in respect of the damage a greater amount than the amount of those damages as so limited or reduced.

This legislation gives a very comprehensive framework. A right to seek a contribution is widely framed (in s.1(1)) and subsists after a party has settled a claim and is no longer liable (s.1(2)) or is no longer able to be sued as being out of time (s.1(3)). Likewise, the right to claim a contribution is granted to a party who has made a settlement and is no longer being sued (s.1(4)). What the precise level of contribution will be is, as s.2 makes clear, inevitably going to be one of fact, depending on all the circumstances. Most significantly, however, is that the contribution right transcends the tort/contract boundary.

Civil Liability (Contribution) Act 1978 s.6(1)

6.(1) A person is liable in respect of any damage for the purposes of this Act if the person who suffered it (or anyone representing his estate or dependants) is entitled to recover compensation from him in respect of that damage (whatever the legal basis of his liability, whether tort, breach of contract, breach of trust or otherwise).

Thus the contribution provisions will work not just as between joint tortfeasors but also where, for instance, A's building is harmed both by B's breach of contract and the negligence of C, his sub-contractor, who is only liable to A in tort.

The practical operation of these provisions is in one respect at least less straightforward than the statutory framework makes clear. In short, the statute assumes that all the relevant parties are there to be sued, and have not become wound up, bankrupt, mad or irretrievably out of the jurisdiction. The common law well understood these various possibilities and a plaintiff injured by the actions of joint tortfeasors (i.e. those who have collectively contributed to the same damage) can sue one for his whole loss, and leave that hapless defendant to seek a contribution from the rest (*Clark* v. *Newsam* (1847) 1 Exch 131 at 140). This means that the defendant bears the risk of absent joint tortfeasors, and not the plaintiff, so that a party only in part responsible may end up bearing 100% of the bill for damages.

This was a particular fear of local authorities after *Anns* v. *Merton*. Builders often became bankrupt and development companies are frequently wound up once the development project is over. Architects and surveyors, at least when *Anns* first burst into the construction world, were quite often not insured against negligence. Local authorities, however, do not go bankrupt and are not wound up but are insured and it was not difficult to envisage cases, like *Anns* itself, where the local authority is left to face the threat of litigation alone, though only partially to blame for the defective state of the premises.

Contributory negligence

In some cases, it is the plaintiff himself who contributes to the damage for example by his own lack of maintenance of his property. This too can be reflected by apportionment of responsibility.

Law Reform (Contributory Negligence) Act 1945 s.1(1)

1.(1)   Where any person suffers damage as the result partly of his own fault and partly of the fault of any other person or persons, a claim in respect of that damage shall not be defeated by reason of the fault of the person suffering the damage, but the damages recoverable in respect thereof shall be reduced to such extent as the court thinks just and equitable having regard to the claimant's share in the responsibility for the damage:

Provided that—
(a) this subsection shall not operate to defeat any defence arising under a contract;
(b) where any contract or enactment providing for the limitation of liability is applicable to the claim, the amount of damages recoverable by the claimant by virtue of this subsection shall not exceed the maximum limit so applicable.

This means that a deduction will take place from the plaintiff's award of damages to reflect his share of the blame. Such deduction cannot reach 100 per cent, because both parties must be at fault to satisfy the terms of the statute (*Pitts* v. *Hunt* [1990] 1 QB 302). Such deduction may take place even where the initial claim is framed in breach of contract, rather than tort, where the contractual obligation is the same i.e. a duty to take reasonable care *Forsikringsaktieselskapet Vesta* v. *Butcher* [1989] AC 852; [1988] 3 WLR 565 CA.

PART C
Liability for Construction

The prime intention of this, the main section of this book, is to examine in turn the liabilities in negligence now faced by each of the different groups involved in the construction process. This will involve separate assessments of the tortious liability of the builder and, associated with him, his sub-contractors and suppliers of materials, as well as of those architects, engineers, etc, involved with danger work and supervision of a construction project. Also, the differing roles of the local authority will fall to be examined, and finally the legal responsibility of the surveying profession will be reviewed.

However, the recent evolution of the law of negligence as it affects the construction industry has not occurred in such a straightforward compartmentalized manner. In particular, the recent rise and even more recent decline in the role of the tort of negligence has occurred in cases concerning both builders and local authorities, and discussion of the two separate groups' liabilities has been intertwined. It is therefore appropriate to examine the course of that development in general before homing in on the liabilities of each group.

Chapter 7

Development of Premises Liability

The logic of *Donoghue* v. *Stevenson* took a long time to be applied to cases involving defective premises. The traditional approach of *caveat emptor* – let the buyer beware – was clearly at the heart of the decision not to apply *Donoghue* in a building case in *Otto* v. *Bolton and Norris* [1936] 2 KB 46 and it was the 1960s before a limited use was made of negligence.

SHARPE v. E.T. SWEETING & SON LTD

[1963] 1 WLR 665   High Court

The plaintiff was injured when a concrete canopy that was part of her council house fell on her. The house was owned by the Middlesbrough Corporation, and built by the defendants, who were found liable in negligence.

NIELD J (at p. 672):

> The Middlesbrough Corporation, owners of the realty, are not parties in the proceedings for the reason that it is agreed that the corporation as owners enjoy an immunity. This principle was explained in *Bottomley* v. *Bannister* [1932] 1 KB 458, and Scrutton LJ said that in the absence of an express contract the landlord of an unfurnished house is not liable to his tenant, or a vendor of real estate to his purchaser, for defects in the house or land rendering it dangerous or unfit for occupation, even if he has constructed the defects himself or is aware of their existence. And Greer LJ quoted *Cavalier* v. *Pope* [1906] AC 428:
>
> > 'It is well established that no duty is, at law, cast upon a landlord not to let a house in a dangerous or dilapidated condition, and further, that if he does let it while in such a condition, he is not thereby rendered liable in damages for injuries which may be sustained by the tenant, his (the tenant's) servants, guests, customers, or others invited by him to enter the premises by reason

of this defective condition. A purchaser of the freehold is, in my judgment, in no better position than a tenant. No case was cited to us in which a tenant or purchaser has ever recovered against a lessor or vendor, either by implied contract or in tort, by establishing the liability of the vendor or landlord for injuries sustained through the house or its fixtures being unsafe at the date of the sale or of the lease.'

This decision was reached by the Court of Appeal before the decision in *Donoghue* v. *Stevenson* [1932] AC 562. However, thereafter another case came before the court: *Otto* v. *Bolton and Norris* [1936] 2 KB 46. That was decided by Atkinson J who said, referring to *Cavalier* v. *Pope* and other cases:

'Now unless the law there laid down has clearly and plainly been declared to be wrong in *Donoghue* v. *Stevenson* it is, of course, binding upon me. That was a case dealing with chattels, and there is not a word in the case from beginning to end which indicates that the law relating to the building and sale of houses is the same as that relating to the manufacture and sale of chattels. *Bottomley* v. *Bannister* is, in fact, referred to by Lord Atkin without any relevant criticism. Clearly, contractually the law is different, in that on the sale of a chattel there may be, and very often is, an implied warranty of fitness, but there is never an implied warranty of fitness in the sale of a house.'

It is not easy to understand entirely the reason for what one may call 'the owner's immunity'. Counsel for the defendants urged the view that perhaps the chief reason was that it was unfair that an owner should be called upon to meet a claim made after a lapse of years. He urged, too, that logically such immunity should be enjoyed also by builders for the same reason. For myself, I doubt the cogency of this reason. It is true that it might be difficult after a lapse of years to know whether a dangerous defect in a house was due to age, or interference, or negligent building. But that difficulty should surely be capable of being met by applying the general principles of the onus of proof under which the plaintiff would be required to prove that the dangerous defect was due to negligent building and not other causes. However, the rule as to the owner's immunity is too deeply embedded in the common law to be capable of disturbance by this court. Nonetheless, I would say that whilst it is contended that it is right to disallow an extension of the principles in *Donoghue* v. *Stevenson*, it is certainly right, in my judgment, to disallow any extension of the principles relating to owner's immunity. I respectfully adopt the reasoning of Lord McDermott, Lord Chief Justice of Ireland in the Northern Ireland case *Gallagher* v. *McDowell Ltd* [1961] NI 26, considering the question of the extension of the principles in *Donoghue* v. *Stevenson*:

'Will its extension to building contractors cause or lead to some mischief which would justify the rejection of such an extension? The attitude that any enlargement of the field of tortious liability is always to be regarded as a step in the right direction is not one to be commended. Some gap between morality and law is inevitable and, if the gap is not too large, may be for the benefit of both codes. On the other hand, the changes to be expected in a progressive society call, from time to time, for such adjustments in the domain of legal responsibility as will promote justice and fair dealing. Those are terms which may mean different things in different epochs, but from the point of view of what is generally regarded as just and fair today I see no reason why a responsibility similar to that of the manufacturers and repairers of chattels should not fall upon the builders of houses, or why, if it does so fall, harm to trade or commerce or some other important facet of the life of the community should be apprehended. The immunities of the landowner seem to me to strengthen rather than weaken these conclusions.'

This left an artificial division between the ordinary builder, who could not be found liable in negligence, and the builder–owner, who remained immune from a negligence action, even in respect of identically defective work. Such an awkward divide could not remain unquestioned for long.

DUTTON v. BOGNOR REGIS URBAN DISTRICT COUNCIL

[1972] 1 QB 373   Court of Appeal

The plaintiff was the second purchaser of a house built by the United Building Company. The foundations were inspected by the local authority under their statutory powers, and approved. The plaintiff, in investigating structural defects that began to appear, discovered that the foundations were inadequate for the site, which was an old rubbish tip. The builder settled the claim against him, and the plaintiff succeeded against the local authority.

LORD DENNING MR (at p. 392):

In this case the significant thing, to my mind, is that the legislature gives the local authority a great deal of *control* over building work and the way it is done. They make byelaws governing every stage of the work. They require plans to be submitted to them for approval. They appoint surveyors and inspectors to visit the work and see if the byelaws are being complied with. In case of any contravention of the byelaws, they can compel the owner to remove the offending work and make it comply with the byelaws. They can also take proceedings for a fine.

In my opinion, the control thus entrusted to the local authority is so extensive that it carries with it a duty. It puts on the council the responsibility of exercising that control properly and with reasonable care. The common law has always held that a right of control over the doing of work carries with it a degree of responsibility in respect of the work. Such has long been the case where an employer has the right to control the way in which an independent contractor does his work: see *Mersey Docks and Harbour Board* v. *Coggins &*

Griffith (Liverpool) Ltd [1947] AC 1. It is also the case when an owner or a local authority exercises control over property for the purpose of doing repairs: see *Mint* v. *Good* [1951] 1 KB 517; *Greene* v. *Chelsea Borough Council* [1954] 2 QB 127; *Brew Brothers Ltd* v. *Snax (Ross) Ltd* [1970] 1 QB 612; or over a grating in a highway—*McFarlane* v. *Gwalter* [1959] 2 QB 332 and *Scott* v. *Green & Sons (A Firm)* [1969] 1 WLR 301. So here, I think, the local authority, having a right of control over the building of a house, have a responsibility in respect of it. They must, I think, take reasonable care to see that the byelaws are complied with. They must appoint building inspectors to examine the work in progress. Those inspectors must be diligent and visit the work as occasion requires. They must carry out their inspection with reasonable care so as to ensure that the byelaws are complied with.

But to whom is that duty owed? And what are the consequences if it is not done?

The Position of the Builder

[Counsel for Bognor Regis Urban District Council] submitted that the inspector owed no duty to a purchaser of the house. He said that on the authorities the builder, Mr Holroyd, owed no duty to a purchaser of the house. The builder was not liable for his negligence in the construction of the house. So also the council's inspector should not be liable for passing the bad work.

I would agree that if the builder is not liable for the bad work the council ought not to be liable for passing it. So I will consider whether or not the builder is liable. Counsel for Bognor Regis Urban District Council relied on *Bottomley* v. *Bannister* [1932] 1 KB 458. That certainly supports his submission. But I do not think it is good law today.

In the 19th century, and the first part of this century, most lawyers believed that no one who was not a party to a contract could sue on it or anything arising out of it. They held that if one of the parties to a contract was negligent in carrying it out, no third person who was injured by that negligence could sue for damages on that account. The reason given was that the only duty of care was that imposed by the contract. It was owed to the other contracting party, and to no one else. Time after time counsel for injured plaintiffs sought to escape from the rigour of this rule. But they were met invariably with the answer given by [counsel for Wright] in *Winterbottom* v. *Wright* (1842) 10 M. & W. 109, 115:

> 'If we were to hold that the plaintiff could sue in such a case, there is no point at which such actions would stop. The only safe rule is to confine the right to recover to those who enter into the contract: if we go one step beyond that, there is no reason why we should not go fifty.'

So the courts confined the right to recover to those who entered into the contract. If the manufacturer or repairer of an article did it negligently, and someone was injured, the injured person could not recover: see *Earl* v. *Lubbock* [1905] 1 KB 253 and *Blacker* v. *Lake and Elliot Ltd* (1912) 106 LT 533. If the landlord of a house contracted with the tenant to repair it and failed to do it – or did it negligently – with the result that someone was injured, the injured person could not recover: see *Cavalier* v. *Pope* [1906] AC 428. If the owner of land built a house on it and sold it to a purchaser, but he did his work so negligently that someone was injured, the injured person could not recover: see *Bottomley* v. *Bannister* [1932] 1 KB 458. Unless in each case he was a party to the contract.

That 19th century doctrine may have been appropriate in the conditions then prevailing. But it was not suited to the 20th century. Accordingly it was done away with in *Donoghue* v. *Stevenson* [1932] AC 562. But the case only dealt with the manufacturer of an article. *Cavalier* v. *Pope* (on landlords) and *Bottomley* v. *Bannister* (on builders) were considered by the House in *Donoghue* v. *Stevenson*, but they were not overruled. It was suggested that they were distinguishable on the ground that they did not deal with chattels but with real property; see by Lord Atkin at p. 598 and by Lord Macmillan at p. 609. Hence they were treated by the courts as being still cases of authority. So much so that in 1936 a judge at first instance held that a builder who builds a house for sale is under no duty to build it carefully. If a person was injured by his negligence, he could not recover: see *Otto* v. *Bolton and Norris* [1936] 2 KB 46.

The distinction between chattels and real property is quite unsustainable. If the manufacturer of an article is liable to a person injured by his negligence, so should the builder of a house be liable. After the lapse of 30 years this was recognised. In *Gallagher* v. *N. McDowell Ltd* [1961] NI 26, Lord MacDermott CJ and his colleagues in the Northern Ireland Court of Appeal held that a contractor who built a house negligently was liable to a person injured by his negligence. This was followed by Nield J in *Sharpe* v. *E T Sweeting & Son Ltd* [1963] 1 WLR 665. But the judges in those cases confined themselves to cases in which the builder was only a contractor and was not the owner of the house itself. When the builder is himself the owner, they assumed that *Bottomley* v. *Bannister* was still authority for exempting him from liability for negligence.

There is no sense in maintaining this distinction. It would mean that a contractor who builds a house on another's land is liable for negligence in constructing it, but that a speculative builder, who buys land and himself builds houses on it for sale, and is just as negligent as the contractor, is not liable. That cannot be right. Each must be under the same duty of care and to the same persons. If a visitor is injured by the negligent construction, the injured person is entitled to sue the builder, alleging that he built the house negligently. The builder cannot defend himself by saying: 'True I was the builder; but I was the owner as well. So I am not liable.' The injured person can reply: 'I do not care whether you were the owner or not, I am suing you in your capacity as builder and that is enough to make you liable.'

We had a similar problem some years ago. The liability of a contractor doing work on land was said to be different from the liability of an occupier doing the selfsame work. We held that each was liable for negligence: see *Billings (A.C.) & Sons* v. *Riden* [1957] 1 QB 46, and our decision was upheld by the House of Lords: [1958] AC 240: see also *Miller* v. *South of Scotland Electricity Board*, [1958] SC 20, 37–8.

I hold, therefore, that a builder is liable for negligence in constructing a house – whereby a visitor is injured – and it is no excuse for him to say that he was the owner of it. In my opinion *Bottomley* v. *Bannister* is no longer authority. Nor is *Otto* v. *Bolton and Norris*. They are both overruled. *Cavalier* v. *Pope* [1906] has gone too.

(and at p. 397):

This case is entirely novel. Never before has a claim been made against a council or its surveyor for negligence in passing a house. The case itself can be brought within the words of Lord Atkin in *Donoghue* v. *Stevenson*: but it is a question whether we should apply them here. In *Dorset Yacht Co. Ltd* v. *Home Office* [1970] AC 1004, Lord Reid said, at p. 1023, that the words of Lord Atkin expressed a principle which ought to apply in general 'unless there is some justification or valid explanation for its exclusion'. So did Lord Pearson at p. 1054. But Lord Diplock spoke differently. He said it was a guide but not a principle of universal application (p. 1060). It seems to me that it is a question of policy which we, as judges, have to decide. The time has come when, in cases of new import, we should decide them according to the reason of the thing.

In previous times, when faced with a new problem, the judges have not openly asked themselves the question: what is the best policy for the law to adopt? But the question has always been there in the background. It has been concealed behind such questions as: Was the defendant under any duty to the plaintiff? Was the relationship between them sufficiently proximate? Was the injury direct or indirect? Was it foreseeable, or not? Was it too remote? And so forth.

Nowadays we direct ourselves to considerations of policy. In *Rondel* v. *Worsley* [1969] 1 AC 191, we thought that if advocates were liable to be sued for negligence they would be hampered in carrying out their duties. In *Dorset Yacht Co. Ltd* v. *Home Office*, we thought that the Home Office ought to pay for damage done by escaping Borstal boys, if the staff was negligent, but we confined it to damage done in the immediate vicinity. In *S.C.M. (United Kingdom) Ltd* v. *W.J. Whittall & Son Ltd* [1971] 1 QB 337, some of us thought that economic loss ought not to be put on one pair of shoulders, but spread among all the sufferers. In *Launchbury* v. *Morgans* [1971] 2 QB 245, we thought that as the owner of the family car was insured she should bear the loss. In short, we look at the relationship of the parties: and then say, as a matter of policy, on whom the loss should fall.

What are the considerations of policy here? I will take them in order.

First, Mrs Dutton has suffered a grievous loss. The house fell down without any fault of hers. She is in no position herself to bear the loss. Who ought in justice to bear it? I should think those who were responsible. Who are they? In the first place, the builder was responsible. It was he who laid the foundations so badly that the house fell down. In the second place, the council's inspector was responsible. It was his job to examine the foundations to see if they would take the load of the house. He failed to do it properly. In the third place, the council should answer for his failure. They were entrusted by Parliament with the task of seeing that houses were properly built. They received public funds for the purpose. The very object was to protect purchasers and occupiers of houses. Yet they failed to protect them. Their shoulders are broad enough to bear the loss.

Next I ask: is there any reason in point of law why the council should not be held liable? Hitherto many lawyers have thought that a builder (who was also the owner) was not liable. If that were truly the law, I would not have thought it fair to make the council liable when the builder was not liable. But I hold that the builder who builds a house badly is liable, even though he is himself the owner. On this footing, there is nothing unfair in holding the council's surveyor also liable.

Then I ask: If liability were imposed on the council, would it have an adverse effect on the work? Would it mean that the council would not inspect at all, rather than risk liability for inspecting badly? Would it mean that inspectors would be harassed in their work or be subject to baseless charges? Would it mean that they would be extra cautious, and hold up work unnecessarily? Such considerations have influenced cases in the past, as in *Rondel* v. *Worsley*. But here I see no danger. If liability is imposed on the council, it would tend, I think, to make them do their work better, rather than worse.

Next, I ask: Is there any economic reason why liability should not be imposed on the council? In some cases the law has drawn the line to prevent recovery of damages. It sets a limit to damages for economic loss, or for shock, or theft by escaping convicts. The reason is that if no limit were set there would be no end to the money payable. But I see no such reason here for limiting damages. In nearly every case the builder will be primarily liable. He will be insured and his insurance company will pay the damages. It will be very rarely that the council will be sued or found liable. If it is, much the greater responsibility will fall on the builder and little on the council.

Finally I ask myself: If we permit this new action, are we opening the door too much? Will it lead to a flood of cases which neither the council nor the courts will be able to handle? Such considerations have sometimes in the past led the courts to reject novel claims. But I see no need to reject this claim on this ground. The injured person will always have his claim against the builder. He will rarely allege – and still less be able to prove – a case against the council.

All these considerations lead me to the conclusion that the policy of the law should be, and is, that the council should be liable for the negligence of their surveyors in passing work as good when in truth it is bad.

With varying degrees of enthusiasm, Sachs LJ and Stamp LJ endorsed this conclusion. This extraordinary case represents a triumph of remorseless logic. The question is whether the local authority is liable, yet in answering that question positively, the Court of Appeal succeeded also in ensnaring all builders, owners or not, into the web of liability. This conclusion was reached irrespective of the very different levels of involvement in the building work of builder and local authority, but the general approach was soon endorsed.

ANNS v. MERTON LONDON BOROUGH COUNCIL

[1978] AC 728   House of Lords

For facts see p. 9, above.

For facts see p. 9, above.

LORD WILBERFORCE (at p. 753):

What then is the extent of the local authority's duty towards these persons? Although, as I have suggested, a situation of 'proximity' existed between the council and owners and occupiers of the houses, I do not think that a description of the council's duty can be based upon the 'neighbourhood' principle alone or upon merely any such factual relationship as 'control' as suggested by the Court of Appeal. So to base it would be to neglect an essential factor which is that the local authority is a public body discharging functions under statute: its powers and duties are definable in terms of public not private law. The problem which this type of action creates, is to define the circumstances in which the law should impose, over and above, or perhaps alongside, these public law powers and duties, a duty in private law towards individuals such that they may sue for damages in a civil court. It is in this context that the distinction sought to be drawn between duties and mere powers has to be examined.

Most, indeed probably all, statutes relating to public authorities or public bodies, contain in them a large area of policy. The courts call this 'discretion' meaning that the decision is one for the authority or body to make, and not for the courts. Many statutes also prescribe or at least presuppose the practical execution of policy decisions: a convenient description of this is to say that in addition to the area of policy or discretion, there is an operational area. Although this distinction between the policy area and the operational area is convenient, and illuminating, it is probably a distinction of degree; many 'operational' powers or duties have in them some element of 'discretion'. It can safely be said that the more 'operational' a power of duty may be, the easier it is to superimpose upon it a common law duty of care.

I do not think that it is right to limit this to a duty to avoid causing extra or additional damage beyond what must be expected to arise from the exercise of the power or duty. That may be correct when the act done under the statute *inherently* must adversely *affect* the interest of individuals. But many other acts can be done without causing any harm to anyone – indeed may be directed to preventing harm from occurring. In these cases the duty is the normal one of taking care to avoid harm to those likely to be affected.

Let us examine the Public Health Act 1936 in the light of this. Undoubtedly it lays out a wide area of policy. It is for the local authority, a public and elected body, to decide upon the scale of resources which it can make available in order to carry out its functions under Part II of the Act – how many inspectors, with what expert qualifications, it should recruit, how often inspections are to be made, what tests are to be carried out, must be for its decision. It is no accident that the Act is drafted in terms of functions and powers rather than in terms of positive duty. As was well said, public authorities have to strike a balance between the claims of efficiency and thrift (du Parcq LJ in *Kent* v. *East Suffolk Rivers Catchment Board* [1940] 1 KB 319, 338): whether they get the balance right can only be decided through the ballot box, not in the courts. It is said – there are reflections of this in the judgments in *Dutton* v. *Bognor Regis Urban District Council* [1972] 1 QB 373 – that the local authority is under no duty to inspect, and this is used as the foundation for an argument, also found in some of the cases, that if it need not inspect at all, it cannot be liable for negligent inspection: if it were to be held so liable, so it is said, councils would simply decide against inspection. I think that this is too crude an argument. It overlooks the fact that local authorities are public bodies operating under statute with a clear responsibility for public health in their area. They must, and in fact do, make their discretionary decisions responsibly and for reasons which accord with the statutory purpose; see *Ayr Harbour Trustees* v. *Oswald* (1883) 8 App Cas 623, 639, *per* Lord Watson:

'the powers which [section 10] confers are discretionary . . . But it is the plain import of the clause that the harbour trustees . . . shall be vested with, and shall avail themselves of, these discretionary powers, whenever and as often as they may be of opinion that the public interest will be promoted by their exercise.'

If they do not exercise their discretion in this way they can be challenged in the courts. Thus, to say that councils are under no duty to inspect, is not a sufficient statement of the position. They are under a duty to give proper consideration to the question whether they should inspect or not. Their immunity from attack, in the event of failure to inspect, in other words, though great is not absolute. And because it is not absolute, the necessary premise for the proposition 'if no duty to inspect, then no duty to take care in inspection' vanishes.

Passing then to the duty as regards inspection, if made. On principle there must surely be a duty to exercise reasonable care. The standard of care must be related to the duty to be performed – namely to ensure compliance with the byelaws. It must be related to the fact that the person responsible for construction in accordance with the byelaws is the builder, and that the inspector's function is supervisory. It must be related to the fact that once the inspector has passed the foundations they will be covered up, with no subsequent opportunity for inspection. But this duty, heavily operational though it may be, is still a duty arising under the statute. There may be a discretionary element in its exercise – discretionary as to the time and manner of inspection, and the techniques to be used. A plaintiff complaining of negligence must prove, the burden being on him, that action taken was not within the limits of a discretion bona fide exercised, before he can begin to rely upon a common law duty of care. But if he can do this, he should, in principle, be able to sue.

Is there, then, authority against the existence of any such duty or any reason to restrict it? It is said that there is an absolute distinction in the law between statutory duty and statutory power – the former giving rise to possible liability, the latter not, or at least not doing so unless the exercise of the power involves some positive act creating some fresh or additional damage.

My Lords, I do not believe that any such absolute rule exists: or perhaps more accurately, that such rules as exist in relation to powers and duties existing under particular statutes, provide sufficient definition of the rights of individuals affected by their exercise, or indeed their non-exercise, unless they take account of the possibility that, parallel with public law duties there may coexist those duties which persons – private or public – are under at common law to avoid causing damage to others in sufficient proximity to them.

He further explored the nature of the liability at p. 758:

There is, in my opinion, no difficulty about this. A reasonable man in the position of the inspector must realise that if the foundations are covered in without adequate depth or strength as required by the byelaws, injury to safety or health may be suffered by owners or occupiers of the house. The duty is owed to them – not of course to a negligent building owner, the source of his own loss. I would leave open the case of users, who might themselves have a remedy against the occupier under the Occupiers' Liability Act 1957. A right of action can only be conferred upon an owner or occupier, who is such when the damage occurs (see below). This disposes of the possible objection that an endless indeterminate class of potential plaintiffs may be called into existence.

This must be related closely to the purpose for which powers of inspection are granted, namely, to secure compliance with the byelaws. The duty is to take reasonable care, no more, no less, to secure that the builder does not cover in foundations which do not comply with byelaw requirements. The allegations in the statements of claim, in so far as they are based upon non-compliance with the plans, are misconceived.

(and at p. 759):

The damages recoverable include all those which foreseeably arise from the breach of the duty of care which, as regards the council, I have held to be a duty to take reasonable care to secure compliance with the byelaws. Subject always to adequate proof of causation, these damages may include damages for personal injury and damage to property. In my opinion they may also include damage to the dwelling house itself; for the whole purpose of the byelaws in requiring foundations to be of a certain standard is to prevent damage arising from weakness of the foundations which is certain to endanger the health or safety of occupants.

To allow recovery for such damage to the house follows, in my opinion, from normal principle. If classification is required, the relevant damage is in my opinion material, physical damage, and what is recoverable is the amount of expenditure necessary to restore the dwelling to a condition in which it is no longer a danger to the health or safety of persons occupying and possibly (depending on the circumstances) expenses arising from necessary displacement. On the question of damages generally I have derived much assistance from the judgment (dissenting on this point, but of strong persuasive force) of Laskin J in the Canadian Supreme Court case of *Rivtow Marine Ltd* v. *Washington Iron Works* [1973] 6 WWR 692, 715 and from the judgments of the New Zealand Court of Appeal (furnished by courtesy of that court) in *Bowen* v. *Paramount Builders (Hamilton) Ltd and McKay* [1975] 2 NZLR 546.

When does the cause of action arise? We can leave aside cases of personal injury or damage to other property as presenting no difficulty. It is only the damage for the house which requires consideration. In my respectful opinion the Court of Appeal was right when, in *Sparham-Souter* v. *Town and Country Developments (Essex) Ltd* [1976] QB 858 it abjured the view that the cause of action arose immediately upon delivery, i.e., conveyance of the defective house. It can only arise when the state of the building is such that there is present or imminent danger to the health or safety of persons occupying it. We are not concerned at this stage with any issue relating to remedial action nor are we called upon to decide upon what the measure of the damages should be; such questions, possibly very difficult in some cases, will be for the court to decide. It is sufficient to say that a cause of action arises at the point I have indicated.

The position of the builder was also considered, at p. 758:

I agree with the majority in the Court of Appeal in thinking that it would be unreasonable to impose liability in respect of defective foundations upon the council, if the builder, whose primary fault it was, should be immune from liability. So it is necessary to consider this point, although it does not directly arise in the present appeal. If there was at one time a supposed rule that the doctrine of *Donoghue* v. *Stevenson* [1932] AC 562 did not apply to reality, there is no doubt under modern authority that a builder of defective premises may be liable in negligence to persons who thereby suffer injury: see *Gallagher* v. *N. McDowell Ltd* [1961] NI 26 *per* Lord MacDermott CJ – a case of personal injury. Similar decisions have been given in regard to architects – (*Clayton* v. *Woodman & Son (Builders) Ltd* [1962] 2 QB 533 and *Clay* v. *A.J. Crump & Sons Ltd* [1964] 1 QB 533). *Gallagher's* case expressly leaves open the question whether the immunity against action of builder owners, established by older authorities (e.g. *Bottomley* v. *Bannister* [1932] 1 KB 458) still survives.

That immunity, as I understand it, rests partly upon the distinction being made between chattels and real property, partly upon the principle of 'caveat emptor' or, in the case where the owner leases the property, on the proposition 'for, fraud apart, there is no law against letting a tumbledown house': see *Robbins* v. *Jones* (1863) 15 CBNS 221, 240 *per* Erle CJ. But leaving aside such cases as arise between contracting parties, when the terms of the contract have to be considered (see *Voli* v. *Inglewood Shire Council* (1963) 110 CLR 74, 85, *per* Windeyer J), I am unable to understand why this principle or proposition should prevent recovery in a suitable case by a person, who has subsequently acquired the house, upon the principle of *Donoghue* v. *Stevenson*: the same rules should apply to all careless acts of a builder: whether he happens also to own the land or not. I agree generally with the conclusions of Lord Denning

MR on this point in *Dutton* v. *Bognor Regis Urban District Council* [1972] 1 QB 373, 392–4. In the alternative, since it is the duty of the builder (owner or not) to comply with the byelaws, I would be of opinion that an action could be brought against him, in effect, for breach of statutory duty by any person for whose benefit or protection the byelaw was made. So I do not think that there is any basis here for arguing from a supposed immunity of the builder to immunity of the council.

Again the same link is made, as Lord Wilberforce accepts that either both the builder and local authority are liable, or neither should be. The careful analysis of how the local authority may be liable is coupled with a clear understanding of the separate position of the builder. This latter aspect was subsequently developed by the Court of Appeal.

BATTY v. METROPOLITAN PROPERTY REALISATIONS

[1978] QB 554   Court of Appeal

The plaintiffs' house was on the edge of a valley. An earth slip on adjacent land caused damage to their garden, and evidence was given that the house itself was doomed. The plaintiffs succeeded against both builder and developer.

MEGAW LJ (at p. 567):

I now turn to the appeal by the second defendants, the builders. [Counsel for Metropolitan Property Realisations] very helpfully and clearly put his argument in the form of six submissions. The first submission was this. A builder should not be taken to be under any duty of care in relation to defects in or observable only upon land which is not available to him in connection with his operations and is neither owned by him nor is in his possession.

[Counsel for Metropolitan Property Realisations], properly and as I think inevitably, conceded that, as he put it, 'in general terms a duty situation can arise between a builder and an occupier with whom the builder is not in privity of contract'. A duty of the *Donoghue* v. *Stevenson* [1932] AC 562 type can arise, it is conceded, in relation to reality. But he contends that the duty extends only to defects – to symptoms of possible instability, for example, affecting properly built foundations – where those defects are in, or observable upon, the actual site on which the house is to be built. If defects exist on neighbouring land which is not in his ownership or possession – or in respect of which he would require someone else's permission to go upon it – there is no duty, it is said, to look for or to observe or to take any action in relation to such symptoms.

[Counsel for Metropolitan Property Realisations] contends that there is a consideration which ought to reduce the scope of the builder's *Donoghue* v.

Stevenson duty to a potential occupier of the house which he builds. That is on the basis of Lord Wilberforce's 'Secondly ... whether there are any considerations which ought to negative, or to reduce or limit the scope of the duty ...' The scope, says [counsel for Metropolitan Property Realisations], ought to be limited by reference to the builder's statutory duty under the Building Regulations. If he builds foundations which comply in all respects with those regulations and any other statutory provisions, and the defects are outside the area of the foundations themselves, then the builder has no further duty.

With all respect to [this] argument, I do not accept it. Of course, the question whether or not there has been a breach of the duty will depend on all relevant considerations going to the question: did the builder act as a competent and careful builder would have acted in what he did or did not do by way of examination and investigation? But I see no reason why, as a matter of law, or by reference to any question of policy considerations, as a matter of the existence of the duty, it should depend on whether or not the symptoms, the observable defects, are on land of which the builder has ownership or in respect of which he has a legal right of entry without requiring some other persons' permission.

I would reject the first submission.

The second submission made by [counsel for Metropolitan Property Realisations] was this. If a builder is under any such duty, it should be limited to defects which are discoverable without subsoil investigation. The way that [he] put his submission was: if there is a duty on the builder, it is not a very heavy one. It would not include subsoil investigation. If a builder knows that what he sees would endanger the stability of the house, it is difficult to say there is not a duty. [Counsel for Metropolitan Property Realisations], on being asked by a member of the court why should that not apply also if he ought to have known, submitted that that would put too high a burden on a builder, and that it is undesirable that the law should put such a duty on him. [He] accepted, as I understand it, that this submission of his must mean that he is submitting that a builder is under no duty to look at adjoining land: if, however, he does know, though he had no duty to look, of something on the adjoining land which indicates a danger without further investigation, he might be under a duty. I see no reason for limiting the duty as a matter of law in the manner in which it is suggested by that submission.

The third submission was: the second defendants were not under any higher duty than that attaching to any other builder who builds for a development company under a building contract. The argument as developed by [counsel for Metropolitan Property Realisations] was that the second defendants here were under no higher obligation than if they had been retained as builders by contractors at arm's length. That is, their duty is no higher because of the particular circumstances here of the relationship between the developers and the builders and what happened between them in relation to the inspection of the site before the building started and their arrangements in regard to the building and what was to happen in relation to it. It may well be that on certain facts a builder would be entitled to rely upon an examination which he knew or reasonably supposed had been made by others on whose competence the builder could properly and reasonably rely. But in this case, on the evidence, the true view is that the decision to build on this site was a joint decision of the builders and the developers, the second defendants and the first defendants. In my judgment, if one is a party to the decision to build on the particular site in circumstances such as were shown in the evidence to exist in this case, the *Donoghue* v. *Stevenson* duty applies. Indeed, I have difficulty in seeing how, as was a part of [counsel's] argument at one stage, the fact that there was here an intervening contract – the contract for the 999 year lease between the first defendants and the plaintiffs – can affect the question whether the second defendants are under a *Donoghue* v. *Stevenson* type of liability. For in *Donoghue* v. *Stevenson* itself the whole question of the manufacturer's liability was considered, and decided to exist, despite the existence of an intervening contract between the retailer and the purchaser of the goods.

I would reject the third submission, on the facts of this case. Again, in my view, it is not a question of the existence of the duty. It is a question whether, in a particular case, on the facts of that case, it has been broken.

The fourth submission was this: if a builder is to be taken to be under any such duty, then no cause of action for breach of it arises (a) in respect of damage to the house itself or (b) alternatively until the house itself is damaged or is in such a state as to present present or imminent danger to the health or safety of the occupier.

This submission, essentially, as I see it, is founded on a question which was raised and the view which was expressed by Stamp LJ in *Dutton* v. *Bognor Regis Urban District Council* [1972] 1 QB 373. Stamp LJ said, at p. 415c:

> 'What causes the difficulty – and it is I think at this point that the court is asked to apply the law of negligence to a new situation – is that whereas the builder had, as I will assume, no duty to the plaintiff not carelessly to build a house with a concealed defect, yet it is sought to impute a not dissimilar duty to the defendant council. At this point I repeat and emphasise the difference between the position of a local authority clothed with the authority of an Act of Parliament to perform the function of making sure that the foundations of a house are secure for the benefit of the subsequent owners of the house and a builder who is concerned to make a profit. So approaching the matter, there is in my judgment nothing illogical or anomalous in fixing the former with a duty to which the latter is not subject. The former by undertaking the task is in my judgment undertaking a responsibility at least as high as that which the defendant in the *Hedley Byrne* case [1964] AC 465 would in the opinion of the majority in the House of Lords have undertaken had he not excluded responsibility . . .'

I pause here to say that obviously this fourth proposition cannot be treated as entirely independent of the first [counsel's] propositions, with which I have already dealt.

The doubt which was raised by that passage in that judgment of Stamp LJ was, as I see it, put at rest by passages in the speech of Lord Wilberforce in *Anns* v. *Merton London Borough Council* [1977] 2 WLR 1024, 1038 and 1039, (quoted above, p. 118–122).

The argument that the local authority should not be liable because it would be unreasonable that it should be held liable when the builder was not held liable was rejected, because Lord Wilberforce sees no reason why the builder should not be held liable. True, he specifically refers to a case where there was a breach of statutory duty – non-compliance with the Building Regulations. But that was the particular question arising in that case. I see no logical or practical reason for so confining it; nor, in my view, did Lord Wilberforce so intend.

As to the question of the nature of the damage which gives rise to the cause of action, it seems to me that an answer given by [counsel for Batty] was a simple and full answer. If it indeed is necessary that it should be shown that there has been physical damage to the property before the action will lie against the builder, in the present case there was physical damage to the property in the landslide, or landslip, of 1974. True, the foundations of the house for the time being remained undisturbed. True, the bricks and mortar of the house, as the judge has found, remained undamaged. But there was physical damage to the garden – a part of the property convened. If physical damage be necessary in order to found the action, there was physical damage. But, apart from what might be regarded as that possibly accidental element here, there is, I think, a wider reason why [the] proposition [by counsel for Metropolitan Property Realisations] should not succeed on the facts of this case.

Again I refer to the speech of Lord Wilberforce in *Anns* v. *Merton London Borough Council* [1977] 2 WLR 1024. Lord Wilberforce is dealing with the question '*When does the cause of action arise?*'; and he uses this sentence, at p. 1039:

'It can only arise when the state of the building is such that there is present or imminent danger to the health or safety of persons occupying it . . .'

Was there not here imminent danger to the health or safety of persons occupying this house, at the time when the action was brought? Indeed, Mrs Batty, one of the plaintiffs, has been awarded damages for the consequences to her health and peace of mind of the foreseen disaster. Why should this not be treated as being a case of imminent danger to the safety and health of people occupying the house? No one knows, or can say with certainty – not even the greatest expert – whether the foundations of the house will move and the house perhaps suddenly tumble tomorrow, or in a year's time, or in three years' time, or in 10 years' time. The law, in my judgment, is not so foolish as to say that a cause of action against the builder does not arise in those circumstances because there is no *imminent* danger. I would reject that submission.

[The fifth submission by counsel for Metropolitan Property Realisations] is one that goes to the facts. It is this. In any event the state of the terrain was not such as to render it careless for the second defendants not to require further investigation prior to the commencement of the building. [He] has properly taken us to passages in the evidence, in addition to the passages to which we had been referred by [counsel] for the first defendants, going to this question as to what would have been seen on the site and the neighbourhood of the site at the relevant time before building started, and as to the evidence as to what that ought to have conveyed to a reasonably careful builder observing those symptoms. Once again, I do not propose to go into the evidence on those matters, helpful though counsel's references to it were. I am quite satisfied that on the evidence the judge was right in his finding that the symptoms were such that investigation was called for by a reasonably careful builder, and that if the investigation which was called for by reason of those symptoms had been made the house would not have been built.

   The sixth and last of [his] submissions is this. The plaintiffs' house has not suffered damage and is not in such a state as to produce present or imminent danger to the plaintiffs' health or safety. [Counsel for Metropolitan Property Realisations] made it clear that he was not challenging the judge's finding that the house was likely to be engulfed or the judge's finding that it is now valueless. This is really related to the fourth submission, with which I have already dealt, and in that I have said all that I think it is necessary to say and all that I am minded to say referring to this proposition.

   I would therefore dismiss the appeals of the first and second defendants; and, by reference to the plaintiffs' cross-notice, I would direct that judgment be entered for the plaintiffs against the first defendants for liability in tort as well as for liability in contract.

Now, the picture is totally different. Before proceeding to examine how the courts have demolished this area of law as quickly as they had built it up, it is right to pause and see what the courts had achieved.

Dutton, Anns and *Batty* together represent a fascinating example of judicial activism. The growing awareness in the 1970s that the tort of negligence was a set of principles, rather than a mere collection of isolated examples, fuelled the remorseless logic of this trio of decisions. If you're liable for products which cause harm, so by analogy you should be liable for premises; if a builder who is not an owner is liable, so should a builder who is an owner; if builders are liable, then so are local authorities. The end result, within a very few years, is a liability across the board for negligently caused present or imminent threats to the health and safety of occupiers.

Such a liability is clearly in the interests of users of buildings. Often prevented from using contractual remedies – against the builder by the doctrine of privity of contract and against the vendor by the still strong *caveat emptor* principle – the plaintiff could find recourse in tort instead. That recourse, however, existed only in a limited form; the threat of injury had to exist and, as ever, the burden of proving that there had been a breach of duty lay on the plaintiff. The courts had done in most senses less to protect the users of buildings than Parliament had done during the same period in the Defective Premises Act 1972 and other legislation (see pp. 169 below).

So what went wrong? The facts of *Batty* form the basis of an initial explanation. As it happens, both person and property were harmed by the failure to build on a suitable site, though only to the extent of damage to Mrs Batty's nerve-endings and a few of her shrubs. It is clear, though, that even if these chance elements were not present, the decision in *Batty* would have been the same, on the evidence of the inevitable and imminent destruction of the property. (As a footnote, it should be recorded that the earth movement subsequently ceased, and years later the house was still standing, and habitable.)

On this view, we have the plaintiff obtaining substantial damages in negligence when no physical damage of any kind has occurred in a claim based on the mere threat of future damage, and that type of damage can easily be regarded as purely economic in nature, with all the restrictions that are invoked in that category of claims.

D & F ESTATES v. CHURCH COMMISSIONERS FOR ENGLAND

[1989] AC 177 House of Lords

For facts, see p. 33 above.

LORD BRIDGE characterized the loss as economic, at p. 206 (see p. 33–34 above). He continued, at p. 207:

It seems to me clear that the cost of replacing the defective plaster itself, either as carried out in 1980 or as intended to be carried out in future, was not an item of damage for which the builder of Chelwood House could possibly be made liable in negligence under the principle of *Donoghue* v. *Stevenson* [1932] AC 562 or any legitimate development of that principle. To make him so liable would be to impose upon him for the benefit of those with whom he had no contractual relationship the obligation of one who warranted the quality of the plaster as regards materials, workmanship and fitness for purpose. I am glad to reach the conclusion that this is not the law, if only for the reason that a conclusion to the opposite effect would mean that the courts, in developing the common law, had gone much farther than the legislature were prepared to go in 1972, after comprehensive examination of the subject by the Law Commission, in making builders liable for defects in the quality of their work to all who subsequently acquire interests in buildings they have erected. The statutory duty imposed by the Act of 1972 was confined to dwelling-houses and limited to defects appearing within six years. The common law duty, if it existed, could not be so confined or so limited. I cannot help feeling that consumer protection is an area of law where legislation is much better left to the legislators.

LORD OLIVER (at p. 211):

It is, I think, clear that the decision of this House in *Anns* v. *Merton London Borough Council* [1978] AC 728 introduced, in relation to the construction of buildings, an entirely new type of product liability, if not, indeed, an entirely novel concept of the tort of negligence. What is not clear is the extent of the liability under this new principle. In the context of the instant appeal, the key passage from the speech of Lord Wilberforce in that case is that which commences at p. 759, and which has already been quoted by my noble and learned friend.

A number of points emerge from this:

(1)   The damage which gives rise to the action may be damage to the person or to property on the ordinary *Donoghue* v. *Stevenson* [1932] AC 562 principle. But it may be damage to the defective structure itself which has, as yet, caused no injury either to person or to other property, but has merely given rise to a risk of injury.

(2)   There may not even be 'damage' to the structure. It may have been inherently defective and dangerous ab initio without any deterioration between the original construction and the perception of risk.

(3)   The damage to or defect in the structure, if it is to give rise to a cause of action, must be damage of a particular kind, i.e. damage or defect likely to cause injury to health or – possibly – injury to other property (an extension arising only by implication from the approval by this House of the decision of the Court of appeal in *Dutton* v. *Bognor Regis Urban District Council* [1972] 1 QB 373).

(4)   The cause of action so arising does not arise on delivery of the defective building or on the occurrence of the damage but upon the damage becoming a 'present or imminent risk' to health or (semble) to property and it is for that risk that compensation is to be awarded.

(5)   The measure of damage is at large but, by implication from the approval of the dissenting judgment in the Canadian case referred to (*Rivtow Marine Ltd* v. *Washington Iron Works* [1973] 6 WWR 692), it must at least include the cost of averting the danger.

These propositions involve a number of entirely novel concepts. In the first place, in no other context has it previously been suggested that a cause of action in tort arises in English law for the defective manufacture of an article which causes no injury other than injury to the defective article itself. If I buy a secondhand car to which there has been fitted a pneumatic tyre which, as a result of carelessness in manufacture, is dangerously defective and which bursts, causing injury to me or to the car, no doubt the negligent manufacturer is liable in tort on the ordinary application of *Donoghue* v. *Stevenson*. But if the tyre bursts without causing any injury other than to itself or if I discover the defect before a burst occurs, I know of no principle upon which I can claim to recover from the manufacturer in tort the cost of making good the defect which, in practice, could only be the cost of supplying and fitting a new tyre.

That would be, in effect, to attach to goods a non-contractual warranty of fitness which would follow the goods into whosoever's hands they came. Such a concept was suggested, obiter, by Lord Denning MR in *Dutton's* case, at p. 396, but it was entirely unsupported by any authority and is, in my opinion, contrary to principle.

The proposition that damages are recoverable in tort for negligent manufacture when the only damage sustained in either an initial defect in or subsequent injury to the very thing that is manufactured is one which is peculiar to the construction of a building and is, I think, logically explicable only to the hypothesis suggested by my noble and learned friend, Lord Bridge of Harwich, that in the case of such a complicated structure the other constituent parts can be treated as separate items of property distinct from that portion of the whole which has given rise to the damage – for instance, in *Anns'* case, treating the defective foundations as something distinct from the remainder of the building. So regarded this would be no more than the ordinary application of the *Donoghue* v. *Stevenson* principle. It is true that in such a case the damages would include, and in some cases might be restricted to, the costs of replacing or making good the defective part, but that would be because such remedial work would be essential to the repair of the property which had been damage by it.

But even so there are anomalies. If that were the correct analysis, then any damage sustained by the building should ground an action in tort from the moment when it occurs. But *Anns'* case tells us – and, at any rate so far as the local authority was concerned, this was a ground of decision and not merely obiter – that the cause of action does not arise until the damage becomes a present or imminent danger to the safety or health of the occupants and the damages recoverable are to be measured, not by the cost of repairing the damage which has been actually caused by the negligence of the builder, but by the (possibly much more limited) cost of putting the building into a state in which it is no longer a danger to the health or safety of the occupants.

It has, therefore, to be recognised that *Anns'* case introduced not only a new principle of a parallel common law duty in a local authority stemming from but existing alongside its statutory duties and conditioned by the purpose of those statutory duties, but also an entirely new concept of the tort of negligence in cases relating to the construction of buildings. The negligent builder is not answerable for all the reasonably foreseeable consequences of his negligence, but only for consequences of a particular type. Moreover, the consequence which triggers the liability is not, in truth, the damage to the building, qua damage, but the creation of the risk or apprehension of damage involving the safety of person or property. Take, for instance, the case of a building carelessly constructed in a manner which makes it inherently defective ab initio but where the defect comes to light only as a result, say, of a structural survey carried out several years later at the instance of a subsequent owner. What gives rise to the action is then not 'damage' in any accepted sense of the word but the perception of possible but avoidable damage in the future. The logic of according the owner a remedy at that stage is illustrated by the dissenting judgment of Laskin J in the Canadian case referred to and it is this: if the plaintiff had been injured the negligent builder would undoubtedly have been liable on *Donoghue* v. *Stevenson* principles. He has not been injured, but he has been put on notice to an extent sufficient to deprive himself of any remedy if he *is* now injured and he therefore suffers, and suffers only, the immediate economic loss entailed in preventing or avoiding the injury and the concomitant liability for it of the negligent builder which his own perception has brought to his attention. It is fair therefore that he should recover this loss, which is as much due to the fault of the builder as would have been the injury if it had occurred. Thus it has to be accepted either that the damage giving rise to the cause of action is pure economic loss not consequential upon injury to person or property – a concept not so far accepted into English law outside the *Hedley Byrne* type of liability (*Hedley Byrne & Co. Ltd* v. *Heller and Partners Ltd* [1964] AC 465) – or that there is a new species of the tort of negligence in which the occurrence of actual damage is no longer the gist of the action but is replaced by the perception of the risk of damage.

I think that it has to be accepted that this involves an entirely new concept of the common law tort of negligence in relation to building cases. Its ambit remains, however, uncertain. So far as *Anns'* case was concerned with liability arising from breach of statutory duty, the liability of the builder was a matter of direct decision. No argument was advanced on behalf of the builder in that case, but it was an essential part of the rationale of the decision in relation to the liability of the local authority that there was a precisely parallel and co-existing liability in the builder. Moreover, it is, I think, now entirely clear that the vendor of a defective building who is also the builder enjoys no immunity from the ordinary consequences of his negligence in the course of constructing the building, but beyond this and so far as the case was concerned with the extent of or limitations on his liability for common law negligence divorced from statutory duty, Lord Wilberforce's observations were, I think, strictly obiter. My Lords, so far as they concern such liability in respect of damage which has actually been caused by the defective structure other than by direct physical damage to persons or to other property, I am bound to say that, with the greatest respect to their source, I find them difficult to reconcile with any conventional analysis of the underlying basis of liability in tort for negligence. A cause of action in negligence at common law which arises only when the sole damage is the mere existence of the defect giving rise to the possibility of damage in the future, which crystallizes only when that damage is imminent, and the damages for which are measured, not by the full amount of the loss attributable to the defect but by the cost of remedying it only to the extent necessary to avert a risk of physical injury, is a novel concept. Regarded as a cause of action arising not from common law negligence but from breach of a statutory duty, there is a logic in so limiting it as to conform with the purpose for which the statutory duty was imposed, that is to say, the protection of the public from injury to health or safety. But there is, on that footing, no logic in extending liability for a breach of statutory duty to cases where the risk of injury is a risk of injury to property only, nor, as it seems to me, is there any logic in importing into a pure common law claim in negligence against a builder the limitations which are directly related only to breach of a particular statutory duty. For my part, therefore, I think the correct analysis, in principle, to be simply that, in a case where no question of breach of statutory duty arises, the builder of a house or other structure is liable at common law for negligence only where actual damage, either to person or to property, results from carelessness on his part in the course of construction. That the liability should embrace damage to the defective article itself is, of course, an anomaly which distinguishes it from liability for the manufacture of a defective chattel but it can, I think, be accounted for on the basis which my noble and learned friend, Lord Bridge of Harwich, suggested, namely that, in the case of a complex structure such as a building, individual parts of the building fall to be treated as separate and distinct items of property. On that footing, damage caused to other parts of the building from, for instance, defective foundations or defective steel-work would ground an action but not damage to the defective part itself except in so far as that part caused other damage, when the damages would include the cost of repair to that part so far as necessary to remedy damage caused to other parts. Thus, to remedy cracking in walls and ceilings caused by defective foundations necessarily involves repairing or replacing the foundations themselves. But, as in the instant case, damage to plaster caused simply by defective fixing of the plaster itself would ground no cause of action apart from contract or under the Defective Premises Act 1972. On what basis, apart from statute, is a builder, in contradistinction to the manufacturer of a chattel, to be made liable beyond this? There is, so far as I am aware, and apart from *Dutton* v. *Bognor Regis Urban District Council* [1972] 1 QB 373, no English authority prior to *Anns* v. *Merton London Borough Council* [1978] AC 728 supporting or even suggesting such a liability. *Dutton's* case was followed by the Court of Appeal in New Zealand in *Bowen* v. *Paramount Builders (Hamilton) Ltd* [1977] 1 NZLR 394 where Richmond P, at p. 410, defined the builder's duty as:

'a duty of care not to create latent sources of physical danger to the person or property of third persons whom he ought reasonably to foresee as likely to be affected thereby'.

He could see no reason why 'if the latent defect causes actual physical damage to the structure of the house' such damage should not give rise to a cause of action. In so holding, the court was clearly influenced by certain United States decisions whose authority has now been much reduced if not destroyed by the Supreme Court decision in *East River Steamship Corporation* v. *Transamerica Delaval Inc.*, (1986) 106 S.Ct. 2295 referred to by my noble and learned friend. The measure of damage in *Bowen's* case went a great deal beyond that suggested in *Anns'* case, for it not only covered the cost of putting the building into a state in which it was no longer dangerous to health or safety but extended to the restoration of its aesthetic appearance and depreciation in value. This really suggests what is, in effect, a transmissible warranty of fitness and, for the reasons already mentioned, I do not for my part think that *Bowen's* case can be supported as an accurate reflection of the law of England. *Rivtow Marine Ltd* v. *Washington Ironworks* [1973] 6 WWR 692, the dissenting judgment in which was, to some extent, relied upon by Lord Wilberforce in *Anns'* case, does not, I think, really assist very much. It is true that it was there held by the majority of the Supreme Court of Canada that the manufacturers and the supplier of defective equipment were liable for the economic loss suffered by the plaintiff as a result of the defective equipment having to be taken out of service, but the basis for the decision was the doctrine of reliance established by the *Hedley Byrne* case which placed upon the defendants a duty to warn of defects of which they were aware. Even on this basis, however, the damages did not exend to the cost of repairing the defective article itself.

Since *Anns'* case there have, of course, been the decision of the Court of Appeal in *Batty* v. *Metropolitan Property Realisations Ltd* [1978] QB 554 and the decision of this House in *Junior Books Ltd* v. *Veitchi Co. Ltd* [1983] 1 AC 520. I do not, for my part, think that the latter is of any help in the present context. As my noble and learned friend, Lord Bridge of Harwich, has mentioned it depends upon so close and unique a relationship with the plaintiff that it is really of no use as an authority on the general duty of care and it rests, in any event, upon the *Hedley Byrne* doctrine of reliance. So far as the general limits of the general duty of care in negligence are concerned, I, too, respectfully adopt what is said in the dissenting speech in that case of Lord Brandon of Oakbrook.

Batty v. *Metropolitan Property Realisations Ltd*, however, is directly in point and it needs to be carefully considered because it is, in my opinion, equally difficult to reconcile with any previously accepted concept of the tort of negligence. The defendant builder in that case had previously owned the land on which the plaintiff's house was built and was working in close conjunction with the plaintiff's vendor, who had bought the land from him. Thus the plaintiffs had a contractual relationship with the vendor, but none with the builder. There was no negligence in the construction of the house as such, nor was there any breach of statutory duty, nor had any damage yet occurred to the house. The negligence consisted solely in not appreciating what the builder ought reasonably to have appreciated, that is to say, that the immediately adjacent land was in such a condition that it would ultimately bring about the subsidence of the plaintiff's land and the consequent destruction of anything built upon it. At the date of the action and of the hearing no actual damage had been occasioned to the house. All that had happened was that a part of the garden had subsided, that being the event which alerted the plaintiffs to the danger which threatened the house. That, however, was not an event in any way attributable to fault on anyone's part but was attributable merely to the natural condition of the adjoining land. So that although there had been physical damage to the garden, it was not physical damage caused by any neglect on the part of the builders. The case is thus, on analysis, one in which the claim was for damages for pure economic loss caused by the putting onto the market of a product which, because defective, would become a danger to health and safety and thus of less value than it was supposed to be. It is not specified in the report of the case how the damages of £13,000 were calculated, but it seems that that sum must have been based on the difference between the market value of the house (which was doomed to destruction and therefore valueless) and the value of an equivalent house built on land not subject to landslips. Thus what the plaintiffs obtained from the builders by way of damages in tort was the sum for which the builders would have been liable if they had given an express contractual warranty of fitness – a sum related directly not to averting the danger created by the builders' negligence but to the replacement of an asset which, by reason of the danger, had lost its value. The decision in *Batty's* case was based upon *Anns'* case, but in fact went one step further because there was not in fact any physical damage resulting from the builders' negligence, although Megaw LJ, at p. 571, appears to have considered that what mattered was the occurrence of physical damage to some property of the plaintiff, however caused. As in *Anns*, the cause of action was related not to damage actually caused by the negligent act but to the creation of the danger of damage, and the case is therefore direct authority for the recovery of damages in negligence for pure economic loss – a proposition now firmly established in New Zealand: see *Mount Albert Borough Council* v. *Johnson* [1979] 2 NZLR 234.

My Lords, I confess to the greatest difficulty in reconciling this with any previously accepted concept of the tort of negligence at common law and I share the doubt expressed by my noble and learned friend, Lord Bridge of Harwich, whether it was correctly decided, at any rate so far as the liability of the builder was concerned. The case was, however, one in which the builder and the developer, with whom the plaintiffs had a direct contractual relationship, were, throughout, acting closely in concern and it may be that the actual decision, although not argued on this ground, can be justified by reference to the principle of reliance established by the decision of this House in *Hedley Byrne & Co. Ltd* v. *Heller & Partners Ltd* [1964] AC 465.

My Lords, I have to confess that the underlying logical basis for and the boundaries of the doctrine emerging from *Anns* v. *Merton London Borough Council* [1978] AC 728 are not entirely clear to me and it is in any event unnecessary for the purposes of the instant appeal to attempt a definitive exposition. This much at least seems clear: that in so far as the case is authority for the proposition that a builder responsible for the construction of a building is liable in tort at common law for damage occurring through his negligence to the very thing which he has constructed, such liability is limited directly to cases where the defect is one which threatens the health or safety of occupants or of third parties and (possibly) other property. In such a case, however, the damages recoverable are limited to expenses necessarily incurred in averting that danger. The case cannot, in my opinion, properly be adapted to support the recovery of damages for pure economic loss going beyond that, and for the reasons given by my noble and learned friend, with whose analysis I respectfully agree, such loss is not in principle recoverable in tort unless the case can be brought within the principle of reliance established by the *Hedley Byrne* case. In the instant case the defective plaster caused no damage to the remainder of the building and in so far as it presented a risk of damage to other property or to the person of any occupant that was remediable simply by the process of removal. I agree, accordingly, for the reasons which my noble and learned friend has given, that the cost of replacing the defective plaster is not an item for which the builder can be held liable in negligence.

What Lord Oliver was seeking to establish is that *Anns* had transgressed by crossing the frontier between physical damage and pure economic loss. Given the cogency of the earlier part of his speech in particular, it might logically be expected that he would have gone on to overrule the decision, but he did not take this drastic step, at this stage. It is clear that he was keen to restrict *Anns* to cases involving 'complex structures' (see pp. 153 below) or one where there was a particularly close relationship between the parties.

Again, the implications of this about-turn need to be examined carefully. There is something wrong with a building that is an imminent threat to its users' health and safety. It may have been built in just the same defective manner as the building next door which has already collapsed. Yet the *D & F* case seeks to distinguish between the two cases. Why? And why here?

Connoisseurs of platitudes will be well aware that the proverbial line has to be drawn somewhere. Given that the policy of the courts has always been to use duty of care to restrict claims, it has to be accepted that there will always be such boundaries and critics of the judges will always seek to denigrate these boundaries by citing troublesome examples. That said, this particular boundary does seem to be placed awkwardly. The tort of negligence has always given greater priority to the compensation of physical injury than to making amends for economic loss. It would therefore seem rational to allow claims where physical damage occurs or is threatened on a generous basis and, as in *Anns*, separate out such cases from those where there is a genuinely qualitative defect which may cost money to make good, but which does not create any threat to life or limb. Such a division would, of course, make no difference to the result in *D & F*, given the (perhaps surprising) finding that falling chunks of plaster were no threat to health and safety. The notion of liability centring on such a threat, borrowed from the statutory context of the local authority's powers, may be surprising in its origin, but reasonable in its effect.

The House of Lords clearly now needed to sort out the basis of its approach to claims for defective premises. The tension between *Anns* and *D & F* was productive of great uncertainty, which it was aimed to resolve by gathering a full House of seven Law Lords to re-examine this whole area in a local authority case.

MURPHY v. BRENTWOOD DISTRICT COUNCIL

[1991] 1 AC 398   House of Lords

For facts, see p. 31, above.

LORD KEITH cited Lord Denning MR in *Dutton* (p. 113, above) and continued (at p. 465).

The jump which is here made from liability under the *Donoghue* v. *Stevenson* principle for damage to person or property caused by a latent defect in a carelessly manufactured article to liability for the cost of rectifying a defect in such an article which is ex hypothesi no longer latent is difficult to accept. As Stamp LJ recognised in the same case, at pp. 414–15, there is no liability in tort upon a manufacturer towards the purchaser from a retailer of an article which turns out to be useless or valueless through defects due to careless manufacture. The loss is economic. It is difficult to draw a distinction in principle between an article which is useless or valueless and one which suffers from a defect which would render it dangerous in use but which is discovered by the purchaser in time to avert any possibility of injury. The purchaser may incur expense in putting right the defect, or, more probably, discard the article. In either case the loss is purely economic. Stamp LJ appears to have taken the view that in the case of a house the builder would not be liable to a purchaser where the defect was discovered in time to prevent injury but that a local authority which had failed to discover the defect by careful inspection during the course of construction was so liable.

Batty v. *Metropolitan Property Realisations Ltd* [1978] QB 554 was a case where a house which suffered no defects of construction had been built on land subject to the danger of slippage. A landslip carried away part of the garden but there was no damage to the house itself. Due to the prospect, however, that at some future time the house might be completely carried away, it was rendered valueless. There was no possibility of remedial works such as might save the house from being carried away. The Court of appeal allowed recovery in tort against the builder of damages based on loss of the value of the house. That again was purely economic loss.

(and at p. 468):

Upon analysis, the nature of the duty held by *Anns* [supra] to be incumbent upon the local authority went very much further than a duty to take reasonable care to prevent injury to safety or health. The duty held to exist may be formulated as one to take reasonable care to avoid putting a future inhabitant owner of a house in a position in which he is threatened, by reason of a defect in the house, with avoidable physical injury to person or health and is obliged, in order to continue to occupy the house without suffering such injury, to expend money for the purpose of rectifying the defect.

The existence of a duty of that nature should not, in my opinion, be affirmed without a careful examination of the implications of such affirmation. To start with, if such a duty is incumbent upon the local authority, a similar duty must necessarily be incumbent also upon the builder of the house. If the builder of the house is to be so subject, there can be no grounds in logic or in principle for not extending liability upon like grounds to the manufacturer of a chattel. That would open up an exceedingly wide field of claims, involving the introduction of something in the nature of a transmissible warranty of quality. The purchaser of an article who discovered that it suffered from a dangerous defect before that defect had caused any damage would be entitled to recover from the manufacturer the cost of rectifying the defect, and presumably, if the article was not capable of economic repair, the amount of loss sustained through discarding it. Then it would be open to question whether there should not also be a right to recovery where the defect renders the article not dangerous but merely useless. The economic loss in either case would be the same. There would also be a problem where the defect causes the destruction of the article itself, without causing any personal injury or damage to other property. A similar problem could arise, if the *Anns* principle is to be treated as confined to real property, where a building collapses when unoccupied.

(and at p. 470):

Liability under the *Anns* decision is postulated upon the existence of a present or imminent danger to health or safety. But considering that the loss involved in incurring expenditure to avert the danger is pure economic loss, there would seem to be no logic in confining the remedy to cases where such danger exists. There is likewise no logic in confining it to cases where some damage (perhaps comparatively slight) has been caused to the building, but refusing it where the existence of the danger has come to light in some other way, for example through a structural survey which happens to have been carried out, or where the danger inherent in some particular component or material has been revealed through failure in some other building. Then there is the question whether the remedy is available where the defect is rectified, not in order to avert danger to an inhabitant occupier himself, but in order to enable an occupier, who may be a corporation, to continue to occupy the building through its employees without putting those employees at risk.

In my opinion it is clear that *Anns* did not proceed upon any basis of established principle, but introduced a new species of liability governed by a principle indeterminate in character but having the potentiality of covering a wide range of situations, involving chattels as well as real property, in which it had never hitherto been thought that the law of negligence had any proper place.

(and at p. 471):

In my opinion there can be no doubt that *Anns* has for long been widely regarded as an unsatisfactory decision. In relation to the scope of the duty owed by a local authority it proceeded upon what must, with due respect to its source, be regarded as a somewhat superficial examination of principle and there has been extreme difficulty, highlighted most recently by the speeches in *D & F Estates* [1989] AC 177, in ascertaining upon exactly what basis of principle it did proceed. I think it must now be recognised that it did not proceed on any basis of principle at all, but constituted a remarkable example of judicial legislation. It has engendered a vast spate of litigation, and each of the cases in the field which have reached this House has been distinguished. Others have been distinguished in the Court of Appeal. The result has been to keep the effect of the decision within reasonable bounds, but that has been achieved only by applying strictly the words of Lord Wilberforce and by refusing to accept the logical implications of the decision itself. These logical implications show that the case properly considered has potentiality for collision with long-established principles regarding liability in the tort of negligence for economic loss. There can be no doubt that to depart from the decision would re-establish a degree of certainty in this field of law which it has done a remarkable amount to upset.

So far as policy considerations are concerned, it is no doubt the case that extending the scope of the tort of negligence may tend to inhibit carelessness and improve standards of manufacture and construction. On the other hand, overkill may present its own disadvantages, as was remarked in *Rowling* v. *Takaro Properties Ltd* [1988] AC 473, 502. There may be room for the view that *Anns*-type liability will tend to encourage owners of buildings found to be dangerous to repair rather than run the risk of injury. The owner may, however, and perhaps quite often does, prefer to sell the building at its diminished value, as happened in the present case.

It must, of course, be kept in mind that the decision has stood for some 13 years. On the other hand, it is not a decision of a type that would seem likely to be taken into account to a significant extent by citizens or indeed local authorities in ordering their affairs. No doubt its existence results in local authorities having to pay increased insurance premiums, but to be relieved of that necessity would be to their advantage, not to their detriment. To overrule it is unlikely to result in significantly increased insurance premiums for householders. It is perhaps of some significance that most litigation involving the decision consists in contests between insurance companies, as is largely the position in the present case. The decision is capable of being regarded as affording a measure of justice, but as against that the impossibility of finding any coherent and logically based doctrine behind it is calculated to put the law of negligence into a state of confusion defying rational analysis. It is also material that *Anns* has the effect of imposing upon builders generally a liability going far beyond that which Parliament thought fit to impose upon house builders alone by the Defective Premises Act 1972, a statute very material to the policy of the decision but not adverted to in it. There is much to be said for the view that in what is essentially a consumer protection field, as was observed by Lord Bridge of Harwich in *D & F Estates* [1989] AC 177, 207, the precise extent and limits of the liabilities which in the public interest should be imposed upon builders and local authorities are best left to the legislature.

My Lords, I would hold that *Anns* was wrongly decided as regards the scope of any private law duty of care resting upon local authorities in relation to their function of taking steps to secure compliance with building byelaws or regulations and should be departed from. It follows that *Dutton* v. *Bognor Regis Urban District Council* [1972] 1 QB 373 should be overruled, as should all cases subsequent to *Anns* which were decided in reliance on it.

LORD BRIDGE (at p. 475):

If a manufacturer negligently puts into circulation a chattel containing a latent defect which renders it dangerous to persons or property, the manufacturer, on the well known principles established by *Donoghue* v. *Stevenson* [1932] AC 562, will be liable in tort for injury to persons or damage to property which the chattel causes. But if a manufacturer produces and sells a chattel which is merely defective in quality, even to the extent that it is valueless for the purpose for which it is intended, the manufacturer's liability at common law arises only under and by reference to the terms of any contract to which he is a party in relation to the chattel; the common law does not impose on him any liability in tort to persons to whom he owes no duty in contract but who, having acquired the chattel, suffer economic loss because the chattel is defective in quality. If a dangerous defect in a chattel is discovered before it causes any personal injury or damage to property, because the danger is now known and the chattel cannot safely be used unless the defect is repaired, the defect becomes merely a defect in quality. The chattel is either capable of repair at economic cost or it is worthless and must be scrapped. In either case the loss sustained by the owner or hirer of the chattel is purely economic. It is recoverable against any party who owes the loser a relevant contractual duty. But it is not recoverable in tort in the absence of a special relationship of proximity imposing on the tortfeasor a duty of care to safeguard the plaintiff from economic loss. There is no such special relationship between the manufacturer of a chattel and a remote owner or hirer.

I believe that these principles are equally applicable to buildings. If a builder erects a structure containing a latent defect which renders it dangerous to persons or property, he will be liable in tort for injury to persons or damage to property resulting from that dangerous defect. But if the defect becomes apparent before any injury or damage has been caused, the loss sustained by the building owner is purely economic. If the defect can be repaired at economic cost, that is the measure of the loss. If the building cannot be repaired, it may have to be abandoned as unfit for occupation and therefore valueless. These economic losses are recoverable if they flow from breach of a relevant contractual duty, but, here again, in the absence of a special relationship of proximity they are not recoverable in tort.

(and at p. 479):

I have so far been considering the potential liability of a builder for negligent defects in the structure of a building to persons to whom he owes no contractual duty. Since the relevant statutory function of the local authority is directed to no other purpose than securing compliance with building byelaws or regulations by the builder, I agree with the view expressed in *Anns* [1978] AC 728 and by the majority of the Court of Appeal in *Dutton* [1972] 1 QB 373 that a negligent performance of that function can attract no greater liability than attaches to the negligence of the builder whose fault was the primary tort giving rise to any relevant damage. I am content for present purposes to assume, though I am by no means satisfied that the assumption is correct, that where the local authority, as in this case or in *Dutton*, have in fact approved the defective plans or inspected the defective foundations and negligently failed to discover the defect, their potential liability in tort is coextensive with that of the builder.

(and at p. 479):

A necessary element in the building owner's cause of action against the negligent local authority, which does not appear to have been contemplated in *Dutton* but which, it is said in *Anns*, must be present before the cause of action accrues, is that the state of the building is such that there is present or imminent danger to the health or safety of persons occupying it. Correspondingly the damages recoverable are said to include the amount of expenditure necessary to restore the building to a condition in which it is no longer such a danger, but presumably not any further expenditure incurred in any merely qualitative restoration. I find these features of the *Anns* doctrine very difficult to understand. The theoretical difficulty of reconciling this aspect of the doctrine with previously accepted legal principle was pointed out by Lord Oliver of Aylmerton in *D & F Estates* [1989] AC 177, 212D–213D. But apart from this there are, as it appears to me, two insuperable difficulties arising from the requirement of imminent danger to health or safety as an ingredient of the cause of action which lead to quite irrational and capricious consequences in the application of the *Anns* doctrine. The first difficulty will arise where the relevant defect in the building, when it is first discovered, is not a present or imminent danger to health or safety. What is the owner to do if he is advised that the building will gradually deteriorate, if not repaired, and will in due course become a danger to health and safety, but that the longer he waits to effect repairs the greater the cost will be? Must he spend £1,000 now on the necessary repairs with no redress against the local authority? Or is he entitled to wait until the building has so far deteriorated that he has a cause of action and then to recover from the local authority the £5,000 which the necessary repairs are now going to cost? I can find no answer to this conundrum. A second difficulty will arise where the latent defect is not discovered until it causes the sudden and total collapse of the building, which occurs when the building is temporarily unoccupied and causes no damage to property except to the building itself. The building is now no longer capable of occupation and hence cannot be a danger to health or safety. It seems a very strange result that the building owner should be without remedy in this situation if he would have been able to recover from the local authority the full cost of repairing the building if only the defect had been discovered before the building fell down.

All these considerations lead inevitably to the conclusion that a building owner can only recover the cost of repairing a defective building on the ground of the authority's negligence in performing its statutory function of approving plans or inspecting buildings in the course of construction if the scope of the authority's duty of care is wide enough to embrace purely economic loss. The House has already held in *D & F Estates* that a builder, in the absence of any contractual duty or of a special relationship of proximity introducing the *Hedley Byrne* principle of reliance, owes no duty of care in tort in respect of the quality of his work. As I pointed out in *D & F Estates*, to hold that the builder owed such a duty of care to any person acquiring an interest in the product of the builder's work would be to impose upon him the obligations of an indefinitely transmissible warranty of quality.

By section 1 of the Defective Premises Act 1972 Parliament has in fact imposed on builders and others undertaking work in the provision of dwellings the obligations of a transmissible warranty of the quality of their work and of the fitness for habitation of the completed dwelling. But besides being limited to dwellings, liability under the Act is subject to a limitation period of six years from the completion of the work and to the exclusion provided for by section 2. It would be remarkable to find that similar obligations in the nature of a transmissible warranty of quality, applicable to buildings of every kind and subject to no such limitations or exclusions as are imposed by the Act of 1972, could be derived from the builder's common law duty or from the duty imposed by building byelaws or regulations. In *Anns* Lord Wilberforce expressed the opinion that a builder could be held liable for a breach of statutory duty in respect of buildings which do not comply with the byelaws. But he cannot, I think, have meant that the statutory obligation to build in conformity with the byelaws by itself gives rise to obligations in the nature of transmissible warranties of quality. If he did mean that, I must respectfully disagree. I find it impossible to suppose that anything less than clear express language such as is used in section 1 of the Act of 1972 would suffice to impose such a statutory obligation.

As I have already said, since the function of a local authority in approving plans or inspecting buildings in course of construction is directed to ensuring that the builder complies with building byelaws or regulations, I cannot see how, in principle, the scope of the liability of the authority for a negligent failure to ensure compliance can exceed that of the liability of the builder for his negligent failure to comply.

There may, of course, be situations where, even in the absence of contract, there is a special relationship of proximity between the builder and building owner which is sufficiently akin to contract to introduce the element of reliance so that the scope of the duty of care owed by the builder to the owner is wide enough to embrace purely economic loss. The decision in *Junior Books Ltd* v. *Veitchi Co. Ltd* [1983] 1 AC 520 can, I believe, only be understood on this basis.

LORD OLIVER (at p. 484):

Despite the categorisation of the damage as 'material, physical damage' (*Anns, per* Lord Wilberforce, at p. 759) it is, I think, incontestable on analysis that what the plaintiffs suffered was pure pecuniary loss and nothing more. If one asks, 'What were the damages to be awarded *for*?' clearly they were not to be awarded for injury to the health or person of the plaintiffs for they had suffered none. But equally clearly, although the 'damage' was described, both in the Court of Appeal in *Dutton* and in this House in *Anns*, as physical or material damage, this simply does not withstand analysis. To begin with, it makes no sort of sense to accord a remedy where the defective nature of the structure has manifested itself by some physical symptom, such as a crack or a fractured pipe, but to deny it where the defect has been brought to light by, for instance, a structural survey in connection with a proposed sale. Moreover, the imminent danger to health or safety which was said to be the essential ground of the action was not the result of the physical manifestations which had appeared but of the inherently defective nature of the structure which they revealed. They were merely the outward signs of a deterioration resulting from the inherently defective condition with which the building had been brought into being from its inception and cannot properly be described as damage caused to the building in any accepted use of the word 'damage'.

(and at p. 485):

The critical question, as was pointed out in the analysis of Brennan J in his judgment in *Council of the Shire of Sutherland* v. *Heyman*, 157 CLR 424, is not the nature of the damage in itself, whether physical or pecuniary, but whether the scope of the duty of care in the circumstances of the case is such as to embrace damage of the kind which the plaintiff claims to have sustained: see *Caparo Industries plc.* v. *Dickman* [1990] 2 AC 605. The essential question which has to be asked in every case, given that damage which is the essential ingredient of the action has occurred, is whether the relationship between the plaintiff and the defendant is such – or, to use the favoured expression, whether it is of sufficient 'proximity' – that it imposes upon the latter a duty to take care to avoid or prevent that loss which has in fact been sustained. That the requisite degree of proximity may be established in circumstances in which the plaintiff's injury results from his reliance upon a statement or advice upon which he was entitled to rely and upon which it was contemplated that he would be likely to rely is clear from *Hedley Byrne* and subsequent cases, but *Anns* [1978] AC 728 was not such a case and neither is the instant case. It is not, however, necessarily to be assumed that the reliance cases form the only possible category of cases in which a duty to take reasonable care to avoid or prevent pecuniary loss can arise. *Morrison Steamship Co. Ltd* v. *Greystoke Castle (Cargo Owners)* [1947] AC 265, for instance, clearly was not a reliance case. Nor indeed was *Ross* v. *Caunters* [1970] Ch 297 so far as the disappointed beneficiary was concerned. Another example may be *Ministry of Housing and Local Government* v. *Sharp* [1970] 2 QB 223, although this may, on analysis, properly be categorised as a reliance case.
    Nor is it self-evident logically where the line is to be drawn. Where, for instance, the defendant's careless conduct results in the interruption of the electricity supply to business premises adjoining the highway, it is not easy to discern the logic in holding that a sufficient relationship of proximity exists between him and a factory owner who has suffered loss because material in the course of manufacture is rendered useless but that none exists between him and the owner of, for instance, an adjoining restaurant who suffers the loss of profit on the meals which he is unable to prepare and sell. In both cases the real loss is pecuniary. The solution to such borderline cases has so far been achieved pragmatically (see *Spartan Steel & Alloys Ltd* v. *Martin & Co. (Contractors) Ltd* [1973] QB 27) not by the application of logic but by the perceived necessity as a matter of policy to place some limits – perhaps

arbitrary limits – to what would otherwise be an endless, cumulative causative chain bounded only by theoretical foreseeability.

I frankly doubt whether, in searching for such limits, the categorisation of the damage as 'material', 'physical', 'pecuniary' or 'economic' provides a particularly useful contribution. Where it does, I think, serve a useful purpose is in identifying those cases in which it is necessary to search for and find something more than the mere reasonable foreseeability of damage which has occurred as providing the degree of 'proximity' necessary to support the action.

(at p. 487):

Anyone, whether he be a professional builder or a do-it-yourself enthusiast, who builds or alters a semi-permanent structure must be taken to contemplate that at some time in the future it will, whether by purchase, gift or inheritance, come to be occupied by another person and that if it is defectively built or altered it may fall down and injure that person or his property or may put him in a position in which, if he wishes to occupy it safely or comfortably, he will have to expend money on rectifying the defect. The case of physical injury to the owner or his licensees or his or their property presents no difficulty. He who was responsible for the defect – and it will be convenient to refer to him compendiously as 'the builder' – is, by the reasonable foreseeability of that injury, in a proximate 'neighbour' relationship with the injured person on ordinary *Donoghue* v. *Stevenson* principles. But when no such injury has occurred and when the defect has been discovered and is therefore no longer latent, whence arises that relationship of proximity required to fix him with responsibility for putting right the defect? Foresight alone is not enough but from what else can the relationship be derived? Apart from contract, the manufacturer of a chattel assumes no responsibility to a third party into whose hands it has come for the cost of putting it into a state in which it can safely continue to be used for the purpose for which it was intended. *Anns*, of course, does not go so far as to hold the builder liable for every latent defect which depreciates the value of the property but limits the recovery, and thus the duty, to the cost of putting it into a state in which it is no longer an imminent threat to the health or safety of the occupant. But it is difficult to see any logical basis for such a distinction. If there is no relationship of proximity such as to create a duty to avoid pecuniary loss resulting from the plaintiff's perception of non-dangerous defects, upon what principle can such a duty arise at the moment when the defect is perceived to be an imminent danger to health? Take the case of an owner–occupier who has inherited the property from a derivative purchaser. He suffers, in fact, no 'loss' save that the property for which he paid nothing is less valuable to him by the amount which it will cost him to repair it if he wishes to continue to live in it. If one assumes the parallel case of one who has come into possession of a defective chattel – for instance, a yacht – which may be a danger if it is used without being repaired, it is impossible to see upon what principle such a person, simply because the chattel has become dangerous, could recover the cost of repair from the original manufacturer.

The suggested distinction between more defect and dangerous defect which underlies the judgment of Laskin J in *Rivtow Marine Ltd* v. *Washington Iron Works* [1973] 6 WWR 692 is, I believe, fallacious. The argument appears to be that because, if the defect had not been discovered and someone had been injured, the defendant would have been liable to pay damages for the resultant physical injury on the principle of *Donoghue* v. *Stevenson* it is absurd to deny liability for the cost of preventing such injury from ever occurring. But once the danger ceases to be latent there never could be any liability. The plaintiff's expenditure is not expenditure incurred in minimising the damage or in preventing the injury from occurring. The injury will not now ever occur unless the plaintiff causes it to do so by courting a danger of which he is aware and his expenditure is incurred not in preventing an otherwise inevitable injury but in order to enable him to continue to use the property of the chattel.

My Lords, for the reasons which I endeavoured to state in the course of my speech in *D & F Estates Ltd* v. *Church Commissioners for England* [1989] AC 177 and which are expounded in more felicitous terms both in the speeches of my noble and learned friends in the instant case and in that of my noble and learned friend, Lord Keith of Kinkel, in *Department of the Environment* v. *Thomas Bates and Son Ltd* [1991] 1 AC 499, I have found it impossible to reconcile the liability of the builder propounded in *Anns* with any previously accepted principles of the tort of negligence and I am able to see no circumstances from which there can be deduced a relationship of proximity such as to render the builder liable in tort for pure pecuniary damage sustained by a derivative owner with whom he has no contractual or other relationship. Whether, as suggested in the speech of my noble and learned friend, Lord Bridge of Harwich, he could be held responsible for the cost necessarily incurred by a building owner in protecting himself from potential liability to third parties is a question upon which I prefer to reserve my opinion until the case arises, although I am not at the moment convinced of the basis for making such a distinction.

If, then, the law imposes upon the person primarily responsible for placing on the market a defective building no liability to a remote purchaser for expenditure incurred in making good defects which, ex hypothesi, have injured nobody, upon what principle is liability in tort to be imposed upon a local authority for failing to exercise its regulatory powers so as to prevent conduct which, on this hypothesis, is not tortious? Or, to put it another way, what is it, apart from the foreseeability that the builder's failure to observe the regulations may create a situation in which expenditure by a remote owner will be required, that creates the relationship of proximity between the authority and the remote purchaser? A possible explanation might, at first sight, seem to be that the relationship arises from the mere existence of the public duty of supervision imposed by the statute. That, I think, must have been the view of Stamp LJ in *Dutton* [1972] 1 QB 373, for he regarded the liability of the local authority as arising quite independently of that of the builder. His was, however, a minority view which derives no support from the reasoning of this House in *Anns* [1978] AC 728 and cannot stand up to analysis except on the basis (a) that the damage sustained was physical damage and (b) that the local authority, by reason of its ability to oversee the operation, was the direct cause of the defective construction. Neither of these propositions in my judgment is tenable.

The instant case is, to an extent, a stronger case than *Anns*, because there the authority was under no duty to carry out an inspection whereas here there was a clear statutory duty to withhold approval of the defective design. This, however, can make no difference in principle and the reasoning of the majority in *Anns*, which clearly links the liability of the local authority to that of the builder, must equally apply. The local authority's duty to future owners of the building to take reasonable care in exercising its supervisory function was expressed in *Anns* to arise 'on principle', but it is not easy to see what the principle was, unless it was simply the foreseeability of possible injury alone, which, it is now clear, is not in itself enough. The only existing principle upon which liability could be based was that propounded in *Dorset Yacht* [1970] AC 1004, that is to say, that the relationship which existed between the authority and the plaintiff was such as to give rise to a positive duty to prevent another person, the builder, from inflicting pecuniary injury. But in a series of decisions in subsequent cases – in particular *Curran* v. *Northern Ireland Co-ownership Housing Association Ltd* [1987] AC 718 and *Hill* v. *Chief Constable of West Yorkshire* – this House has been unable to find in the case of other regulatory agencies with powers as wide as or wider than those under the Public Health Acts such a relationship between the regulatory authority and members of the public for whose protection the statutory powers were conferred: see also *Yuen Kun-Yeu* v. *Attorney-General of Hong Kong* [1988] AC 175.

My Lords, I can see no reason why a local authority, by reason of its statutory powers under the Public Health Acts or its duties under the building regulations, should be in any different case. Ex hypothesi there is nothing in the terms or purpose of the statutory provisions which supports the creation of a private law right of action for breach of statutory duty. There is equally nothing in the statutory provisions which even suggests that the purpose of the statute was to protect owners of buildings from economic loss. Nor is there any easily discernible reason why the existence of the statutory duties, in contradistinction to those existing in the case of other regulatory agencies, should be held in the case of a local authority to create a special relationship imposing a private law duty to members of the public to prevent the conduct of another person which is not itself tortious. Take the simple example of the builder who builds a house with inadequate foundations and presents it to his son and daughter-in-law as a wedding present. It would be manifestly absurd, if the son spends money on rectifying the defect which has come to light, to hold him entitled to recover the expenditure from his father because the gift turns out to be less advantageous than he at first supposed. It seems to me no less absurd to hold that nevertheless there exists between the authority which failed properly to inspect and the donee of the property a relationship entitling the latter to recover from the authority the expenditure which he cannot recover from the donor. Yet that must be the logical result of the application of *Anns*, unless one is to say that the necessary relationship of proximity exists, not between the authority and all subsequent owners and occupiers, but only between the authority and the owners and occupiers who have acquired a property for value. With the greatest deference to the high authority of the opinions expressed in *Anns* and in *Dutton*, I cannot see, once it is recognised, as I think that it has to be, that the only damage sustained by discovery of the defective condition of the structure is pure pecuniary loss, how those decisions can be sustained as either an application or a permissible extension of existing principle.

The question that I have found most difficult is whether, having regard to the time which has elapsed and the enormous amount of litigation which has been instituted in reliance upon *Anns*, it is right that this House should now depart from it. In his speech in *Dorset Yacht* [1970] AC 1004, Lord Diplock observed, at p. 1064:

> 'As any proposition which relates to the duty of controlling another man to prevent his doing damage to a third deals with a category of civil wrongs of which the English courts have hitherto had little experience it would not be consistent with the methodology of the development of the law by judicial decision that any new proposition should be stated in wider terms than are necessary for the determination of the present appeal. Public policy may call for the immediate recognition of a new sub-category of relations which are the source of a duty of this nature additional to the sub-category described in the established proposition, but further experience of actual cases would be needed before the time became ripe for the coalescence of sub-categories into a broader category of relations giving rise to the duty, such as was effected with respect to the duty of care of a manufacturer of products in *Donoghue* v. *Stevenson* [1932] AC 562. Nevertheless, any new sub-category will form part of the English law of civil wrongs and must be consistent with its general principles.'

For the reasons which I have endeavoured to express I do not think that *Anns* can be regarded as consistent with those general principles. Nor do I think that it can properly be left to stand as a peculiar doctrine applicable simply to defective buildings, for I do not think that its logical consequences can be contained within so confined a compass. It may be said that to hold local authorities liable in damages for failure effectively to perform their regulatory functions serves a useful social purpose by providing what is, in effect, an insurance fund from which those who are unfortunate enough to have acquired defective premises can recover part at least of the expense to which they have been put or the loss of value which they have sustained. One cannot but have sympathy with such a view although I am not sure that I see why the burden should fall on the community at large rather than be left to be covered by private insurance. But, in any event, like my noble and learned friends, I think that the achievement of beneficial social purposes by the creation of entirely new liabilities is a matter which properly falls within the province of the legislature and within that province alone. At the date when *Anns* was decided the Defective Premises Act 1972, enacted after most careful consideration by the Law Commission, had shown clearly the limits within which Parliament had thought it right to superimpose additional liabilities upon those previously existing at common law and it is one of the curious features of the case that no mention even of the existence of this important measure, let alone of its provisions – and in particular the provision regarding the accrual of the cause of action – appears in any of the speeches or in the summary in the Law Reports of the argument of counsel.

There may be very sound social and political reasons for imposing upon local authorities the burden of acting, in effect, as insurers that buildings erected in their areas have been properly constructed in accordance with the relevant building regulations. Statute may so provide. It has not done so and I do not, for my part, think that it is right for the courts not simply to expand existing principles but to create at large new principles in order to fulfil a social need in an area of consumer protection which has already been perceived by the legislature but for which, presumably advisedly, it has not thought it necessary to provide. I would accordingly allow the appeal.

This blast against the whole series of 1970s developments has shaken the whole of the law of tort to its roots, and the repercussions of it are at their strongest in cases of liability for defective premises. It is therefore especially important to examine the case very carefully, so as to see what precisely the House of Lords has achieved.

First and foremost is the overruling of *Anns, Dutton* and *Batty*, and all other cases based thereupon. This cuts a huge swathe through the previous corpus of law. Note, though, that the Lords are clear (e.g. Lord Mackay at p. 32, above; Lord Keith at p. 137 above), that this over-ruling is only in the context of the liability of local authorities and, it is presumed in the light of the clear linkage in both law and fact between the two categories of defendant, in the area of builders' liability too. So some of the more general comments in *Anns* may survive, albeit now in much modified form after *Yuen Kun-Yeu* (see p. 16, above).

A second comment is to note the factual background of the *Murphy* case. No injury occurred to the occupants of the damaged house, and so the case is only authority for cases where there is no actual injury, but only the threat thereof. Indeed, on the facts, the threat of injury seemed remote; the plaintiff sold the house to a builder who, we are told, lived happily with his family without carrying out any remedial work, the subsidence having stabilized. Case-law subsequent to *Anns* (e.g. *Bluett* v. *Woodspring District Council* [1983] JPL 242) appeared to have established that subsidence that had stabilized was no longer a threat to health and safety. True, in *Murphy*, a gas pipe had fractured as a result of the subsidence but this was easily repaired and did not of itself indicate an ongoing threat of damage (see Ralph Gibson LJ in the Court of Appeal: [1991] 1 AC 398 at 415). So *Murphy*, like *D & F* before it, was not very satisfactory territory to apply the *Anns* formula. It is hard to see that this point will have much relevance, however, given the force of the attack on *Anns* and all its works.

The implications of leaving the Murphys of this world without a tort remedy should not be overlooked. Accepting that there was a threat of harm to the Murphy family caused by the prospect of a gas explosion, the response of the House of Lords is, to say the least, unsympathetic. Given that their loss arises from expenditure not just to make the house nicer but to make it safe, it seems harsh to describe this as a merely qualitative defect. A clear distinction appears in fact to be between the clear necessity to do work to prevent damage to life and the mere desirability of doing work which would improve or restore the quality of the property. It is suggested here, as in *D & F*, that deserving complainants are lumped in indiscriminately with less meritorious claims, thus causing potential injustice to them. After all, in one sense all loss is economic in that even a broken leg can only be compensated by money.

We should not dwell unduly on the destruction of *Anns*. Rather, we need to see what is left and try to construct a picture of the liability of builders, local authorities and others in the new post-*Murphy* world.

Chapter 8

Liability of the Builder

In considering the negligence liability of the 'builder', it should be borne in mind that that term covers a very wide variety of different people and different operations, from multinational firms working on major industrial projects to a local odd-job man building a new boundary wall. The general principles of the tort of negligence apply equally whatever the situation, although the exigencies of litigation mean that most of the reported cases are in the former category.

It is also fair to say that the modern construction project involves much teamwork and collaboration between the many different trades and professions concerned. It is therefore perhaps artificial to divide the building team up into its components but there is a clear distinction to be drawn between the supervisors and the supervised, and this chapter focuses on the latter category.

Builders' duties of care

(a) Duty to avoid threats to health and safety

This duty has been removed by *Murphy*. Lord Jauncey's speech neatly explains why.

MURPHY v. BRENTWOOD DISTRICT COUNCIL

[1991] 1 AC 398 House of Lords.

LORD JAUNCEY (at p. 497):

> My Lords, if, as I believe, the decision in *Anns* cannot be reconciled with the principle of *Donoghue* v. *Stevenson* upon the basis of the complex structure theory, is there any other established principle upon which it could be

justified? When Lord Wilberforce said that the damages recoverable might include those for damage to the house itself, it is clear that he was referring to damage separate from but caused by the defective foundations. However, the measure of such damages would be limited to what was necessary to remove the danger to the health or safety of the occupants, which might well include the cost of repairing the initial defect but might equally well be less than that required to repair all the damage. Furthermore, the cause of action would only arise when there was present or imminent danger to the occupants. Thus the two prerequisites to an action based on *Anns* were (1) the existence of material physical damage resulting from the original defect and (2) the presence or imminence of danger associated with that damage. These prerequisites give rise to a number of difficulties. In the first place, if the basis of the duty is that persons should not be placed in a position of danger it is difficult to draw a logical distinction between danger which manifests itself because of physical damage and danger which is discovered fortuitously, for example, by a survey or inspection. Why, it might be asked, should the houseowner in the latter case have no right of action if he takes steps to remove the danger before physical damage has occurred but have such a right if he waits until damage has occurred when remedial costs may very well be much higher? In the second place, the concept of imminent danger gives rise to considerable practical difficulties. Is a danger imminent when it is bound to occur, albeit not for some time, or is it imminent only if it is likely to occur in the immediate future? Different persons will have different views as to what constitutes imminence and plaintiffs will be in doubt as to when their causes of action accrue. If the house collapses without any warning and injures nobody any danger inherent in its construction has been removed. It would be a very strange result that the owner should have no remedy in such an event but should have a remedy if the danger had manifested itself before collapse.

Instead we now have to look at the damage and decide whether it will fall into the category of economic loss (which most threats to safety, as in *Murphy* itself, will) or into the category of physical injury, and analyse the relevant duties accordingly.

(b)   Duty not to cause economic loss

A good example of the way in which a case that in the past may have fallen within the *Anns* principle now comes into this category is provided by the House of Lords in a case argued before *Murphy* but the decision in which was given on the same day.

DEPARTMENT OF THE ENVIRONMENT v.
THOMAS BATES & SON

[1991] 1 AC 499   House of Lords

The plaintiffs leased office premises from EMI. The defendants were
responsible for the construction work under contract to EMI; the work
was close to completion when the plaintiffs acquired their interest in the
property. After eight years, leaks began to affect the property and
investigations showed that some concrete pillars were weak, and could
not support the design load of the building. The plaintiffs unsuccessfully
claimed for the cost of repair works.

LORD KEITH (at p. 519):

> The foundation of the plaintiffs' case is *Anns* v. *Merton London Borough
> Council* [1978] AC 728. That decision was concerned directly only with the
> liability in negligence of a local authority in respect of its functions in regard
> to securing compliance with building byelaws and regulations. The position of
> the builder as regards liability towards a remote purchaser of a building which
> suffered from defects due to carelessness in construction was touched on very
> briefly. However, it has since been generally accepted that similar principles
> govern the liability both of the local authority and of the builder.
>
> It has been held by this House in *Murphy* v. *Brentwood District Council*
> [1991] 1 AC 398 that *Anns* was wrongly decided and should be departed from,
> by reason of the erroneous views there expressed as to the scope of any duty
> of care owed to purchasers of houses by local authorities when exercising the
> powers conferred upon them for the purpose of securing compliance with
> building regulations. The process of reasoning by which the House reached its
> conclusion necessarily included close examination of the position of the
> builder who was primarily responsible, through lack of care in the construc-
> tion process, for the presence of defects in the building. It was the unanimous
> view that, while the builder would be liable under the principle of *Donoghue* v.
> *Stevenson* [1932] AC 562 in the event of the defect, before it had been
> discovered, causing physical injury to persons or damage to property other
> than the building itself, there was no sound basis in principle for holding him
> liable for the pure economic loss suffered by a purchaser who discovered the
> defect, however such discovery might come about, and required to expend
> money in order to make the building safe and suitable for its intended
> purpose.
>
> In the present case it is clear that the loss suffered by the plaintiffs is pure
> economic loss. At the time the plaintiffs carried out the remedial work on the
> concrete pillars the building was not unsafe by reason of the defective
> construction of these pillars. It did, however, suffer from a defect of quality
> which made the plaintiffs' lease less valuable than it would otherwise have
> been, in respect that the building could not be loaded up to its design capacity
> unless any occupier who wished so to load it had incurred the expenditure

necessary for the strengthening of the pillars. It was wholly uncertain whether during the currency of their lease the plaintiffs themselves would ever be likely to require to load the building up to its design capacity, but a purchaser from them might well have wanted to do so. Such a purchaser, faced with the need to strengthen the pillars, would obviously have paid less for the lease than if they had been sound. This underlines the purely economic character of the plaintiffs' loss. To hold in favour of the plaintiffs would involve a very significant extension of the doctrine of *Anns* so as to cover the situation where there existed no damage to the building and no imminent danger to personal safety or health. If *Anns* were correctly decided, such an extension could reasonably be regarded as entirely logical. The undesirability of such an extension, for the reason stated in *Murphy* v. *Brentwood District Council*, formed an important part of the grounds which led to the conclusion that *Anns* was not correctly decided. That conclusion must lead inevitably to the result that the plaintiffs' claim fails.

In effect, the House of Lords are of the view that there is no threat to health and safety here. Such a threat could only arise if the building was used right up to its full design load. The resulting inability so to use the building is merely a qualitative defect, affecting its value, and as such is economic loss. Even if there had been a more real threat to health and safety still, according to *Murphy*, the fact of it being known removes the danger since remedial work can be carried out; this too simply affects the value of the building and so it too is economic loss.

This has bizarre results. The good householder, seeing the first signs of cracks developing, will rush to carry out repair works. For this, he will get little reward from the law of negligence, which will count this as economic loss and which, usually, will be unclaimable. The idle householder, however, who lazily watches the cracks expand until the building collapses on him, has now suffered personal injury; this seems to form the basis of a valid claim (see p. 152, below) and so he is rewarded for his idleness (unless the court has the last laugh and holds that, in not repairing at an early stage, he has failed adequately to mitigate his losses).

However, this is not to say that no claim can be entertained in respect of economic loss. Remedies do exist, albeit highly restrictive ones, for economic loss caused by negligent acts. The source of a plaintiff's right of action remains *Junior Books* v. *Veitchi* (see p. 22 above). It will be recalled that this decision allows such a claim in situations of especially close proximity. More surprisingly, this much-criticized decision, flayed by judge and commentator alike, has survived *Murphy*, which might have been expected to overrule it but was content with claiming the scalp of *Anns* instead. Indeed, in *Murphy, Junior Books* is given modest support by for example, Lord Bridge (at p. 140 above).

In what circumstances might a builder be sued under the *Junior Books* principle? Lord Bridge sums up the case as revolving around:

(i)   a special relationship of proximity;
(ii)   that incorporates reliance; and
(iii)   which is akin to contract.

These criteria are not easy to satisfy in a case, like *Murphy*, involving an ordinary dwelling-house. There, although the householder is in fact relying on the skill of the contractors involved, they do not know of him or his requirements. This does not seem akin to contract, or to connote any especially close proximity. In the *Department of Environment* case, too, there would be no hope of employing a *Junior Books* argument successfully, since the building work was carried out for EMI on a speculative basis, and the plaintiff only came in to the picture when the building work was almost over. Again, far from having a special relationship, they have virtually no relationship at all.

However, it would be different if there had been prior dealings between the parties. If the office premises were known in advance to be occupied by the Department, their needs and concerns, including their intended use of the full design load, would be likely to become known to the contractors. Then the parties, although not necessarily privy to the same contract, would be in a relationship where reliance may well be placed by the Department on the contractors' judgment, and possible *Junior Books* liability might arise. The other situation where such a claim might well succeed is on similar facts to *Junior Books* itself i.e. where a sub-contractor is nominated by the employer and is closely involved with the project from its early stages – not exactly a rare situation. It is rash to predict anything in relation to negligence these days, but it may well be that plaintiffs, denied their valuable *Anns* remedy, will produce a modest revival in the fortunes of *Junior Books*.

(c)   Duty not to cause injury to person or property

Most buildings do not collapse. Even those which may will usually show signs of deterioration and their responsible owners will act to remedy the defects. Just occasionally, though, buildings do cause damage to their occupiers and their property. For example, defects in construction may remain hidden until triggered by an extraneous event, such as the gas explosion at Ronan Point, then fail catastrophically. Defective drainage work, or, more likely, defective foundations work damaging drains, is an obvious potential threat of personal injury in the form of disease. Is the builder liable for such damage?

In principle the answer should be that he is. A duty would seem to be owed, the future occupiers of a building being a small and definable group in relation to whom the builder could be said to be in proximity. The only difficulty comes from *Murphy*; the question of a duty on the local authority in such circumstances is regarded as an open question (see p. 222, below). Given the now longstanding acceptance of linkage between builders' liability and that of local authorities, it would seem likely that *if* a future court were to hold that local authorities owed no duty in respect of physical injury, then it would very much open up the question of whether the builder also should be so liable. That said, it would be difficult to justify a policy of non-compensation for the injured, and the discussion in *Murphy* of complex structures (see p. 154 below) suggests that there will be a duty to this category of plaintiffs.

(d)   Other possible duties

One of the less attractive features of the recent leading decisions in this area is the lack of restraint the senior judges show in making throwaway remarks that raise great issues, but do not answer them. This leads to great uncertainty and two such areas currently exist.

The first of these is 'complex structure' theory. As we have seen, where a building is damaged in itself, this is now seen as a purely economic loss. On the other hand, where an outside force, such as a defendant's lorry, harms the building, this is physical damage. Most buildings are very complicated structures and one part may harm another. The foundations, if defective, can harm the walls above; central heating pipes can burst and floodwater may damage the fabric of the building. Suggestions have been made that such cases are really instances of one part of the building harming another, and that therefore such cases should be seen as being ones of physical damage to the building.

D & F ESTATES v. CHURCH COMMISSIONERS FOR ENGLAND

[1989] AC 177 House of Lords

For facts, see p. 33, above.

LORD BRIDGE (at p. 206):

> I can see that more difficult questions may arise in relation to a more complex structure like a dwelling-house. One view would be that such a structure should be treated in law as a single indivisible unit. On this basis, if the unit becomes a potential source of danger when a hitherto hidden defect in construction manifests itself, the builder, as in the case of the garden wall, should not in principle be liable for the cost of remedying the defect. It is for

this reason that I now question the result, as against the builder, of the decision in *Batty* v. *Metropolitan Property Realisations Ltd* [1978] QB 554.

However, I can see that it may well be arguable that in the case of complex structures, as indeed possibly in the case of complex chattels, one element of the structure should be regarded for the purpose of the application of the principles under discussion as distinct from another element, so that damage to one part of the structure caused by a hidden defect in another part may qualify to be treated as damage to 'other property', and whether the argument should prevail may depend on the circumstances of the case. It would be unwise and it is unnecessary for the purpose of deciding the present appeal to attempt to offer authoritative solutions to these difficult problems in the abstract. I should wish to hear fuller argument.

See also Lord Oliver at p. 214, quoted at p. 131, above.

Though tentative in character, these remarks attracted instant attention, and much abuse. They had clear potential for widespread use, if even something as relatively straightforward as a dwelling-house were to be seen as a complex structure. They also had clear potential for chaos and confusion, however, since the new formulation was clearly inherently vague, and in need of major development and definition by the courts. Their Lordships have subsequently tried to put the lid back on this particular Pandora's Box.

MURPHY v. BRENTWOOD DISTRICT COUNCIL

[1991] 1 AC 398   House of Lords

For facts, see p. 31 above.

LORD KEITH (at p. 470):

In *D. & F. Estates Ltd* v. *Church Commissioners for England* [1989] AC 177 both Lord Bridge of Harwich and Lord Oliver of Aylmerton expressed themselves as having difficulty in reconciling the decision in *Anns* with pre-existing principle and as being uncertain as to the nature and scope of such new principle as it introduced. Lord Bridge, at p. 206, suggested that in the case of a complex structure such as a building one element of the structure might be regarded for *Donoghue* v. *Stevenson* purposes as distinct from another element, so that damage to one part of the structure caused by a hidden defect in another part might qualify to be treated as damage to 'other property'. I think that it would be unrealistic to take this view as regards a building the whole of which had been erected and equipped by the same contractor. In that situation the whole package provided by the contractor would, in my opinion, fall to be regarded as one unit rendered unsound as such by a defect in the particular part. On the other hand where, for example, the electric wiring had been installed by a sub-contractor and due to a defect caused by lack of care a fire occurred which destroyed the building, it might not be stretching ordinary principles too far to hold the electrical sub-contractor liable for the damage.

LORD BRIDGE (at p. 478):

The reality is that the structural elements in any building form a single indivisible unit of which the different parts are essentially interdependent. To the extent that there is any defect in one part of the structure it must to a greater or lesser degree necessarily affect all other parts of the structure. Therefore any defect in the structure is a defect in the quality of the whole and it is quite artificial, in order to impose a legal liability which the law would not otherwise impose, to treat a defect in an integral structure, so far as it weakens the structure, as a dangerous defect liable to cause damage to 'other property'.

A critical distinction must be drawn here between some part of a complex structure which is said to be a 'danger' only because it does not perform its proper function in sustaining the other parts and some distinct item incorporated in the structure which positively malfunctions so as to inflict positive damage on the structure in which it is incorporated. Thus, if a defective central heating boiler explodes and damages a house or a defective electrical installation malfunctions and sets the house on fire, I see no reason to doubt that the owner of the house, if he can prove that the damage was due to the negligence of the boiler manufacturer in the one case or the electrical contractor on the other, can recover damages in tort on *Donoghue* v. *Stevenson* [1932] AC 562 principles. But the position in law is entirely different where, by reason of the inadequacy of the foundations of the building to support the weight of the superstructure, differential settlement and consequent cracking occurs. Here, once the first cracks appear, the structure as a whole is seen to be defective and the nature of the defect is known.

LORD OLIVER (at p. 484):

In the speech of my noble and learned friend, Lord Bridge of Harwich, and in my own speech in *D. & F. Estates Ltd* v. *Church Commissioners for England* [1989] AC 177 there was canvassed what has been called 'the complex structure theory'. This has been rightly criticised by academic writers although I confess that I thought that both my noble and learned friend and I had made it clear that it was a theory which was not embraced with any enthusiasm but was advanced as the only logically possible explanation of the categorisation of the damage in *Anns* as 'material, physical damage'. My noble and learned friend has, in the course of his speech in the present case, amply demonstrated the artificiality of the theory and, for the reasons which he has given, it must be rejected as a viable explanation of the underlying basis for the decision in *Anns*.

LORD JAUNCEY (at p. 497):

> My Lords, I agree with the views of my noble and learned friend, Lord Bridge of Harwich, in this appeal that to apply the complex structure theory to a house so that each part of the entire structure is treated as a separate piece of property is quite unrealistic. A builder who builds a house from foundations upwards is creating a single integrated unit of which the individual components are interdependent. To treat the foundations as a piece of property separate from the walls or the floors is a wholly artificial exercise. If the foundations are inadequate the whole house is affected. Furthermore, if the complex structure theory is tenable there is no reason in principle why it should not also be applied to chattels consisting of integrated parts such as a ship or a piece of machinery. The consequences of such an application would be far-reaching. It seems to me that the only context for the complex structure theory in the case of a building would be where one integral component of the structure was built by a separate contractor and where a defect in such a component had caused damage to other parts of the structure, e.g. a steel frame erected by a specialist contractor which failed to give adequate support to floors or walls. Defects in such ancillary equipment as central heating boilers or electrical installations would be subject to the normal *Donoghue* v. *Stevenson* principle if such defects gave rise to damage to other parts of the building.

So, farewell then to the old complex structure theory. And welcome to its new, equally vague but more limited, successor. The complex structure theory remains with us if the separate components are installed by separate contractors, if Lords Keith and Jauncey are to be believed. Lord Bridge, however, prefers to look at the defective component and enquire as to whether it is part of the structure itself or whether it is an ancillary feature, such as heating or electrical works – in the latter case, this he would regard as a case of physical damage.

The new approach to complex structures succeeds in combining inequity and uncertainty. It is inequitable in that it creates yet another set of artificial divides. The owner of a property with defects installed by a sub-contractor curiously is placed in a better position than the owner of an identical property with identical defects installed by the main contractor. Uncertainty too is created. Even if their Lordships could agree on what the new approach is, there is still going to be the need for much litigation to delineate the situations where there is sufficient distinction between the cause of the damage and the damage itself to allow the claim to be treated as physical in character.

The second area of uncertainty created by *Murphy* is in relation to a particular category of threats to health and safety. We have seen that a threat to the householder and his visitors alone falls to be treated as economic loss. However, put simply, buildings can fall outwards as well as inwards. What duty, if any, is the builder under towards an occupier who carries out repair work to prevent a threat to third parties?

MURPHY v. BRENTWOOD DISTRICT COUNCIL

[1991] 1 AC 398   House of Lords

For facts, see p. 31, above.

LORD BRIDGE (at p. 475):

> The only qualification I would make to this is that, if a building stands so close to the boundary of the building owner's land that after discovery of the dangerous defect it remains a potential source of injury to persons or property on neighbouring land or on the highway, the building owner ought, in principle, to be entitled to recover in tort from the negligent builder the cost of obviating the danger, whether by repair or by demolition, so far as that cost is necessarily incurred in order to protect himself from potential liability to third parties.

Lord Oliver also raised this possibility but stated that he was not 'convinced of the basis for making such a distinction'.

Can the distinction be justified? Lord Bridge appears to see the rationale for his proposal as being the plaintiff acting reasonably to stop him being sued by the third parties. But is this a real threat? The answer seems to be in the tort of nuisance. Even though not responsible for the creation of the nuisance, the occupier will be strictly liable to the users of the highway if harmed by the property (*Wringe* v. *Cohen* [1940] 1 KB 229), while he will also be liable to his neighbours in private nuisance for damage to their property even if he has done nothing about the nuisance himself (*Sedleigh-Denfield* v. *O'Callaghan* [1940] AC 880).

So, it would be harsh on the occupier if he could not in these circumstances claim against the builder for the cost of remedial work necessary to avoid the actions against him. That said, however, it is very difficult to see how averting threats to third parties is any different from averting threats to oneself and one's visitors; both do look very much like economic loss as currently defined. Similarly, the pressure on the occupier to carry out repairs to prevent him being sued by his neighbours is, if anything, less forceful than the pressure on him to carry out repairs to prevent harm to himself and his family. Given the clear decision in *Murphy*, it would seem invidious to create an anomalous exception of the kind Lord Bridge suggests.

Builders' standard of care

Once the plaintiff has succeeded in getting within the ambit of the various duties of care that builders still owe, the next question to arise is whether the builder has fallen below the standard of care to be expected from the reasonable builder. This raises the problem of applying the one test to the range of different builders and types of building work.

WORLOCK v. S.A.W.S.

(1982) 265 EG 774   Court of Appeal

The defendants built a bungalow for the plaintiff. The defendants only provided labour; the plaintiff was responsible for materials and insurance. The defendants argued that this affected their duty of care, but were held liable in both contract and tort.

ROBERT GOFF LJ (at p. 76):

> I shall consider first the appeal of the builder. For him [counsel for S.A.W.S], while accepting that a builder who undertook the whole work of construction of a bungalow, whether or not with the supervision of an architect, would be subject to the standard of care which the learned judge held in this case was imposed on the builder, both in contract and in tort, nevertheless submitted that there were special circumstances of the present case which made it inappropriate for the builder to be judged by the standards suitable for a builder who was in overall control of the works. In this connection, he relied on two particular features of the contract – first, that the plaintiff as employer was responsible for supplying most of the materials for the work, so that to a large extent the builder was employed on a labour-only basis – and second, the provision in the contract set out in [the defendant's] quotation, that 'insurance, public and employers' was to be supplied by the plaintiff.
>
> Like the learned judge, I feel considerable sympathy for [the defendant.] It is indeed very sad that he should now be in his present position. But I feel bound to say that, as a matter of law, I entertain no doubt that the learned judge was right in the conclusion which he reached as to the builder's liability. I cannot see that the two features of the contract, on which [counsel for S.A.W.S.] relies, affect the builder's responsibility in law for the work which he carried out under the contract. It is true that the relevant part of the Specification requires the builder to 'excavate and level over the site of all buildings, footpaths, etc., trenches for foundations and services, all as shown'; that 'as shown' must have meant as shown on the plans: and that against the section on the relevant plan it was simply stated, in respect of the slab (I omit the damp-proof membrane) – 'Floor – 2″ concrete screed', beneath that '4″ min. concrete', and beneath that '4″ min. hardcore'. Even so, provisions of the

contract show that it was contemplated that difficulties might be encountered in relation to the soil: I refer not only to the words '4″ min. hardcore' on the plan, and the passage in the quotation which contemplated that further brickwork might well be required, but also to paragraph 2 of Section 1 of the Specification, which required the contractor to visit the site before tendering and make himself thoroughly acquainted with (*inter alia*) the general nature of the soil and sub-soil. Furthermore if, as I am satisfied was the case, the builder was required in law to exercise in relation to the work of laying the hardcore the standard of care which was expected of a reasonably competent building contractor, then in the exercise of that duty the builder should, on the learned judge's findings, have recognised the type of soil in the area where the hardcore was to be laid, have considered whether it was good practice to leave soil of that type under the hardcore, and have concluded that it was not. The fact that the builder had contracted in relation (*inter alia*) to this part of the work, to provide labour only did not, in my judgment, detract from the duty imposed upon him under the contract. It simply meant that the plaintiff was to supply the materials, and that the builder was relieved from that responsibility; with the materials supplied by the plaintiff the builder still had to carry out the work, and he must in law have been required to exercise the ordinary standard of care in so doing. He was not a navvy employed to dig foundations: he was a building contractor engaged to build a house, albeit with materials to a large extent supplied by his employer. Nor, in my judgment, did it make any difference that the plaintiff was to provide 'insurance, public and employers'. That meant no more than that, under the contract, the plaintiff had to obtain the relevant cover and pay the premium: it could, in my judgment, have no impact upon the standard of care to be imposed upon the builder in respect of the work he undertook under the contract. I can find no fault with the reasoning or the conclusion of the learned judge on this point. It follows that, in my judgment, the builder's appeal must be dismissed.

The same approach applies not only in the context of different contracts, but at each end of the spectrum of skills.

HONE v. BENSON

(1978) 248 EG 1013   High Court

The plaintiffs purchased a restaurant from the defendants. It was alleged that the premises had been badly converted and had a defective hot water and central heating system. Most of this work had been carried out by the defendants themselves, with enthusiasm rather than skill. It was held that a duty of care could exist in such circumstances.

JUDGE EDGAR FAY QC (at p. 1014):

> It is clear, I think, that in a case where there has been building work done by
> a builder and done negligently he is liable to persons who have suffered injury
> in consequence of his work. The peculiarity of the present case is that, unlike
> the vast majority of cases, the building work was not conducted by a building
> contractor but by building owners who, so far as we know from the pleadings,
> were amateurs, although assisted by a skilled plumber it is said, namely, the
> person whose name appears on the pleadings.
>
> In the first place, in response to the suggestion [by counsel for Hone] that
> liability is thus established, [counsel for Benson] says that I must consider
> carefully whether there is a duty of care lying not in a case such as *Anns* or
> *Dutton* upon the professional builder but upon an amateur who in these do-
> it-yourself days does work on his own account, I find myself unable to
> appreciate any distinction in law between the two. A person who takes on
> skilled work holds himself out to be judged by the standard of skill of those
> able and qualified to do such work. I can see no reason in law or justice –
> indeed, I can see reasons for the contrary – why those who, it may be
> unskilfully, do do-it-yourself building work should be held just as liable as a
> professional builder. It seems to me that a purchaser from the defendants in
> this case is the neighbour (as that word was used in *Donoghue* v. *Stevenson*, and
> to him (or them in this case) the duty of care was in fact owed.

GEORGE WIMPEY & CO. LTD v. D.V. POOLE

(1984) 27 BLR 58   High Court

Wimpey designed and built a quay wall for Vosper Thorneycroft. The
wall proved to be unsatisfactory as a result of design defects. Wimpey
carried out various remedial works and then sought to establish that their
design work had been negligent. This unusual argument was because they
were insured against losses arising from negligent work. Wimpey's
argument was unsuccessful.

WEBSTER J (at p. 76):

> It was common ground that, ordinarily, the test of negligence to be applied to
> a professional man is that stated by McNair J in *Bolam* v. *Friern Hospital
> Management Committee* [1957] 1 WLR 582, at p. 586, subsequently approved
> by the Privy Council in *Chin Keow* v. *Government of Malaysia* [1967] 1 WLR
> 813 and by the House of Lords in *Whitehouse* v. *Jordan* [1981] 1 WLR 246.
> That well known test reads:
>
> > 'But where you get a situation which involves the use of some special skill
> > or competence, then the test as to whether there has been negligence or not

is not the test of the man on the top of a Clapham omnibus, because he has not got this special skill. The test is the standard of the ordinary skilled man exercising and professing to have that special skill.'

An error of judgment may or may not be negligent; whether it constitutes a negligent error of judgment is to be determined by applying that same test: see *Whitehouse* v. *Jordan.*

[Counsel] on behalf of Wimpeys sought to put two glosses on the test for the purposes of this case. The first is that, as he submits, the test is not 'the standard of the ordinary skilled man exercising and professing to have that special skill' if the client deliberately obtains and pays for someone with specially high skills.

Megarry J, as he then was, considered but did not decide the question in the *Duchess of Argyll* v. *Beuselinck* [1972] 2 Lloyd's Rep 172, a claim for negligence against a solicitor. At pp. 183–4 Megarry J said:

'One question that arose during the argument was that of the standard of care required of a solicitor; and although counsel did their best to assist me, the question remained obscure. It was common ground that, at any rate in normal cases, an action for negligence by a solicitor is an action in contract: see *Groom* v. *Crocker* [1939] 1 KB 194. At one stage, [counsel] asserted that this was of importance only in regard to limitation; but I think that later he accepted that this was too restricted a view. I can see that in actions in tort, the standard of care to be applied will normally be that of the reasonable man: those lacking in care and skill fail to observe the standards of the reasonable man at their peril, and the unusually careful and highly skilled are not held liable for falling below their own high standards if they nevertheless do all that a reasonable man would have done. But to say that in tort the standard of care is uniform does not necessarily carry the point in circumstances where the action is for a breach of an implied duty of care in a contract whereby a client retains a solicitor. No doubt the inexperienced solicitor is liable if he fails to attain the standard of a reasonably competent solicitor. But if the client employs a solicitor of high standard and great experience, will an action for negligence fail if it appears that the solicitor did not exercise the care and skill to be expected of him, though he did not fall below the standard of a reasonably competent solicitor? If the client engages an expert, and doubtless expects to pay commensurate fees, is he not entitled to expect something more than the standard of the reasonably competent? I am speaking not merely of those expert in a particular branch of the law, as contrasted with a general practitioner, but also of those of long experience and great skill as contrasted with those practising in the same field of the law but being of a more ordinary calibre and having less experience. The essence of the contract of retainer, it may be said, is that the client is retaining the particular solicitor or firm in question, and he is therefore entitled to expect from that solicitor or firm a standard of care and skill commensurate with the skill and experience which that solicitor or firm has. The uniform standard of care postulated for the world at large in tort hardly seems appropriate when the duty is not one imposed by the law of tort but arises from a contractual obligation existing between the client and the particular solicitor or firm in question. If, as is usual, the retainer contains no express terms as to the solicitor's duty of care, and the matter rests upon an implied term, what is that term in the case of a solicitor of long experience or special skill? Is it that he will put at his client's disposal the care and skill of an average solicitor, or the care and skill that he has? I may say that [counsel] advanced no contention that it was the latter standard that was to be applied; but I wish to make it clear that I have not overlooked the point, which one day may require further consideration.'

According to the researches of counsel, that question has not yet received further consideration. Oliver J, as he then was, referred to the *Duchess of Argyll* in *Midland Bank Trust Co. Ltd* v. *Hett, Stubbs & Kemp* [1979] 1 Ch 384, at p. 403, but without, apparently, modifying the conventional test and for my part, if the question be material, I feel constrained by the clear words of the test as expressed by McNair J, and by the approval of that test without qualification by the Privy Council and the House of Lords, to treat it as unqualified. Since the hearing ended I have considered the judgment of Kilner-Brown J in *Greaves & Co. (Contractors) Ltd* v. *Baynham Meikle and Partners* [1974] 1 WLR 1261 where a similar point was considered. The decision in that case, however, rested on 'special circumstances' (see p. 1269 C–D) and is not, in my view, inconsistent with the conclusion I have reached.

The second gloss which [counsel for Wimpey] sought to put upon the test was that it is the duty of a professional man to exercise reasonable care in the light of his actual knowledge and that the question whether he exercised reasonable care cannot be answered by reference to a lesser degree of knowledge than he had, on the grounds that the ordinarily competent practitioner would only have had that lesser degree of knowledge.

I accept [the] submission [by counsel for Wimpey]; but I do not regard it as a gloss upon the test of negligence as applied to a professional man. As it seems to me, that test is only to be applied where the professional man causes damages because he lacks some knowledge or awareness. The test establishes the degree of knowledge or awareness which he ought to have in that context. Where, however, a professional man has knowledge, and acts or fails to act in a way which, having that knowledge, he ought reasonably to foresee would cause damage, then, if the other aspects of duty are present, he would be liable in negligence by virtue of the direct application of Lord Atkins' original test in *McAlister (or Donoghue) (a Pauper)* v. *Stevenson* [1932] AC 562, at 580:

'You must take reasonable care to avoid acts or omissions which you can reasonably foresee would be likely to injure your neighbour.'

Lastly, while considering the test of negligence, I should advert to the necessity of avoiding hindsight and of judging the conduct of the designer or designers of the quay wall by the standards of the time at which it was designed, not by any later standards, still less the standards of today.

That this approach is correct appears to be confirmed by *Wilsher* v. *Essex Area Health Authority* (p. 45, above).

Contractors and sub-contractors

The respective rights and responsibilities of contractors and sub-contractors have continued to be a fertile area for legal dispute. We have already seen (Chapter 5, above) that the Courts are generally willing to consider the overall contractual setting in determining whether a duty of care arises and what its extent should be. The next case provides a good example of how this can work.

WELSH TECHNICAL SERVICE ORGANIZATION v. HADEN YOUNG

(1987) 37 BLR 130   High Court

The defendants were nominated sub-contractors involved with building work at the plaintiff's hospital. They negligently set fire to the building, but were held not to be liable in tort.

MACPHERSON J (at p. 136):

My task is to resolve a preliminary issue ordered to be tried by Master Hodgson in the following terms:

'Whether any provisions of the tender documents the main contract or the sub-contract alone or in combination limit or exclude liability of the defendants to the plaintiff in respect of damage by fire caused by negligence of the servant or agent of the defendant.'

[Counsel's] argument [for Welsh Technical Service Organization] goes, in summary, as follows. He says:

(1) There is no privity of contract between Welsh Health and Haden Young (save in the limited terms already mentioned).
(2) Haden Young owes a duty of care to Welsh Health because there is between the alleged wrongdoer and the person who has suffered damage a sufficient relationship of proximity or neighbourhood such that in the reasonable contemplation of the former, carelessness on his part may be likely to cause damage to the latter, in which case a *prima facie* duty of care arises (see *per* Lord Wilberforce in *Anns* v. *Merton London Borough Council* [1978] AC 729, at 751; 5BLR 1 at 14).
(3) No exclusion clause between the parties excludes liability for such

negligence, nor is there any restriction of liability of Haden Young to Welsh Health.

(4) Haden Young has no shield against Welsh Health's claim in tort.
(5) In particular the tender 'arrangements' have no contractual or other force restricting Welsh Health's claim or preventing Haden Young being sued in tort.
(6) Haden Young cannot construct any kind of contract from the tender documents, which are simply the precursors of the eventual sub-contract and which set the scene or set up the sub-contract price. In any event the architects had no authority so to contract.
(7) The scope of Haden Young's duty or liability should not be cut down or restricted. What is there here (says [counsel for Welsh Technical Service Organization]) so unusual or which points specifically to the inference that everybody proceeded upon the proposition that it was impolitic for Welsh Health's action to proceed against Haden Young?

[Counsel] for IDC stands substantially aloof. He points out that in analysis the risks defined in the documents are consistent, in the sense that fire risk goes up the chain while general risks go downwards. But he does say (in support of Haden Young) that there is no reason why parties should not be free to allocate risks as they choose, and that effect can be and should be given to that which parties are plainly attempting to do. Whatever the exact course of the tender arrangements may be, [he] says, effect ought to be given to the intentions of the parties (as seen from the documents) which is here reflected in the contractual arrangements which were made, and in particular in clauses 18 and 20 of the main contract and clauses 3–5 of the sub-contract.

[Counsel] for Haden Young has set out a summary of his submissions in a document which will be filed with the court's papers. In essence he looks fixedly at the tender arrangements, since he accepts at once the basic problems caused by lack of privity of contract.

But, says [counsel for Haden Young], there is upon the particular facts of this case a contract between Welsh Health and Haden Young, or at least a collateral warranty or an estoppel which allows Haden Young to say that Welsh Health unreservedly took upon themselves the sole risk of fire (within the meaning of clause 20).

[He] further argues that the conditions of tender and/or clause 20 were incorporated into the sub-contract itself.

Finally, he says that any duty of care or liability for breach of duty was limited by the terms on which Haden Young were invited to tender and by the effect of clause 20B of the main contract. In this regard [he] invokes the second part of Lord Wilberforce's statement in *Anns* where these words are used:

'Secondly, if the first question (as to *prima facie* duty) is answered affirmatively it is necessary to consider whether there are any considerations which ought to negative or to reduce or limit the scope of the duty or the class of persons to whom it is owed or the damages to which a breach of it may give rise . . .'

In seeking to resolve this case, I make the following observations, and refer to some of the cases to which counsel have referred me.

(1) There is no problem in my judgment as to the basic right of Welsh Health to sue Haden Young in negligence. The relationship between the parties was certainly as close as it could be short of actual privity of contract. Haden Young did owe a *prima facie* duty to avoid damage to the works themselves and to persons and property other than the subject matter of the works and a duty to avoid causing pure economic loss consequential upon defects in their work or acts of negligence perpetrated by their men. See *Junior Books Ltd* v. *Veitchi Ltd* [1983] AC 520; 21 BLR 66. In particular I refer to the speech of Lord Roskill at [1983] AC 541–6; 21 BLR 82–90.

(2) It is perfectly correct that no exclusion clause as such protects Haden Young. There is, however, nothing to prevent a building owner undertaking or agreeing to bear the whole risk of damage by fire. And it has been said in terms that where this is done through clauses 18 and 20C of the standard form such risks will include fire caused by the negligence of the contractors or sub-contractors (see *per* Lord Keith of Kinkel in *Scottish Housing Association* v. *Wimpey UK Ltd* [1986] 1 WLR 995 at p. 999; 34 BLR 1 at 6–7.

An agreement in this form is thus undoubtedly wide enough and clear enough to include the risk of damage by fire caused by negligence. It is thus an exclusion clause in its effect, since the building owner positively takes on the whole risk.

[Counsel for Welsh Technical Service Organization] accepts without reservation that clauses 18–20 would prevent any action by Welsh Health against IDC. None of the problems about exclusion clauses and their ambit or extent so far as negligence is concerned arise.

(3) I do not find much assistance from two older cases to which he referred me, namely *Hampton* v. *Glamorgan County Coucil* [1917] AC 13, and *Vigers Sons & Co.* v. *Swindell* [1939] 3 All ER 590.

The first concerned primarily questions of agency. The second dealt primarily with questions of authority. Each case does, of course, highlight the principles which are accepted by both sides as to the general rules as to lack of privity between employer and sub-contractor.

But, as [counsel for Haden Young] says, each case (and in particular *Hampton's* case) stresses the need to focus upon the particular facts of a particular case. As Viscount Haldane said at p. 20):

> 'There is only one safe way when you are construing a document like this, and that is to take the principle which must govern all similar cases, to bear it firmly in mind, and in the light of that principle to read through the contract.'

In parenthesis, I find no support for any argument in the instant case based upon lack of authority. Such a plea is not made, and upon the documents I see no basis for such an argument here.

(4) It is of considerable importance in my judgment to bear in mind always the commercial reality of the whole situation.

This is not a case, such as, for example, where an unconscionable bargain is imposed upon an unadvised, unsuspecting and perhaps guileless customer.

Here experienced business men are considering large-scale contracts with every aspect of the matter open and identified. In particular, it must be necessary for all the parties, including the tenderers to consider questions of insurance and its cost throughout. Indeed the contractor and sub-contractor are obliged between themselves to approve each other's arrangements for insurance in respect of their respective liabilities (clause 4). And clause 5 indicates to the sub-contractor the need for him to ascertain the position under the main contract.

In its tender Haden Young must be able therefore to establish what the position is to be. And, of course, all the tender documents came from Welsh Health, and are not allowed to be altered at the risk of invalidating the tender. This is provided by condition 1 of the conditions of tender put forward on behalf of Welsh Health.

At one time I wondered whether or not there was in this context some relevance in the actual arrangements made by Haden Young for insurance. But in the end I do not believe that there is. The question is what bound or should control the parties, and this should not be decided what Haden Young may or may not have decided to effect by way of cover.

In my judgment it is an important aspect of the case that Haden Young's commercial position as to price (which is perhaps the main object of the tender procedure) must be much affected if not governed by the insurance position which Welsh Health set out for their consideration in the tender documents.

Primarily, these were conditions 10 and 11.5 of the conditions of tender.

But I stress also in this context the full specification, and its express reference to the contract conditions which are to prevail (including clauses 18–20) and the specific paragraph, No. 27, dealing with fire insurance which states as follows:

'The sub-contractor is to make due allowance and take all risk for, and be solely and entirely responsible for the insurance of all plant tools and equipment under the sub-contractor's control.'

I am convinced that a common sense reading of all the tender documents and of the contracts in this case would lead a reasonable commercial person to conclude that the building owner was to bear the sole risk of fire in this case.

Conclusion

Particularly in the light of that last conviction, it will be no surprise that in this case I prefer the arguments advanced for Haden Young to those advanced by [counsel for Welsh Technical Service Organization].

Indeed in my judgment it would be unjust that the party which set out so clearly at the outset its own intention to bear the risk of fire should be able to pass that risk back to the sub-contractor, in spite of the fact that an employee of the sub-contractor has acted negligently.

Primarily, I base my decision upon acceptance of the argument that there was indeed here a contract between Welsh Health and Haden Young at the outset and upon acceptance of the tender.

In many cases a tender may simply show an informal willingness to bargain. And in many cases the tender arrangements may simply set up the price and be purely preliminary steps on the way to a sub-contract.

But in the present case I am persuaded that both parties intended to be bound and were bound contractually. The combined effect of the tender form and the standard form is in my judgment clearly to form an offer capable of acceptance. The wording in particular of the tender form can only be interpreted as being an offer to do the work for a certain sum, and an undertaking to enter into the sub-contract with the building owner. The document in my judgment contemplates contractual relations, and insofar as the matter could be said to have been conditional there was a binding contract when both the offer was accepted (as it plainly was by letter) and when the condition was fulfilled by execution of the sub-contract.

As to the conditions of tender, I have already indicated my views as to their commercial meaning and interpretation. And in my judgment the plain conclusion to be drawn is that Welsh Health were prepared to agree and did agree at the outset that once they accepted Haden Young's price they did so upon the conditions set out, and in particular on condition that Welsh Health would bear sole risk for loss or damage by fire.

Looking at the matter the other way round, I am convinced that if Haden Young had not entered into its sub-contract Welsh Health would in this case have been at once able to sue for any loss incurred in having to accept a higher offer than that put forward by Haden Young. In my judgment the language of their own documents is contractual, and their meaning is plain.

Once that is established Welsh Health cannot ask for damages, since an agreement to bear sole risk is in my judgment just that, and is (as I have already said) equivalent in force to an exclusion clause.

Furthermore, although it is not strictly necessary so to decide in view of my decision on this basic issue, I see strength in the argument that the conditions of tender became upon the facts of this case incorporated into the sub-contract, with the result that the plaintiffs' claim is (in shorthand terms) bad for circuity. These arrangements were made in the context of contractual relations which would involve all three parties. IDC were already involved in connection with the hospital itself, and it seems plain that each of the three parties knew what was happening in connection with the other two. It is, of course, true that the sub-contract does not in terms expressly incorporate the conditions of tender, but I see strength in the argument that they are to be impliedly incorporated, in view of the whole circumstances and (vis-à-vis the sub-contractor) the express provision of condition 2 of the conditions of tender requiring their incorporation.

If I were not convinced that the matter is contractual in its simplest form, I would certainly have been disposed to accept the alternative argument that there was here a collateral warranty given by Welsh Health to bear sole risk in consideration of Haden Young agreeing to tender at their price and to sub-contract to IDC. The effect would be the same in my judgment, although I prefer the earlier interpretation of the documents and their force.

I say nothing about estoppel, save that the argument as to estoppel seems to me to be subsidiary and may lack a plea of detriment. Insofar as the case alleges promissory estoppel nothing is added by the plea in view of my earlier conclusions as to contract or warranty.

I next turn shortly to the second part of Lord Wilberforce's words in *Anns* to which reference has already been made.

Once again it is necessary to look at the matter as it developed, and to bear in mind that Haden Young's duty and, in particular, their liability for breach of that duty must be looked at in the light of all the 'arrangements' which were made between all three parties, most particularly bearing in mind the statements repeatedly made that Welsh Health would bear the sole risk of damage by fire.

In principle in my judgment such a situation is a 'consideration which ought to negative, or to reduce, or limit the scope of the duty . . . or the damages to which a breach of it may give rise.'

Lord Roskill in *Junior Books* at [1983] 1 AC 546 E; 21 BLR 90 envisaged that an exclusion clause 'according to the manner in which it was worded might in some circumstances limit the duty of care . . .' and in my judgment it is fair to conclude that the statements made and contracts concluded in this case could well have negatived the damages to which Haden Young's breach of duty might give rise.

His Honour Judge Hawser QC, in *Twins Transport Ltd* v. *Patrick and Another* (1983) 25 BLR 65 also envisaged such a situation – although in that case he rejected the submission based upon it because of the lack of knowledge or consent to the terms of the exclusion clause by the relevant party.

His Honour Judge Fay QC also considered the matter in *Rumbelows Ltd* v. *AMK and Another* (1980) 19 BLR 25. He referred to the case of *Morris* v. *C.W. Martin & Sons Ltd* [1966] 1 QB 716 and stated his belief that:

'if the plaintiffs had not only known of but also assented to the second defendants' conditions, then having regard to *Morris* v. *Martin & Sons* I think the implied term as between the plaintiff and first defendant would indeed have been modified so as to limit the first defendant's liability to conform with the limits on the second defendant's liability.'

Upon the facts of that case Judge Fay found that 'the evidence did not establish such a state of affairs'.

Finally, the matter has also been considered by His Honour Judge Smout QC in *Southern Water Authority* v. *Lewis and Duvivier* (1984) 27 BLR 116; [1985] 2 All ER 1077. At 27 BLR 24–5; [1985] 2 All ER 1085–6 the judge considered both the *prima facie* duty and the possible considerations which might cut it down.

The contract and its terms were, of course, different in that case, and markedly so were the facts in the case. But the learned official referee saw and accepted that the 'contractual setting' might well define the area of risk (see 27

BLR 124 [1985] 2 All ER 1086B). And he decided upon the facts of that case that the scope of the sub-contractor's liability could be limited because of an earlier choice so to limit that particular liability by the plaintiff's predecessor.

Again, in view of my firm conclusion as to the first point in this case, these matters may not strictly arise. But it seemed to me right to say where I stand in relation to these arguments in the instant case.

My conclusion is simply that in the light of all that Welsh Health put forward and insisted upon by way of conditions of tender, and in view of the whole 'contractual setting', it would indeed be right to negative their *prima facie* right to damages. Such matters being ideally 'considerations' within the second part of Lord Wilberforce's words. I am at least heartened to some extent by the approach of the three official referees whose experience and wisdom in these fields is much greater than mine, even if these cases are not directly parallel to the present case.

My judgment means that in all the circumstances Haden Young do not have to bear the cost of their own palpable negligence. I do not believe that there is anything unjust about that, since Welsh Health knew full well what they were proposing to all those involved. The fact that they bear their own losses is neither here nor there.

The case is firstly notable for the judge's finding that the parties were in a sufficiently proximate relationship for liability under the *Junior Books* principle to arise. This historic finding was however not to help the plaintiff. The contractual setting, established by the plaintiff, was enough to indicate that no tortious liability was intended and that, in a commercial case, is reason enough for there not being such a liability even as between parties as close as these.

Other tortious liabilities of builders

Two statutory provisions appear also to add to the liabilities in tort faced by the builder.

Defective Premises Act 1972 ss. 1(1); 1(2)

1.(1) A person taking on work for or in connection with the provision of a dwelling (whether the dwelling is provided by the erection or by the conversion or enlargement of a building) owes a duty—
   (a) if the dwelling is provided to the order of any person, to that person; and
   (b) without prejudice to paragraph (a) above, to every person who acquires an interest (whether legal or equitable) in the dwelling;
   to see that the work which he takes on is done in a workmanlike or, as the case may be, professional manner, with proper materials and so

that as regards that work the dwelling will be fit for habitation when completed.

(2) A person who takes on any such work for another on terms that he is to do it in accordance with instructions given by or on behalf of that other shall, to the extent to which he does it properly in accordance with those instructions, be treated for the purposes of this section as discharging the duty imposed on him by subsection (1) above except where he owes a duty to that other to warn him of any defects in the instructions and fails to discharge that duty.

This appears to be an important and wide-ranging duty. It is placed on not just builders but on anyone who takes on work in connection with the provision of homes, and appears to provide a stricter form of liability: the work must be done in a professional way and the completed building 'will be' fit for habitation. However, its utility has in the part been severely reduced by s.2 of the Act, which precludes the Act from operating where there is 'an approved scheme' in operation. This is a reference to the NHBC (National House-Builders Council) scheme which provides contractual guarantees to occupiers of new homes and which extends to subsequent purchasers too. The scheme extends to virtually all new homes, so the 1972 Act only applied to a tiny minority of dwellings (non-residential premises are outside the scope of the Act).

However, it has recently been agreed between the NHBC and the Government that the NHBC scheme no longer will be approved under s.2, thus, at a stroke, enabling most purchasers to avail themselves of the Act's protection, as demonstrated by *Andrews* v. *Schooling, The Times,* 21.3.91.

Building Act 1984 s.38(1)

38.(1) Subject to this section—
   (a) breach of a duty imposed by building regulations, so far as it causes damage, is actionable, except in so far as the regulations provide otherwise, and
   (b) as regards such a duty, building regulations may provide for a prescribed defence to be available in an action for breach of that duty brought by virtue of this subsection.

This piece of legislation had its origins in the Health and Safety at Work Act 1974 and on its face would seem an ideal and straightforward way of placing liability on the builder if his work falls below the objective standard of the Building Regulations. However, the Government has never brought this section (or its predecessor) into force, and seems unlikely so to do in the future. It is thus of no effect.

Liability for supply of materials

So far concentration has been placed on the work carried out by the builder. However, the builder is usually also involved in supply of the materials used, and is under an implied contractual warranty to ensure that the materials are fit for their intended purpose and will be of good quality. As ever, as buildings are transferred from party to party, contractual rights are lost and tortious remedies come to the centre of the stage, either against the builder or, in turn, the original producer or supplier of the materials. The application of negligence principles to such product liability cases began with *Donoghue* v. *Stevenson* itself, but in recent years has taken a very different form. It should be emphasized that the legislation concerns itself with fixtures, and not with the actual buildings themselves.

Consumer Protection Act 1987 s. 2(1)

> **2.**(1) Subject to the following provisions of this Part, where any damage is caused wholly or partly by a defect in a product, every person to whom subsection (2) below applies shall be liable for the damage.

This created a strict liability for defective products – in other words there is no need to prove a breach of duty. But on whom is this liability imposed?

Consumer Protection Act 1987 s.2(2)

> (2) This subsection applies to—
>   (a) the producer of the product;
>   (b) any person who, by putting his name on the product or using a trade mark or other distinguishing mark in relation to the product, has held himself out to be the producer of the product;
>   (c) Any person who has imported the product into a member State from a place outside the member States in order, in the course of any business of his, to supply it to another.

The main liability is imposed by s.2(2) on the producer i.e. the manufacturer of the article. This liability is shared by 'apparent producers', who hold themselves out as producers though in fact they are not – brand names such as St Michael are a common example – and also by importers into the European Community. The builder, unless actually engaged in the production of the materials himself, will not be liable under s.2(2), and will be able to pass the liability on.

Consumer Protection Act 1987 s.2(3)

(3) Subject as aforesaid, where any damage is caused wholly or partly by a defect in a product, any person who supplied the product (whether to the person who suffered the damage, to the producer of any product in which the product in question is comprised or to any other person) shall be liable for the damage if—

(a) the person who suffered the damage requests the supplier to identify one or more of the persons (whether still in existence or not) to whom subsection (2) above applies in relation to the product;

(b) that request is made within a reasonable period after the damage occurs and at a time when it is not reasonably practicable for the person making the request to identify all those persons; and

(c) the supplier fails, within a reasonable period after receiving the request, either to comply with the request or to identify the person who supplied the product to him.

(4) Neither subsection (2) nor subsection (3) above shall apply to a person in respect of any defect in any game or agricultural produce if the only supply of the game or produce by that person to another was at a time when it had not undergone an industrial process.

(5) Where two or more persons are liable by virtue of this Part for the same damage, their liability shall be joint and several.

(6) This section shall be without prejudice to any liability arising otherwise than by virtue of this Part.

Here the builder will count as a 'supplier' of the materials, and will therefore be at risk under this subsection if unable or unwilling to identify the producer. This suggests that, at the very least, the builder should take care to retain good records of the sources of his materials.

Consumer Protection Act 1987 s.3

3.(1) Subject to the following provisions of this section, there is a defect in a product for the purposes of this Part if the safety of the product is not such as persons generally are entitled to expect; and for those purposes 'safety', in relation to a product, shall include safety with respect to products comprised in that product and safety in the context of risks of damage to property, as well as in the context of risks of death or personal injury.

(2) In determining for the purposes of subsection (1) above what persons generally are entitled to expect in relation to a product all the circumstances shall be taken into account, including—

(a) the manner in which, and purposes for which, the product has been marketed, its get-up, the use of any mark in relation to the product and any instructions for, or warnings with respect to, doing or refraining from doing anything with or in relation to the product;

(b) what might reasonably be expected to be done with or in relation to the product; and

(c) the time when the product was supplied by its producer to another; and nothing in this section shall require a defect to be inferred from the fact alone that the safety of a product which is supplied after that time is greater than the safety of the product in question.

This reinforces s.2(1), by making clear that products will be regarded as defective if they are unsafe to their users, irrespective of why they have come into that condition.

Section 4 provides some defences, of which the most important is s.4(1)(e) which helps defendants using high technology since it is a defence if the state of the art is such that other producers would not have discovered the defect.

Consumer Protection Act 1987 ss.5(1)–5(4)

5.(1) Subject to the following provisions of this section, in this Part 'damage' means death or personal injury or any loss of or damage to any property (including land).

(2) A person shall not be liable under section 2 above in respect of any defect in a product for the loss of or any damage to the product itself or for the loss of or any damage to the whole or any part of any product which has been supplied with the product in question comprised in it.

(3) A person shall not be liable under section 2 above for any loss of or damage to any property which, at the time it is lost or damaged, is not—

(a) of a description of property ordinarily intended for private use, occupation or consumption; and

(b) intended by the person suffering the loss or damage mainly for his own private use, occupation or consumption.

(4) No damages shall be awarded to any person by virtue of this Part in respect of any loss of or damage to any property if the amount which would fall to be so awarded to that person, apart from this subsection and any liability for interest, does not exceed £275.

The Act here draws the same distinction between claimable and non-claimable forms of loss as does the House of Lords in *D & F* and in *Murphy*. By s.5(2), no claim can be made under the Act from damages to the product itself, or to that in which the product in question is a component. Such claims will fall into the province of the tort of negligence, and in particular within the category of economic loss and the difficulties of making such claims are just as severe in product cases as in any other type.

SIMAAN GENERAL CONTRACTING CO. v. PILKINGTON GLASS LTD

[1988] QB 758   Court of Appeal

Simaan were the main contractors for a building project in Abu Dhabi. The defendants were the manufacturers of double-glazing units which turned out to be the wrong colour and were unacceptable to the employer's architect. Payments to Simaan were withheld, and they claimed unsuccessfully against the defendants.

BINGHAM LJ (at p. 780):

It was submitted for the defendants (in brief summary):

(1) that the plaintiffs could not have succeeded against the defendants before the *Junior Books* case;
(2) that the *Junior Books* case should be regarded as a decision on its own facts;
(3) that the assumed facts of the *Junior Books* case included crucial facts not present here, including in particular substantial damage to the subject property, ownership and occupation of the subject property by the pursuers, and close proximity based, inter alia, on nomination of the defenders as sub-contractors by the pursuers and reliance on the defendants' skill and judgment;
(4) that the plaintiffs could not show physical damage to property;
(5) that the plaintiffs could not show physical damage to any property in which they had any legal or possessory interest;
(6) that the plaintiffs could not perfect their cause of action on proof of economic loss alone;
(7) that the glass units had not suffered damage;
(8) that if the glass units had suffered damage the plaintiffs had had no interest in them at the time of the damage;
(9) that there was no express or implied assumption of responsibility by the defendants towards the plaintiffs;
(10) that reasons of policy should deny the plaintiffs a right of recovery against the defendants on the facts assumed here.

In a very carefully thought out and well presented argument for the plaintiffs, [counsel] sought to counter these arguments and to escape the toils of authority by relying on the *Hedley Byrne* case [1964] AC 465 and the principles there laid down and adopted in the *Junior Books* case [1983] 1 AC 520. [He] put forward his claim as one not dependent on physical damage and suggested that the *Junior Books* case also rested on a much wider basis. But if the plaintiffs had to show damage to goods in which they had an interest he contended that they could do so on the ground that the panels, on rejection by the architect, reverted or passed into the plaintiffs' ownership.

I can, I think, state my conclusions fairly shortly.

(1) I accept without reservation that a claim may lie in negligence for recovery of economic loss alone. Were that not so the *Hedley Byrne* case [1964] AC 465 could not have been decided as it was.

(2) I am quite sure that the defendants owed the plaintiffs a conventional *Donoghue* v. *Stevenson* [1932] AC 562 duty of care to avoid physical injury or damage to person or property. Suppose (however improbably) that the defendants manufactured the units so carelessly that they were liable to explode on exposure to strong sunlight and that one of the units did so explode, blinding an employee of the plaintiffs working in the building. I cannot conceive that such employee would fail in a personal injury action against the defendants for failure to prove a duty of care.

(3) There is no meaningful sense in which the plaintiffs can be said to have relied on the defendants. No doubt the plaintiffs hoped and expected that the defendants would supply good quality goods conforming with the contract specification. But the plaintiffs required Feal to buy these units from the defendants for one reason only, namely, that they were contractually obliged to do so and had no choice in the matter. There was no technical discussion of the product between the plaintiffs and the defendants.

(4) Where a specialist sub-contractor is vetted, selected and nominated by a building owner it may be possible to conclude (as in the *Junior Books* case [1983] 1 AC 520) that the nominated sub-contractor has assumed a direct responsibility to the building owner. On that reasoning it might be said that the defendants owed a duty to the Sheikh in tort as well as to Feal in contract. I do not, however, see any basis on which the defendants could be said to have assumed a direct responsibility for the quality of the goods to the plaintiffs: such a responsibility is, I think, inconsistent with the structure of the contract the parties have chosen to make.

(5) The *Junior Books* case has been interpreted as a case arising from physical damage. I doubt if that interpretation accords with Lord Roskill's intention, but it is binding upon us. There is in my view no physical damage in this case. The units are as good as ever they were and will not deteriorate. I bridle somewhat at the assumption of defects which we are asked to make because what we have here are not, in my view, defects but failures to comply with Sale of Goods Act conditions of correspondence with description or sample, merchantability or (perhaps) fitness for purpose. It would, I think, be an abuse of language to describe these units as damaged. The contrast with the floor in the *Junior Books* case is obvious.

(6) I do not accept that the *Hedley Byrne* case [1964] AC 465, and such
      authorities as *Ross* v. *Caunters* [1980] Ch 297, establish a general rule that
      claims in negligence may succeed on proof of foreseeable economic loss
      caused by the defendant even where no damage to property and no
      proprietary or possessory interest are shown. If there were such a general
      rule, the plaintiffs in the *Candlewood* case [1986] AC 1 and *Leigh and
      Sillavan Ltd* v. *Aliakmon Shipping Co. Ltd* [1986] AC 785 would not have
      failed on the ground they did and the causes of action in the *Pirelli* case
      [1983] 2 AC 1 and *London Congregational Union Inc.* v. *Harriss and
      Harriss* [1988] 1 All ER 15 would have been complete at an earlier date.
      However attractive it may theoretically be to postulate a single principle
      capable of embracing every kind of case, that is not how the law has
      developed. It would of course be unsatisfactory if (say) doctors and
      dentists owed their patients a different duty of care. I do not, however,
      think it unsatisfactory or surprising if, as I think, a banker's duty towards
      the recipient of a credit reference and an industrial glass manufacturer's
      duty towards a main contractor, in the absence of any contract between
      them, differ. Here, the plaintiffs' real (and understandable) complaint is
      that the defendants' failure to supply goods in conformity with the
      specification has rendered their main contract less profitable. This is a
      type of claim against which, if laid in tort, the law has consistently set its
      face.

(7) If, contrary to my view, these units can be regarded as damaged at all, the
      damage (or the defects) occurred at the time of manufacture when they
      were the defendants' property. I therefore think that the plaintiffs fail to
      show any interest in the goods at the time when damage occurred. I very
      much doubt if there was any time on site, whether in course of erection
      or after rejection, when the plaintiffs had a proprietary or possessory
      interest in the units, but I do not think it useful to pursue this, since
      neither was the time at which, if at all, physical damage occurred.

(8) I do not think it just and reasonable to impose on the defendants a duty
      of care towards the plaintiffs of the scope contended for. (a) Just as
      equity remedied the inadequacies of the common law, so has the law of
      torts filled gaps left by other causes of action where the interests of justice
      so required. I see no such gap here, because there is no reason why claims
      beginning with the Sheikh should not be pursued down the contractual
      chain, subject to any short-cut which may be agreed upon, ending up
      with a contractual claim against the defendants. That is the usual
      procedure. It must be what the parties contemplated when they made
      their contracts. I see no reason for departing from it. (b) Although the
      defendants did not sell subject to exempting conditions, I fully share the
      difficulty which others have envisaged where there were such conditions.
      Even as it is, the defendants' sale may well have been subject to terms and
      conditions imported by the Sale of Goods Act 1979. Some of those are
      beneficial to the seller. If such terms are to circumscribe a duty which
      would be otherwise owed to a party not a party to the contract and
      unaware of its terms, then that could be unfair to him. But if the duty is

unaffected by the conditions on which the seller supplied the goods, it is in my view unfair to him and makes a mockery of contractual negotiation.

DILLON LJ (at p. 784):

There has consequently been a good deal of discussion in argument about the controversial decision in *Junior Books Ltd* v. *Veitchi Co. Ltd* [1983] 1 AC 520. My own view of the *Junior Books* case is that the speeches of their Lordships have been the subject of so much analysis and discussion with differing explanations of the basis of the case that the case cannot now be regarded as a useful pointer to any development of the law, whatever Lord Roskill may have had in mind when he delivered his speech. Indeed I find it difficult to see that future citation from the *Junior Books* case can ever serve any useful purpose.

Again, though, it should be remembered that, notwithstanding the view so firmly expressed by Dillon LJ, the House of Lords in *Murphy* have sustained the *Junior Books* ruling, albeit on restrictive grounds (see p. 151 above) and that a different conclusion would be reached if the relationship between Simaan and Pilkington had been a closer one, with the two parties dealing together albeit outside the formal framework of a contract. This would be clearly so too if the builder makes a careless statement recommending materials, thus falling within the *Hedley Byrne* v. *Heller* [1964] AC 465 principle of liability.

Builders – summary

The builder does not escape liability as a result of *Murphy*. The builder will be liable, it is likely, for personal injury and some specialist contractors seem likely to be unlucky enough to fall within the parameters of the restrictive *Junior Books* liability for economic loss. Some difficult cases remain to be sorted out on the borderline between these two categories, and separate liabilities arise in connection with the supply of materials.

Thus the builder faces a considerable risk of liability, but this in turn will overlap with, and may be shared by, other members of the building team.

Chapter 9

Liability for Design and Supervision

The builder does not work in a vacuum. He is, in theory at least, under the continuous control of the architects (or consulting engineers) who have carried out the design work on the project and continue to be involved with the work as it is carried out, as supervisors of the construction work. These two roles, of design and supervision, give rise to very different obligations. The other complication that exists in this area is that the precise nature of all the obligations in play need to be carefully examined. What are the respective responsibilities of the architect and the builder to the employer and to subsequent users of the building? And does the architect owe the contractor duties of care too? Of course nowadays many large firms offer both design and building services but still the law would look carefully at the respective roles of each function.

Liability in negligence

Two brief matters of general relevance to the liability of architects can be mentioned. Firstly, the architect is involved in two legally different ways. He both performs acts e.g. while supervising works and also makes statements, as to suitability of materials or contractors and many other matters. It should be examined carefully in which role the architect is alleged to have been negligent, since the law of tort regards liability for negligent statements as arising only on a highly limited basis (see *Hedley Byrne* v. *Heller*, p. 19, above).

The other preliminary matter which needs to be raised is the question of choice of action between tort and contract on the part of the architect's employer. Earlier discussion of the *Tai Hing* and subsequent cases (p. 70, above) has indicated that this is at present an area of considerable uncertainty. However, it is suggested that the area of architect's liability is a classic one where the plaintiff should normally enjoy the choice of mode of action, since the tortious duty to take reasonable care in the performance of the work complements the contractual duties, which generally are also to take reasonable care but may be agreed to be more onerous. This is not an area where the law of tort would conflict with or contradict the normal contractual arrangements, and there therefore seems no reason why it should not normally be available to the plaintiff.

Duties to employer and subsequent owners

Design

The main issue here in recent years has been the extent to which the architect's design duty has become more onerous. The traditional approach has always been that the duty stemming from the employment of an architect has only been one to take reasonable care, but there are some suggestions that this is changing.

GREAVES & CO. (CONTRACTORS) v. BAYNHAM MEIKLE & PARTNERS

[1975] 1 WLR 1095   Court of Appeal

This was a claim by contractors seeking an indemnity from the defendants, who were consultant structural engineers, against their own liability to the owner arising from the unsuitability of the premises for its intended use, the storage of oil drums and moving of them by forklift trucks. The defendants were held to be liable to indemnify the contractors.

LORD DENNING MR:

> It has often been stated that the law will only imply a term when it is reasonable and necessary to do so in order to give business efficacy to the transaction; and, indeed, so obvious that both parties must have intended it. But those statements must be taken with considerable qualification. In the great majority of cases it is no use looking for the intention of both parties. If you asked the parties what they intended, they would say that they never gave it a thought: or, if they did, the one would say that he intended something different from the other. So the courts imply – or, as I would say, impose – a term such as is just and reasonable in the circumstances. Take some of the most familiar of implied terms in the authorities cited to us. Such as the implied condition of fitness on a sale of goods at first implied by the common law and afterwards embodied in the Sale of Goods Act 1893. Or the implied warranty of fitness on a contract for work and materials: *Young & Marten Ltd* v. *McManus Childs Ltd* [1969] 1 AC 454, (9 BLR 77). Or the implied warranty

that a house should be reasonably fit for human habitation: see *Hancock* v. *B.W. Brazier* [1966] 1 WLR 1317. And dozens of other implied terms. If you should read the discussions in the cases, you will find that the judges are not looking for the intention of both parties; nor are they considering what the parties would answer to an officious bystander. They are only seeking to do what is 'in all the circumstances reasonable'. That is how Lord Reid put it in *Young & Marten Ltd* v. *McManus Childs Ltd*; and Lord Upjohn said quite clearly that the implied warranty is 'imposed by law'.

Apply this to the employment of a professional man. The law does not usually imply a warranty that he will achieve the desired result, but only a term that he will use reasonable care and skill. The surgeon does not warrant that he will cure the patient. Nor does the solicitor warrant that he will win the case. But, when a dentist agrees to make a set of false teeth for a patient, there is an implied warranty that they will fit his gums; see *Samuels* v. *Davis* [1943] KB 526.

What then is the position when an architect or an engineer is employed to design a house or a bridge? Is he under an implied warranty that, if the work is carried out to his design, it will be reasonably fit for the purpose? Or is he only under a duty to use reasonable care and skill? This question may require to be answered some day as matter of law. But in the present case I do not think we need answer it. For the evidence shows that both parties were of one mind on the matter. Their common intention was that the engineer should design a warehouse which would be fit for the purpose for which it was required. That common intention gives rise to a term implied *in fact*.

The other judges, Browne and Geoffrey Lane LJ, both went out of their way to emphasize that this was very much a decision on its own special facts and circumstances, in particular the specific knowledge on the part of the defendant of the plaintiff, and the expectations created by a 'package-deal' contract i.e. one where the contractors were themselves responsible to the owner for the works. However, this type of contract is quite common in major projects.

IBA v. EMI AND BICC

(1980) 14 BLR 1   House of Lords

There the plaintiffs' TV mast on Emley Moor in Yorkshire collapsed. The main project contractors were EMI, and BICC were nominated sub-contractors responsible for the design and erection of the mast itself. Both were held liable for its collapse.

LORD FRASER (at p. 34):

The duty of BICC was, as they fully accept, to use the care and skill to be expected of ordinarily competent structural engineers undertaking work of the class in question. O'Connor J held that they had failed in that duty because they omitted to take account of, or to provide for, stresses likely to be created by asymmetric ice loading on the stays of the mast, and to relate those stresses to the separate stresses likely to be caused by vortex shedding on the structure of this type of mast. The phenomena of asymmetric ice loading and vortex shedding have been sufficiently described by my noble and learned friend Lord Dilhorne. The basic assumptions from which to calculate the effect of vortex shedding were not known when the mast was designed, but the existence of stresses due to vortex shedding and their general nature was known to BICC at that time. So was the theoretical possibility of 'lock-on' due to the coincidence of the natural mode of oscillation of a structure with further oscillation caused by vortex shedding. But the occurrence of 'lock-on' in a guyed structure such as this mast was considered by most experts before the accident to be impossible. It was not foreseen by BICC and I do not think it was reasonably foreseeable by them on the information that was then available. Nevertheless, the fact that vortex shedding would be likely to create great stresses at certain wind speeds was known to BICC and they very properly sought advice on the matter from Dr Flint [a consultant]. His advice given in what he called a 'preliminary report' dated 20 November 1963, was that oscillation of an amplitude of up to four feet on each side of the vertical might occur when the wind was blowing at certain speeds. BICC made no changes in their designs for any of the masts in consequence of this report, and although the report related only to the design of the Winter Hill mast, they treated it as giving an all-clear for the design of the Emley Moor mast also. They made no allowance for 'lock-on'.

The other source of stress, asymmetric ice loading on the stays, is much easier for the non-scientist to understand than vortex shedding. The mast had three lanes of stays which were at an angle of 120 degrees to each other. Number 3 lane was approximately on the north side of the mast and numbers 1 and 2 lanes were approximately on the east south-east and west south-west sides respectively. The collapse occurred after four days of very cold weather with a steady breeze of about 20 mph from the east, which caused numbers 1 and 3 lanes of stays to become heavily loaded with ice of a dense and heavy type, while lane 2 remained relatively free from ice. The result, as the learned judge held, was that the ice laden lanes of stays tended to pull the mast over towards the north and east and thus imposed tension on the opposite side. This tension, when added to the tension caused by oscillation, caused one of the welded joints in the lattice-work at a height of 1,027 ft to break with the result that the mast fell. BICC admit that they had not made any allowance for tension arising from asymmetric ice loading of the stays. The reason why they did not do so was that they relied on the code of practice then current in the United Kingdom and designed the mast to withstand the pressures that would be caused by a wind of 80 mph at height of 40 ft above the ground combined with half-inch radial thickness of ice on all structures of the mast above 500 ft, but made no further allowance for ice on the stays. Experience had shown that masts designed in accordance with that code of practice were able to withstand any wind likely to occur in the United Kingdom. But experience in the United Kingdom related exclusively to masts of lattice-work construction, and the fundamental question on negligence is whether BICC ought to have foreseen that it might not be safely applied to a cylindrical mast. O'Connor J found as a fact that:

> 'the evidence is all one way that before the end of the 1960s nobody considered asymmetric ice loading a factor to be taken into account when designing a mast . . . The reason for this also emerges quite clearly from the evidence, it was well known that masts became iced up, but they were designed to withstand very high wind speeds and the assumption was, and indeed the experience was, that the ice would be blown off or shaken off long before the critical wind speeds were reached. It is said that the coming of the cylindrical mast introduced a new dimension into the calculations, namely, that the structure would be subject to stresses induced by vortex shedding at low wind speeds, that is, at a time when the ice could not be expected to have fallen off.'

In my opinion this was indeed a new dimension, and it ought to have been appreciated by BICC. Their fatal error was in failing to recognise that, because the greatest dynamic stresses on a cylindrical mast would occur at low wind speeds, of around 14 knots at ground level, they might occur at the same time as significant stresses due to asymmetric ice loading, because any ice on the stay would not have been shaken or blown off. The error arose not from difficulty of calculation but from the omission of what seems to me to have been a simple piece of reasoning about known facts. I have reminded myself of the risk of being wise after the event and of demanding precautions against damage that were not reasonably foreseeable but I have reached a firm conclusion that BICC failed in their duty of care when they applied the code of practice that had been found appropriate for lattice masts to the new cylindrical mast at Emley Moor without noticing that the reason for disregarding ice on the stays was not applicable to a cylindrical mast. They were therefore negligent in their design.

(and at p. 44):

> If the terms of the contract alone had left room for doubt about that, I think that in a contract of this nature a condition would have been implied to the effect that EMI had accepted some responsibility for the quality of the mast, including its design, and possibly also for its fitness for the purpose for which it was intended. The extent of the responsibility was not fully explored in argument, and, having regard to the decision on negligence in the design, it does not required to be decided. It is now well recognised that in a building contract for work and materials a term is normally implied that the main contractor will accept responsibility to his employer for materials provided by nominated sub-contractors. The reason for the presumption is the practical convenience of having a chain of contractual liability from the employer to the main contractor and from the main contractor to the sub-contractor – see *Young & Marten Ltd* v. *McManus Childs Ltd* [1969] 1 AC 454 (9 BLR 77). Of

course, as Lord Reid pointed out in that case at 465G, 'No warranty ought to be implied in a contract unless it is in all circumstances reasonable.' (9 BLR 87). In most cases the implication will work reasonably because if the main contractor is liable to the employer for defective material, he will generally have a right of redress against the person from whom he bought the material. In the present case it is accepted by BICC that, if EMI are liable in damages to IBA for the design of the mast, then BICC will be liable in turn to EMI. Accordingly, the principle that was applied in *Young & Marten Ltd* in respect of materials, ought in my opinion to be applied here in respect of the complete structure, including its design. Although EMI had no specialist knowledge of mast design, and although IBA knew that and did not rely on their skill to any extent for the design, I see nothing unreasonable in holding that EMI are responsible to IBA for the design seeing that they can in turn recover from BICC who did the actual designing. On the other hand it would seem to be very improbable that IBA would have entered into a contract of this magnitude and this degree of risk without providing for some right of recourse against the principal contractor or the sub-contractors for defects of design.

LORD SCARMAN (at p. 47):

The finding of negligence in the design of the mast makes it unnecessary for the House to determine the extent of the contractual responsibility for the design assumed by EMI to IBA (and by BICC to EMI). As my Lord Fraser of Tullybelton observes:

'it must at the very least have been to ensure that the design would not be made negligently.'

But I would not wish it to be thought that I accept that this is the extent of the design obligation assumed in this case. The extent of the obligation is, of course, to be determined as a matter of construction of the contract. But, in the absence of a clear, contractual indication to the contrary, I see no reason why one who in the course of his business contracts to design, supply, and erect a television aerial mast is not under any obligation to ensure that it is reasonably fit for the purpose for which he knows it is intended to be used. The Court of Appeal held that this was the contractual obligation in this case, and I agree with them. The critical question of fact is whether he for whom the mast was designed relied upon the skill of the supplier (i.e. his or his sub-contractor's skill) to design and supply a mast fit for the known purpose for which it was required.

Counsel for the appellants, however, submitted that, where a design, as in this case, requires the exercise of professional skill, the obligation is no more than to exercise the care and skill of the ordinarily competent member of the profession. Although it might be negligence to-day for a constructional engineer not to realise the danger to a cylindrical mast of the combined forces of vortex shedding (with lock-on) and asymmetric ice loading of the stays, he submitted that it could not have been negligence before the collapse of this mast: for the danger was not then appreciated by the profession. For the purpose of the argument, I will assume (contrary to my view) that there was no negligence in the design of the mast, in that the profession was at that time unaware of the danger. However, I do not accept that the design obligation of the supplier of an article is to be equated with the obligation of a professional man in the practice of his profession. In *Samuels* v. *Davis* [1943] KB 526, the Court of Appeal held that, where a dentist undertakes for reward to make a denture for a patient, it is an implied term of the contract that the denture will be reasonably fit for its intended purpose. I would quote two passages from the judgment of du Parcq LJ. At p. 529 he said (omitting immaterial words):

> '. . . if someone goes to a professional man . . . and says: "Will you make me something which will fit a particular part of my body?" . . . and the professional gentleman says: "Yes," without qualification, he is then warranting that when he has made the article it will fit the part of the body in question.'

And at p. 530 he added:

> 'If a dentist takes out a tooth or a surgeon removes an appendix, he is bound to take reasonable care and to show such skill as may be expected from a qualified practitioner. The case is entirely different where a chattel is ultimately to be delivered.'

> I believe the distinction drawn by du Parcq LJ to be a sound one. In the absence of any terms (express or to be implied) negativing the obligation, one who contracts to design an article for a purpose made known to him undertakes that the design is reasonably fit for the purpose.

The IBA case well illustrates the onerous burden on designers. Not only were BICC held to be liable in negligence for their design work (though the case does reveal that they were not up to the demands of working on the limits of design technology), but once again the main contractor was also held to be liable as part of the 'package-deal' in contract.

The fact that, in these types of contract, the main contractor takes on a contractual responsibility for the design work actually performed by the sub-contractor is likely to be reflected in tortious claims, for example by subsequent owners of the premises. Duties of care arise from situations where the defendant is responsible for something, but fails to live up to the burden of that responsibility. The courts have held that there is no duty when a defendant is not able to take any effective steps (*King* v. *Liverpool City Council* [1986] 1 WLR 890); here the corollary applies, the broad contractual responsibility placed on the defendants giving them the right to take steps to prevent the negligence.

Assessing the place of the *IBA* case today is not easy. As far as the tort liability of the nominated sub-contractors is concerned, this would now presumably be seen as a claim for economic loss since, on the analysis given in *D & F* and in *Murphy*, the mast simply became less valuable by reason of its disintegration. Put another way, no extrinsic factor caused damage to the mast; it merely damaged itself.

However, this is just the type of case where the *Junior Books* principle allowing tortious claims for such economic losses would still seem to apply. Indeed both *IBA* and *Junior Books* have a common background of claims against nominated sub-contractors. Equally statements about the suitability of the mast design are just the sort of statements that, if negligent, will give rise to *Hedley Byrne* liability for negligent misstatement (see p. 194, below).

As far as the importance of the *IBA* case on the scope of the main contractor's liability is concerned, it is clear that the courts will be slow to extend these onerous responsibilities too widely.

GEORGE HAWKINS v.
CHRYSLER (UK) LTD AND BURNE ASSOCIATES

(1986) 38 BLR 36   Court of Appeal

Chrysler wanted new showers at their factory. Hawkins was their employee, injured by slipping on the floor of the shower room. Burne were the engineers who designed the facilities and supervised their installation.

Fox LJ (at p. 46):

It is not in doubt that Burne Associates were under an obligation to exercise reasonable skill and care. The judge's finding as to the negligence issue was as follows:

'I have seen and heard Mr Burne. It may be that he made the wrong decision. Having seen him and having heard him I came to the conclusion that he was exercising a degree of prudence and competence which his professional position required him to exercise and I do not find he was negligent in his approach to this particular problem. He made an inquiry of a sub-contractor who was a specialist in this field and that was the basis upon which it was done. In other circumstances other considerations might have applied, but that was the course he took and having seen him and heard him I do not think I should fault him for it.'

I see no reason to disturb the judge's conclusion upon the negligence issue. I think that there was ample evidence upon which he could so conclude, and I see no reason to suppose that he misdirected himself in his analysis of the evidence before him.

(and at p. 48):

> In my view *Greaves* is a different case from the present, and although the principles there stated are applicable, it seems to me that comparisons with the actual form of words used in the evidence are not of assistance.
>
> In my view the passage in the cross-examination of Mr Burne, to which I have referred, does not justify the inference in fact of a warranty.
>
> This was not a 'package deal' whereby Burne Associates would provide advice and would lay this floor; in my view they were advisers only.
>
> As appears from the question in the cross-examination of Mr Burne to which I have referred [at p. 48] he was being asked whether he was required to use his expertise as an engineer. He was advising as a professional man. Professional men do not normally give warranties. Their obligation is to use reasonable skill and care.

After referring to Lord Scarman's views in the *IBA* case (cited above), he stated (at p. 50):

> I do not think that the present case falls within the principle stated by Lord Scarman in the earlier part of the speech which I have just read. Burne Associates were not supplying anything; it is true that they are described as 'designing' the shower area, but so far as the floor is concerned, the word 'design' seems to me to need some further explanation. They did not provide the floor, or lay it. What they did was to give professional advice as to its suitability. The distinction is, I think, apparent from the latter part of the passage from the speech of Lord Scarman which I have read, and in particular, the reference to the judgment of du Parcq LJ in *Samuels* v. *Davies*.
>
> Here Burne Associates, as it seems to me, were not providing a chattel. Their function was purely advisory. They were professional men, and were acting as professional men.

After further citation of the *Greaves* case (cited above) Fox JL continued (at p. 51):

> All those judicial observations seem to me to accept that, *prima facie*, a professional man, in the exercise of his profession, is normally obliged only to use reasonable care and skill. That is reflected in the standard conditions of employment of architects in the RIBA conditions, and in the standard conditions of engagement for design of engineering projects which is the ACE document – both of which stipulate for the use of reasonable care and skill.
>
> It may be that there are, in consequence, anomalies between the position of contractors and sub-contractors on the one hand, and the position of professional men on the other. Be that as it may, I can see nothing in the present case which justifies me in saying that, in order to give the agreement between the company and Burne Associates business efficacy, the alleged warranty must be implied. I think that it can be given business efficacy, quite

adequate for its purpose by the ordinary obligation of a professional man to use reasonable care and skill in the execution of his work. That obligation was not breached by Burne Associates.

This therefore highlights the exceptional nature of the decisions in both *Greaves* and the *IBA* case. However, the courts are still prepared to follow them in an appropriate case.

Thus in *Holland Hannen & Cubitts (Northern) Ltd* v. *Welsh Health Technical Services Organisation* (1985) 35 BLR 1, nominated sub-contractors were found to have in effect warranted the fitness for purpose of hospital floors and were found in breach of duty for persisting with the design.

Personal duty of architect

An oft-emphasized aspect of the architect's obligation is that it is a non-delegable duty which remains his own responsibility.

MORESK CLEANERS LTD v. THOMAS HENWOOD AND HICKS

[1966] 2 Lloyd's Rep 338   High Court, Official Referee's Business

The plaintiffs asked the defendants to design an extension to their laundry premises. The design proved inadequate and investigation revealed that the defendants had allowed the contractors to prepare the design.

SIR WALKER CARTER QC (at p. 342):

> In my opinion he [the architect] had no implied authority to employ the contractor to design the building. If he wished to take that course, it was essential that he should obtain the permission of the building owner before that was done ... If the defendant was not able, because this form of reinforced concrete was a comparatively new form of construction, to design it himself, he had three courses open to him. One was to say 'this is not my field'. The second was to go to the client, the building owner, and say: 'this reinforced concrete is out of my line. I would like you to employ a structural engineer to deal with this aspect of the matter.' Or he can, while retaining responsibility for the design, himself seek the advice and assistance of a structural engineer, paying for his service out of his own pocket but having at any rate the satisfaction of knowing that if he acts upon that advice and it turns out to be wrong the person whom he employed to give the advice will owe the same duty to him as he, the architect, owes to the building owner.

This was a clearly inappropriate act of delegation to a contractor whose interests were potentially in conflict with those of the employer. But, as the above quote makes clear, there are situations where the architect should involve an independent expert.

INVESTORS IN INDUSTRY COMMERCIAL PROPERTIES v. SOUTH BEDFORDSHIRE DISTRICT COUNCIL

[1986] 1 All ER 787   Court of Appeal

The plaintiffs acquired an infill site for commercial development. Four warehouses were built but their foundations turned out to be defective. No duty was owed by the local authority (see p. 229, below), but the court took the opportunity to consider the council's third party claim against the architects who had relied on the expert advice of consulting engineers. (This section is not reported in the official Law Report: [1986] QB 1034.)

SLADE LJ (at p. 807):

> The only authority to which we have been referred concerning the liability of an architect who has delegated specialised tasks to qualified contractors is *Moresk Cleaners Ltd* v. *Hicks* [1966] 2 Lloyd's Rep 338. That, however, was a case very different from the present, since it concerned the design of a building and it was held that the defendant had no authority to delegate the function of design. The effect of the RIBA conditions does not appear there to have been in issue.
>
> Whether or not in any given instance these conditions apply, it must generally be the duty of an architect to exercise reasonable care in the work which he is engaged to perform. However, [clauses] 1.20, 1.22 and 1.23 of the RIBA conditions, in our judgment, clearly contemplate that, where a particular part of the work involved in a building contract involves specialist knowledge or skill beyond that which an architect of ordinary competence may reasonably be expected to possess, the architect is at liberty to recommend to his client that a reputable independent consultant, who appears to have the relevant specialist knowledge or skill, shall be appointed by the client to perform this task. If following such a recommendation a consultant with these qualifications is appointed, the architect will normally carry no legal responsibility for the work to be done by the expert which is beyond the capability of an architect of ordinary competence, in relation to the work allotted to the expert, the architect's legal responsibility will normally be confined to directing and co-ordinating the expert's work in the whole. However, this is subject to one important qualification. If any danger or problem arises in connection with the work allotted to the expert, of which an architect of ordinary competence reasonably ought to be aware and reasonably could be expected to warn the client, despite the employment of

the expert, and despite what the expert says or does about it, it is in our judgment the duty of the architect to warn the client. In such a contingency he is not entitled to rely blindly on the expert, with no mind of his own, on matters which must or should have been apparent to him.

One particular duty which falls on the architect is in relation to recommending building contractors to the employer. Recent case law has highlighted this area.

PRATT v. HILL ASSOCIATES

(1987) 38 BLR 25   Court of Appeal

The plaintiff intended to demolish a building and replace it with a modern bungalow. Her architect advised her that Swanmore Builders had been very reliable. In fact they were inept and subsequently become insolvent.

PURCHAS LJ (at p. 33):

> In the notice of appeal [counsel] for the appellant in this case sets out the findings of the learned recorder in the first paragraph of the first extract to which I have referred and her findings concerning the chain of causation. She submits that in the face of those findings the recorder was wrong in refusing to award the damages in respect of the interim payments and the costs of the arbitration. These were incurred by the plaintiff as a foreseeable result of her reliance upon the defendant's misrepresentation. She further submits that the recorder incorrectly held that the plaintiff must be taken to have assumed the risk of the builder's insolvency.
>
> There has been some argument at the Bar as to whether or not the insolvency of the builder was to be included in the descripton held out by the architect as 'very reliable'. For my part, I consider that the question of insolvency is irrelevant and that, in being drawn into this issue, the learned recorder fell into error. With respect to him, it is not strictly a relevant feature in the chain of causation, which he satisfactorily described as being established in the first of the two extracts which I have cited. Solvency may well be an incident in determining whether or not damages properly recoverable as a result of a breach of duty can effectively be recovered, but it cannot itself reflect upon that breach of duty. In this case, on the findings of the learned recorder, the architect was in breach of his duty to recommend a suitable reliable builder, and his lack of suitability and reliability led to the disastrous execution of the works. It did not lead to failure to be able to recover damages for the builder's failure on his part to execute the work; nor indeed did insolvency cause the arbitration. The arbitration flowed directly from the failure of the builder to carry out his contract to erect the building. So I find that the learned recorder fell into error in concentrating on insolvency as being

in some way relevant to considering the chain of events which flowed from the established breach to which the learned recorder himself referred and which I have already described in this judgment.

I pause only to record that in a respondent's notice it was asserted that the insolvency was not foreseeable and that the learned recorder was wrong to find that as a fact. In view of what I have said, it is not necessary for me to consider that matter, and indeed I think [he], to do him justice, did not press the point upon us. What he urged in responding to the appeal in this case was that it was not correct to take a global duty and look at the breach, but that each particular aspect of damage flowing had to sound in its own duty and its own breach; and in that way he submitted (I hope I do not do him an injustice) that the builder's insolvency did not give rise to any duty upon which the plaintiff relied. Therefore any breach, even if it was foreseeable, was not relevant and did not sound in damages. This is not quite what the learned recorder himself was discussing in the second paragraph in the second extract of his judgment to which I have referred. He is talking about a duty of care which was negated because he found as a fact that Miss Pratt knew as much about financially risky builders probably as Mr Hill did, and therefore the learned recorder finds that there was no duty in that respect, and, if that was the cause of the damage related to the costs of the arbitration or the making of the payments, those payments would not be recoverable. As I have already said, the basic breach which led to all this loss was the failure of the builders to carry out the work properly and that stemmed directly from the misrepresentation negligently given by the architect which in turn caused Miss Pratt to make a contract with these highly unreliable builders.

For these reasons I would be minded to allow this appeal.

Ongoing duties

Recent cases have frequently sought to draw analogies between liability for defective products and for defective premises. However, obvious differences do exist between the two forms of liability, especially in terms of both the time taken to make the finished article and the time for which it then endures. The question thus arises as to the extent to which the initial design, at the very beginning of the construction process, needs to be continually monitored during the years of building work and the decades of the life of the building that follow.

BRICKFIELD PROPERTIES LTD v. NEWTON

[1971] 1 WLR 862   Court of Appeal

The plaintiffs were suing the defendant architect in relation to their design and supervision of building works. A writ was issued referring to the negligent supervision of the works but then went on later to allege negligent design. The defendant unsuccessfully sought to strike out the allegations of negligence in design as being contrary to the Rules of the Supreme Court.

SACHS LJ (at p. 873):

> Where there are found in completed buildings serious defects of the type here under review the facts relating to design, execution and superintendence are inextricably entangled until such time as the court succeeds in elucidating the position through evidence. The design has inevitably to be closely examined even if the only claim relates to superintendence, and all the more so if the designs are, as is alleged here, experimental or such as need amplification as the construction progresses. The architect is under a continuing duty to check that his design will work in practice and to correct any errors which may emerge. It savours of the ridiculous for the architect to be able to say, as it was here suggested that he could say: 'true, my design was faulty, but, of course, I saw to it that the contractors followed it faithfully' and be enabled on that ground to succeed in the action.
>
> The same – or substantially the same – set of facts falls to be investigated in relation to the design claim and the superintendence claim. The plans and specifications and ancillary documents are relevant to the superintendence claim as well as to the design claim: hence the inability of the defendant to allege prejudice with regard to the preparation of his defence if this appeal is allowed. Accordingly, the 'new cause of action' falls within the ambit of RSC Ord. 20, r. 5 (5), and is one which the court has jurisdiction to permit to be pursued.

This is clearly a sensible decision, reflecting well the professional approach that is rightly expected from the designer of a building through the lifetime of the project. However, recently a court attempted to extend this obligation to new limits.

ECKERSLEY v. BINNIE, NUTTALL AND NORTH WEST WATER AUTHORITY

(1987) LEXIS transcript　　　　　　　High Court, Court of Appeal

This case arose out of the Abbeystead pumping station disaster, when an explosion of methane gas occurred while a party of visitors were inspecting a pumping station, killing and injuring many of them. The design work on this major project had been done at a time when the hazards of accumulations of methane were not known, but the presence and hazards of the methane became known subsequently.

ROSE J:

The designers owed a duty, in relation to the design, construction and likely operation of the system to all those who might reasonably be expected to be affected by the permanent works – the designers' responsibility did not end on handover: there was a continuing duty to warn the Water Authority if the designers became aware of potential danger prior to the explosion. The standard of care required was that of reasonably competent engineers specialising in the design of water transfer systems, including tunnels, applying the standards appropriate at the time of design, construction and operation.

The designers showed a lack of care at the design stage and they were negligent in their failure to reconsider the design in the light of what was and ought to have been discovered during construction. Design was their sole responsibility. They were negligent at the construction stage when the primary responsibility was theirs. They were negligent in failing to warn the Water Authority of the methane danger; also, to a certain extent, they were negligent in not keeping abreast with, and passing on to their clients knowledge which had developed about methane between handover and 1984.

However, the Court of Appeal doubted the existence of such a startling extension to traditional ideas of duty and it is likely that the duty does not extend beyond the time of the architect's final departure from the site.

Scope of duty

Hitherto, it has been assumed that architects and others involved in the design of buildings owe duties equally to their employers and to the subsequent owners and occupiers of the building. *Eames* is an example which demonstrates this.

EAMES LONDON ESTATES LTD v. NORTH HERTFORDSHIRE DISTRICT COUNCIL

(1980) 259 EG 491   High Court (Official Referee's Business)

The plaintiffs purchased two factory units from developers. The units were built on a site that was part old railway embankment, and part infill to bring the rest of the land to the same height. Settlement of the filling took place causing damage to the buildings. Successful claims were brought against the architect as well as the builder and local authority.

JUDGE EDGAR FAY QC (at p. 495):

Who then were the negligent parties? First and foremost I fear I must place the architect, Mr Hyde. He was entrusted with the designing of this building. This includes designing the foundations. I have before me the written evidence of Mr D.C. Hudson RIBA, who states:

> I consider it normal practice for an architect to draw his client's attention to the need for ground conditions to be investigated. Also, that the client be advised of the possible need for a qualified structural engineer to be employed to carry out a detailed site investigation, if the architect was uncertain in any way of the type and bearing capacity of the ground.

Mr Hyde knew he was on made ground. I am astonished that he, with his local connections and knowledge, thought it was all old railway embankment. Accepting that he did, he should nevertheless have satisfied himself about the land's bearing capacity. The 25-in Ordnance Survey maps are part of the furnishings of local architects' offices and a glance at the then current edition would have shown him that the railway had only crossed a part of the site. Indeed the location plan sent by Mr Hyde with the planning application was based on a tracing from the Ordnance Survey showing the railway line.

An architect cannot shed his responsibility for foundations by ascertaining what will get by the local authority as this architect seems to have done. Three-quarters of a ton per sq ft was a loading sufficiently light to show that he had some appreciation of the inferior nature of the ground, that is that he was on inquiry and in my judgment he was negligent in specifying this loading for the piers without any attempt to ascertain for himself whether the ground was suitable for this or any other loading. He was negligent, too, in putting aside the query when it was properly raised by some practical man on the spot. I reject his suggestion that his design was satisfactory for all but the extreme climatic conditions of 1976: the events of that year merely accelerated the steady settlement of the relevant parts.

Though this decision is useful, we must be careful to place this judgment in its proper legal context. The nature of the architect's duty is not really explored. It is clearly tortious in character, since he was employed not by the plaintiffs but by the developers. However, the precise nature of the architect's duty is less clear. Design work involves a blend of acts and statements – the act of drawing up the plans accompanied by explicit or implicit assurances to the employer. The nature of the loss likely to be suffered is, on modern definitions, economic in character. Now claims for negligent acts causing economic loss depend on the especially close proximity akin to contract derived from *Junior Books*, and it is unlikely that this will be satisfied in a case like *Eames* where the plaintiff is the subsequent purchaser of what appears to have been a speculative development, and thus has no personal connection whatsoever with the architect. Plaintiffs are only likely to succeed if they are the employer, subject of course to the *Tai Hing* point, or if they, though not the employer, are actively involved in discussions with the designers from an early stage, as in the *IBA* case above.

Equally if a claim against an architect is framed in negligent misstatement, similar restrictions apply as the plaintiff must show that he is within a *Hedley Byrne* type relationship with the defendant and active reliance appears to be an element within this, as the House of Lords has recently emphasized.

CAPARO INDUSTRIES plc v. DICKMAN

[1990] 2 AC 605   House of Lords

For facts, see p. 27 above.

LORD BRIDGE (at p. 620) reviewed the authorities on negligent misstatement and continued:

> The salient feature of all these cases is that the defendant giving advice or information was fully aware of the nature of the transaction which the plaintiff had in contemplation, knew that the advice or information would be communicated to him directly or indirectly and knew that it was very likely that the plaintiff would rely on that advice or information in deciding whether or not to engage in the transaction in contemplation. In these circumstances the defendant could clearly be expected, subject always to the effect of any disclaimer of responsibility, specifically to anticipate that the plaintiff would rely on the advice or information given by the defendant for the very purpose for which he did in the event rely on it. So also the plaintiff, subject again to the effect of any disclaimer, would in that situation reasonably suppose that he was entitled to rely on the advice or information communicated to him for the very purpose for which he required it. The situation is entirely different where a statement is put into more or less general circulation and may foreseeably be relied on by strangers to the maker of the statement for any one of a variety of different purposes which the maker of the statement has no specific reason to anticipate. To hold the maker of the statement to be under a duty of care in respect of the accuracy of the statement to all and sundry for any purpose for which they may choose to rely on it is not only to subject him, in the classic words of Cardozo CJ to 'liability in an indeterminate amount for an indeterminate time to an indeterminate class': see *Ultramares Corporation v. Touche* (1931) 174 NE 441, 444; it is also to confer on the world at large a quite unwarranted entitlement to appropriate for their own purposes the benefit of the expert knowledge or professional expertise attributed to the maker of the statement. Hence, looking only at the circumstances of these decided cases where a duty of care in respect of negligent statements has been held to exist, I should expect to find that the 'limit or control mechanism . . . imposed upon the liability of a wrongdoer towards those who have suffered economic damage in consequence of his negligence' rested in the necessity to prove, in this category of the tort of negligence, as an essential ingredient of the

'proximity' between the plaintiff and the defendant, that the defendant knew that his statement would be communicated to the plaintiff, either as an individual or as a member of an identifiable class, specifically in connection with a particular transaction or transactions of a particular kind (e.g. in a prospectus inviting investment) and that the plaintiff would be very likely to rely on it for the purpose of deciding whether or not to enter upon that transaction or upon a transaction of that kind.

LORD OLIVER (at p. 638):

What can be deduced from the *Hedley Byrne* case, therefore, is that the necessary relationship between the maker of a statement or giver of advice ('the adviser') and the recipient who acts in reliance upon it ('the advisee') may typically be held to exist where:

(1) the advice is required for a purpose, whether particularly specified or generally described, which is made known, either actually or inferentially, to the adviser at the time when the advice is given;
(2) the adviser knows, either actually or inferentially, that his advice will be communicated to the advisee, either specifically or as a member of an ascertainable class, in order that it should be used by the advisee for that purpose;
(3) it is known either actually or inferentially that the advice so communicated is likely to be acted upon by the advisee for that purpose without independent inquiry; and
(4) it is so acted upon by the advisee to his detriment. That is not, of course, to suggest that these conditions are either conclusive or exclusive, but merely that the actual decision in the case does not warrant any broader propositions.

Caparo represents a clear indication that the courts are now keen to restrict liability for negligent misstatements. The architect sued by a plaintiff with whom, albeit extra-contractually, he has a working relationship will fall within the ambit of these remarks but a subsequent purchaser will be in the same position as the incoming investors in *Caparo* itself, and will form part of a class of unascertained persons to whom, it would seem, no duty of care is now owed. There is an obvious distinction with *Caparo* in that there the class of potential investors is very large whereas here the building is only likely to come into the hands of a very small class of subsequent purchasers. The only way in which subsequent purchasers will be able to avail themselves of an action against the architect is if they come within the definition of what Lord Bridge describes as 'a member of an identifiable class'; this, it is suggested, does not extend to subsequent purchasers who are better seen as members of a small but, to the architect when giving advice, unidentifiable class.

Of course in the less usual event of the building causing physical injury, the victims will be able to bring an action in mainstream negligence based on the decision of the Court of Appeal in *Clay* v. *A.J. Crump & Sons Ltd* [1964] 1 QB 533 (see p. 211 below); claims for physical damage caused by negligent misstatements do not seem to depend on a special relationship, according to this case, though the impact of the then recent *Hedley Byrne* decision was not explored in detail in *Clay*, where the court seems happy to rely on *Donoghue* v. *Stevenson* itself.

Overall, then, subsequent purchasers will have problems in taking action against architects unless the courts take a relatively flexible view of the special relationship. In the recent cases involving building society surveyors (see p. 244 below) the plaintiffs were successful in claiming against surveyors whose work had been commissioned not by them but by the mortgagee, the House of Lords did uphold a claim based in negligent misstatement, but these cases do seem distinguishable here, on the basis of the ready identifiability of the plaintiffs and the fact that they were in fact paying for the survey work to take place.

Standard of care

As ever, establishing the existence of a duty of care is only a first step, though now a difficult one in the context of design work. The plaintiff then has to battle on to establish a breach of that duty, which is never easy against a professional defendant (see Chapter 2, above) given the implications of the *Bolam* test which clearly applies to architects.

WORBOYS v. ACME INVESTMENTS LTD

(1969) 210 EG 335   Court of Appeal

Acme retained Worboys to plan a housing development of up-market properties. Acme sought to withhold some of the fees due on the grounds, *inter alia*, that Worboys were negligent in not including downstairs toilets in these luxury houses. This aspect of the claim failed.

SACHS LJ (at p. 335):

> Lastly, Mr Worboys was said to have been negligent in making no provision for downstairs toilets on the ground floor. He gave no warranty that houses without downstairs toilets would be saleable at between £7,500 and £8,000. No professional evidence was called to show that the omission on an architect's plan of downstairs toilets in houses selling at those prices in the particular area was not in accordance with standard practice or otherwise indicated

incompetence. It was contended, however, that this was a class of case where the court could find breach of professional duty without evidence of lack of competence. There might well be cases where an omission on a plan was so glaring as to require no evidence; for example, where a house was designed without provision for a staircase. But it would be grossly unfair to architects if in respect of other omissions the court could condemn them in negligence in the absence of professional evidence that they had failed to exercise due care. Moreover, Mr Berry, who was a property developer of 25 years' experience, had approved the plans on more than one occasion knowing that they showed no provision for downstairs toilets. The defendants, therefore, could not complain of lack of downstairs toilets which had made the houses less saleable. Accordingly, Mr Worboys's appeal against the finding of negligence in respect of the toilets should be allowed.

NYE SAUNDERS v. BRISTOW

(1987) 37 BLR 92   Court of Appeal

The plaintiffs here were architects retained by the defendant in connection with the proposed renovation of his mansion. The plaintiffs advised that the estimated cost of the proposed work was £238,000. However, this included no provision for contingencies nor for inflation which at this time (1973) was substantial. Within months, the true figure was estimated at £440,000. The plaintiffs unsuccessfully sued for their fees, which the defendant had refused to pay.

STEPHEN BROWN LJ (at p. 102):

It is not disputed on behalf of the appellants that Mr Bristow was entitled to terminate the engagement. It is submitted that nevertheless Mr Nye was entitled to be paid his fees for the work which he had carried out up until the termination in September. There is no doubt that Mr Nye had carried out a very great deal of detailed work. There is no criticism of any kind of the quality of his work, and it has not been suggested in these proceedings, either here or below, that Mr Nye was other than an architect of integrity, experience and high competence. Mr Bristow refused to pay his fees because he said that he had been grossly misled. The legal formulation of that situation is that, in giving the estimate in February 1974 of £238,000 and then revealing in September of the same year that the ultimate cost would be very greatly in excess of that sum – in particular because of the inflation factor – Mr Nye had failed in his duty to take due care in providing a reliable approximate estimate in February 1974. It is claimed that he was in breach of what may conveniently be termed a *Hedley Byrne* duty to Mr Bristow.

The case for the plaintiffs, as presented to the learned deputy official referee, was that the work had been properly carried out by Mr Nye after his engagement as architect for the project and that he had not been at fault in any way in misleading Mr Bristow as to the cost. On behalf of Mr Bristow it was contended that the estimate in February 1974 was given by Mr Nye to Mr Bristow in answer to Mr Bristow's request for the approximate ultimate cost of the work. It was submitted on behalf of Mr Bristow that Mr Nye's failure, in particular to draw attention to the fact that inflation would inevitably increase that estimated figure, amounted to a breach of duty on the part of Mr Nye which led directly to Mr Bristow embarking on the project and then being unable to proceed with it. It is submitted that this constituted negligence on the part of Mr Nye which disentitled him to fees for the work done following his engagement after his estimate had been acted upon.

The issues before the learned deputy official referee were clear cut. The chronology of events was not in dispute. Some matters of detail were disputed, but essentially the basic facts of this unhappy story were not in issue. The duty and standard of care to be expected from Mr Nye was accepted as being that which applied to any profession or calling which required special skill, knowledge or experience. The test is that formulated in a medical negligence case – *Bolam* v. *Friern Hospital Management Committee* [1957] 1 WLR 582. Where there is a conflict as to whether he has discharged that duty, the courts approach the matter upon the basis of considering whether there was evidence that at the time a responsible body of architects would have taken the view that the way in which the subject of enquiry had carried out his duties was an appropriate way of carrying out the duty, and would not hold him guilty of negligence merely because there was a body of competent professional opinion which held that he was at fault. The onus of proving negligence, of course, rests firmly upon the person who alleges it – in this case Mr Bristow.

(and at p. 106):

In his judgment, in assessing the evidence of the experts, including of course the evidence of Mr Nye, who was himself an experienced architect and who had said in terms that it was not the practice in February 1974 to do other than give an estimate based on current prices and not to warn about inflation, even in the most general of terms, the learned judge said:

'I find that there was not a practice accepted as proper by a responsible body of architects in February 1974, that no warning as to inflation need be given when providing an approximation of the cost of the then current outline proposals.'

It seems to me that on the totality of the evidence the judge was fully justified in coming to that conclusion. He was entitled to take the view that the evidence of [the expert witnesses] did not constitute evidence of a responsible body of architects accepting as a proper practice that no warning as to inflation need be given when providing an approximate estimate of the cost of proposed works. It seems to me that the learned judge had ample evidence before him which entitled him to find that there was a failure on the part of Mr Nye to draw the attention of the client to the fact that inflation was a factor which should be taken into account when considering the ultimate cost and that that failure constituted a breach of the *Hedley Byrne* type duty to the defendant. It is right to bear in mind that, although the term 'negligence' has a nasty ring (particularly when it is applied to a professional man), one is not dealing with what might be termed ordinary carelessness. This case concerns a duty which the law says is placed upon somebody who undertakes the responsibility of providing an answer such as Mr Nye had undertaken to do in the circumstances previously outlined and which are not disputed.

Both these cases give clear and straightforward examples of the way the *Bolam* test is applied. Both give great weight to the evidence of standard professional practice and assess the conduct of the defendant alongside it albeit to differing conclusions.

Certification

We now move from the area of design where the architect is, as in many ways, out on his own assessing the requirements of the employer and the constraints of the site, to those areas where the relationship between architect and the builder comes under scrutiny. The first of these are the various formal certification procedures that exist under the JCT and other standard form contracts. These certificates are issued by the architect (or engineer) to the contractors and testify that work has been completed in accordance with the provisions of the contract. This, in turn, enables the contractor to demand final payment for the work from the employer. Obviously, then, any wrongful certification by the architect can lead quickly to losses being suffered by the employer.

The architect as arbitrator

Over the years, however, particular problems have existed in this particular type of case that have precluded what would otherwise seem to be a fertile ground for negligence actions against architects. The problem has been that, over the years, the argument has been put successfully that the quasi-judicial activity being performed here by architects means that they should enjoy the same immunity from suit in negligence that judges themselves enjoy. The House of Lords has now twice had the opportunity to review the position.

SUTCLIFFE v. THACKRAH

[1974] AC 727   House of Lords

The defendants were architects who designed a house for the plaintiff and remained engaged to act on his behalf during its construction. In the course of this, interim certificates were issued which were in respect of work which was subsequently discovered either to have not been done or to have been done, but incompetently. It was held that the defendants had been negligent, and enjoyed no immunity from suit.

LORD REID (at p. 735):

The argument for the respondents starts from the undoubted rule, based on public policy, that a judge is not liable in damages for negligence in performing his judicial duties. The next step is that those employed to perform duties of a judicial character are not liable to their employers for negligence. This rule has been applied to arbitrators for a very long time. It is firmly established and could not now be questioned by your Lordships. It must be founded on public policy but I am not aware of any authoritative statement of the reason for it. I think it is right but it is hardly self-evident. There is a general rule that a person employed to perform duties of a professional character is liable in damages if he causes loss to his employer by failure to take due care or to exercise reasonable professional skill in carrying out his duties. So why should he not be liable if the duties which he is employed to perform are of a judicial character?

The reason must, I think, be derived at least in part from the peculiar nature of duties of a judicial character. In this country judicial duties do not involve investigation. They do not arise until there is a dispute. The parties to a dispute agree to submit the dispute for decision. Each party to it submits his evidence and contention in one form or another. It is then the function of the arbitrator to form a judgment and reach a decision.

In other forms of professional activity the professional man is generally left to make his own investigation. In the end he must make a decision but it is a different kind of decision. He is not determining a dispute: he is deciding what to do in all the circumstances. He may go wrong because he has at some stage failed to take due care and that may not be difficult to prove. But coming to a wrong but honest decision on material submitted for adjudication is rarely due to negligence or lack of care, and it is seldom due to such gross failure to exercise professional skill as would amount to negligence. It is in the vast majority of cases due to error of judgment and there is so much room for differences of opinion in reaching a decision of a judicial character that even the most skilled and experienced arbitrator or other person acting in a judicial capacity may not infrequently reach a decision which others think is plainly wrong.

But a party against whom a decision has been given that is generally thought to be wrong may often think that it has been given negligently, and I think that the immunity of arbitrators from liability for negligence must be based on the belief – probably well founded – that without such immunity arbitrators would be harassed by actions which would have very little chance of success. And it may also have been thought that an arbitrator might be influenced by the thought that he was more likely to be sued if his decision went one way than if it went the other way, or that in some way the immunity put him in a more independent position to reach the decision which he thought right.

But whatever be the grounds of public policy which have given rise to this immunity of persons acting in a judicial capacity, I do not think that they have anything like the same force when applied to professional men when they are not fulfilling a judicial function.

The point can perhaps be most clearly illustrated by considering the case of a skilled man engaged to value some property or object. The circumstances may vary very much. The owner may wish to sell or insure the property and want to know its market value. No one doubts that in that case the valuer may be sued for negligence if his negligent valuation has caused loss to the owner. Or the owner may have reason to believe that a particular person A would buy the property from him and would accept a valuation by a skilled man. Or he may have agreed with A to sell at a price to be fixed by a skilled valuer, or by this particular valuer. And he may or may not have told the valuer about this when engaging him.

There is modern authority to the effect that if the valuer knows that his valuation will affect or bind another person besides his client, the owner, then he can claim an arbitrator's immunity. But why should that be? The valuer is in each case engaged by only one party and he has exactly the same task to perform in all these cases. He must, to the best of his ability, estimate the market price of the property. I do not believe that a professional man would approach his task in any different spirit or be influenced in any significant way because he knew that the interests of some other person besides his employer would be affected by the conclusion which he reached.

On the other hand, the valuer could be engaged by both parties as an arbitrator if there is a dispute about the value of certain property. The dispute would be submitted to him for decision and the parties would put their contentions before him. Then he would have to judge between them and have an arbitrator's immunity.

Now I can come to the position of an architect. He is employed by the building owner but has no contract with the contractor. We do not in this case have occasion to consider whether nevertheless he may have some duty to the contractor: I do not think that a consideration of that matter would help in the present case. The RIBA form of contract sets out the architect's functions in great detail. It has often been said, I think rightly, that the architect has two different types of function to perform. In many matters he is bound to act on his client's instructions, whether he agrees with them or not; but in many other matters requiring professional skill he must form and act on his own opinion.

Many matters may arise in the course of the execution of a building contract where a decision has to be made which will affect the amount of money which the contractor gets. Under the RIBA contract many such decisions have to be made by the architect and the parties agree to accept his decisions. For example, he decides whether the contractor should be reimbursed for loss under clause 11 (variation), clause 24 (disturbance) or clause 34 (antiquities); whether he should be allowed extra time (clause 23); or when work ought reasonably to have been completed (clause 22). And, perhaps most important, he has to decide whether work is defective. These decisions will be reflected in the amounts contained in certificates issued by the architect.

The building owner and the contractor make their contract on the understanding that in all such matters the architect will act in a fair and unbiased manner and it must therefore be implicit in the owner's contract with the architect that he shall not only exercise due care and skill but also reach such decisions fairly, holding the balance between his client and the contractor.

For some reason not clear to me a theory has developed and is reflected in many decided cases to the effect that where the architect has agreed or is required to act fairly he becomes what has often been called a quasi-arbitrator. And then it is said that he is entitled to an arbitrator's immunity from actions for negligence. Others of your Lordships have dealt with the older authorities and I shall not say more about them than that they are difficult to reconcile and often unconvincing. They are not confined by any means to cases involving architects and one view of them has recently, in *Arenson* v. *Arenson* [1973] Ch 346, 370, been succinctly expressed by Buckley L.J.:

'In my judgment, these authorities establish in a manner binding upon us in this court that, where a third party undertakes the role of deciding as between two other parties a question, the determination of which requires the third party to hold the scales fairly between the opposing interests of the two parties, the third party is immune from an action for negligence in respect of anything done in that role.'

I can see no good grounds for this view. If there is any validity in my conjecture as to the reason of public policy giving rise to the immunity of arbitrators, those reasons do not apply to this situation. Persons who undertake to act fairly have often been called 'quasi-arbitrators'. One might almost suppose that to be based on the completely illogical argument: all persons carrying out judicial functions must act fairly, therefore all persons who must act fairly are carrying out judicial functions. There is nothing judicial about an architect's function in determining whether certain work is defective. There is no dispute. He is not jointly engaged by the parties. They do not submit evidence as contentious to him. He makes his own investigations and comes to a decision. It would be taking a very low view to suppose that without his being put in a special position his employer would wish him to act unfairly or that a professional man would be willing to depart from the ordinary honourable standard of professional conduct.

LORD SALMON (at p. 758):

Since arbitrators are in much the same position as judges, in that they carry out more or less the same functions, the law has for generations recognised that public policy requires that they too shall be accorded the immunity to which I have referred. The question is: does this immunity extend beyond arbitrators properly so called, and if so, what are its limits?

It is well established that, in general, persons such as doctors, accountants, barristers (acting in an advisory capacity), valuers and architects owe their clients a duty to exercise reasonable care and skill in rendering the services for which they are engaged. If they commit a breach of this duty which causes their client damage, then they are liable to compensate him for the loss which their negligence has caused him. This is obviously just. The heresy (as it seems to me) has, however, grown up that if a person engaged to act for a client ought to act fairly and impartially towards the person with whom his client is dealing, then he is immune from being sued by his client, however negligent he may have been. In short, liability to compensate your client for the damage you have caused him solely by your own negligence is excluded because of your obligations to act fairly and impartially towards someone else.

May I give your Lordships some examples of the astonishing results to which this heresy leads?

A well-known dealer in 18th century English paintings is brought an 18th century English painting to value for a handsome fee. He is not told why his client requires the valuation. It may be because he intends to sell it or insure it, or perhaps just out of curiosity. The dealer values the picture (entirely honestly but wrongly) at £500. Relying on this valuation, the client asks £500 for the picture, and sells it for that sum. it is subsequently established that the picture was worth £50,000 and that this should have been obvious to anyone in the dealer's position who had exercised reasonable care and the skill which he professed. In such circumstances, the client would have an unanswerable claim against the dealer in negligence. Now suppose exactly the same facts, save that when the client brought the picture to the dealer he told him that the valuation was wanted because he was going to sell the picture to a friend, and the friend had agreed to buy the picture for the value which the dealer put upon it, providing he could afford to do so. It would appear on the authority of certain cases, to which I will refer later, that the dealer would then be immune from being sued by his client because of his additional duty to act impartially and fairly towards his client's friend. It is said that this factor, of itself, puts the dealer in the same position as if he were performing the functions of a judge or arbitrator, and accordingly, so the argument runs, public policy requires that he should have complete immunity in respect of his undoubted negligence, which had admittedly caused his client a loss. I am afraid that I can find no sensible basis for such an astonishing proposition.

Take another example, an architect who has been engaged by a building owner to look after his interests in relation to the construction of a building is (as in the present case) the architect nominated in an RIBA form of contract entered into between the building owner and the contractor. Under that contract to which the architect is not a party, but of which he, of course, has knowledge, the parties agree, amongst other things, that the architect shall issue interim certificates stating the amount due to the contractor in respect of work properly executed, and that, within a specified period after the date of each certificate, the building owner shall pay the amount therein certified (clause 30). In the event of a dispute arising in relation to any such certificate, the contract provides that the parties may at any time submit the dispute to arbitration (clause 35).

No one denies that the architect owes a duty to his client to use proper care and skill in supervising the work and in protecting his client's interests. That, indeed, is what he is paid to do. Nevertheless, it is suggested that because, in issuing the certificates, he must act fairly and impartially as between his client and the contractor, he is immune from being sued by his client if, owing to his negligent supervision or (as in the present case) other negligent conduct, he issues a certificate for far more than the proper amount, and thereby causes his client a serious loss.

As in the case of the valuer, it is said that the architect is performing much the same functions and must, therefore, be regarded as being in the same position as a judge or arbitrator and must accordingly be accorded the same immunity. I confess that I can see no more reason for regarding the architect as being in the same position as a judge or arbitrator than there is for so regarding the valuer. No reason has ever been suggested. I suspect that this is because none exists. The descriptions 'quasi-arbitrator' and 'quasi-judicial functions' have been invoked but never defined. They cannot mean more than in much the same position as an arbitrator or judge. In reality, however, there are the most striking differences between the roles of the valuer and architect in the circumstances to which I have referred and the role of a judge or arbitrator. Judges and arbitrators have disputes submitted to them for decision. The evidence and the contentions of the parties are put before them for their examination and consideration. They then give their decision. None of this is true about the valuer or the architect who were merely carrying out their ordinary business activities. Indeed, their functions do not seem to me even remotely to resemble those of a judge or arbitrator. Moreover, in the case of the architect, the contract provided that the certificate was not binding and that, in the event of any dispute arising in relation to it, that dispute could be submitted to arbitration for decision. Like my noble and learned friend, Lord Reid, I suspect that the heresy that such valuers and architects are to be regarded as being in the same position as judges and arbitrators rests on the fallacy that since all judges and arbitrators must be impartial and fair, anyone who has to be impartial and fair must be treated as a judge or an arbitrator.

These arguments seemed clear and conclusive, but the then recent Court of Appeal case of *Arenson*, mentioned in Lord Reid's speech, also came on appeal to the Lords.

ARENSON v. CASSON BECKMAN RUTLEY & CO.

[1977] AC 405   House of Lords

The defendants in this case were auditors who were entrusted with the task of assessing the fair value of shares to be transferred to his uncle if the plaintiff left his employment with the family's business. The defendants were duly called in and valued the shares at a price one sixth of their value a few months later when the business went public. The plaintiff sued in negligence and won.

LORD SIMON OF GLAISDALE (at p. 419):

> [The argument] starts with the immunity conferred on the arbitrator for reasons of public policy. But in my judgment this is a secondary and subordinate consideration of public policy. There is a primary and anterior consideration of public policy, which should be the starting point. That is that, where there is a duty to act with care with regard to another person and there is a breach of such duty causing damage to the other person, public policy in general demands that such damage should be made good to the party to whom the duty is owed by the person owing the duty. There may be a supervening and secondary public policy which demands, nevertheless, immunity from suit in the particular circumstances (see Lord Morris of Borth-y-Gest in *Sutcliffe* v. *Thackrah* [1974] AC 727, 752). But that the former public policy is primary can be seen from the jealousy with which the law allows any derogation from it. Thus a barrister enjoys immunity, but only in respect of his forensic conduct (since his duty to the court may conflict with and transcend his duty to his client): *Rondel* v. *Worsley* [1969] 1 AC 191.

(and at p. 424):

> But in my view the essential prerequisite for him to claim immunity as an arbitrator is that, by the time the matter is submitted to him for decision, there should be a formulated dispute between at least two parties which his decision is required to resolve. It is not enough that parties who may be affected by the decision have opposed interests – still less that the decision is on a matter which is not agreed between them.

It will be clearly seen that the mere fact that the architect is making an assessment or a valuation between two parties is not enough to make him into an arbitrator and thus potentially immune. There will need to be a clearly formulated legal dispute between the parties before any question of such immunity arises, for example if they continue to dispute the architect's certificate or valuation and decide to take formal legal action to resolve the matter. Of course, if such a dispute arises and there is dissatisfaction with the outcome, this may be capable of resolution by an appeal to the High Court on a point of law (Arbitration Act 1979 s.1(2)).

The problem that would today come to the fore in a case like *Sutcliffe* is the nature of the loss suffered. Sutcliffe paid out too much for the quality of the house he actually received, and thus suffered a purely economic loss. On the other hand, it could be argued that the triangular relationship linking employer, builder and architect may be sufficiently close in character to enable even an economic loss claim to succeed. Of course just as over-payment, such as in *Sutcliffe* will affect the employer, so underpayment will harm the contractor's interest.

Certification and liability to the contractors

Two recent cases have allowed the courts to explore this tricky area.

LUBENHAM FIDELITIES INVESTMENTS CO. LTD v. SOUTH PEMBROKESHIRE DISTRICT COUNCIL

(1986) 33 BLR 1　　　　　　　　　　　　　　　　Court of Appeal

Lubenham were working as contractors for the council. The council's architects deducted an amount wrongfully from an interim payment. As a result of this, the contract ended abruptly. Complex legal actions ensued, including one by the contractors against the architects, which was unsuccessful.

MAY LJ (at p. 55):

> Whatever the cause of the undervaluation the proper remedy available to the contractor is, in our opinion, to request the architect to make the appropriate adjustment in another certificate or if he declines to do so, to take the dispute to arbitration under clause 35. In default of arbitration or a new certificate the conditions themselves give the contractor no right to sue for the higher sum.

(and at p. 71):

> In our judgment Lubenhams broke the chain of causation by persisting in the suspension of the works. They alone were responsible for what followed and for the termination of the two contracts. Accordingly, they have no claim . . .

What this (edited) account amounts to is this. Lubenham's prime remedy is against the architect's incorrect certificate in what the terms of the contractor made clear was a binding arbitration. The court seems to accept that, in tort, a duty of care may be owed – this did not seem to be objected to by council (see 33 BLR 68/69) but found that the issue of causation was of itself enough to deny a remedy to Lubenhams, the authors of their own misfortune. The issue of duty of care was faced head-on in the next case.

MICHAEL SALLISS & CO. LTD v. CALIL

(1988) Const LJ 125　　　　High Court (Official Referee's Business)

This was a claim by the builders, Salliss, and against their employers, Calil. Subsequently, the builders joined the architects (Newman) as third parties and claimed that they owed a duty of care to them in relation to, amongst other things, the preparation of workable drawings for the contractor to work from and in authorizing formal extensions to the contract period as permitted by the contract if authorized as reasonable by the architect. The architect was held to owe a limited set of duties of care towards the contractor.

JUDGE FOX-ANDREWS QC (at p. 130):

> But it is self-evident that a contractor who is a party to a JCT contract looks to the architect/supervising officer to act fairly as between him and the building employer in matters such as certificates and extensions of time. Without a confident belief that that reliance will be justified, in an industry where cash flow is so important to the contractor, contracting could be a hazardous operation. If the architect unfairly promotes the building employer's interest by low certification or merely fails properly to exercise reasonable care and skill in his certification it is reasonable that the contractor should not only have the right as against the owner to have the certificate reviewed in arbitration but also should have the right to recover damages against the unfair architect.
>
> I find that to the extent that the plaintiffs are able to establish damage resulting from the architect's unfairness in respect of matters in which under the contract the architects were required to act impartially damages are recoverable and are not too remote.
>
> But in many respects an architect in circumstances such as these owes no duty to the contractors. He owes no duty of care to contractors in respect of the preparation of plans and specifications or in deciding matters such as whether or not he should cause a survey to be carried out. He owes no duty of care to a contractor whether he should order a variation. Once, however, he has ordered a variation he has to act fairly in pricing it.
>
> The reason why the architects owe no duty in the kind of circumstances I have just set out is because the architects are not required to act fairly. In many cases, e.g. variations, he will be acting on his client's instructions. It would be bizarre if, for example, having been instructed by a client not to incur the expense of a survey the architect was found to have breached some duty to the contractor in not having done so.

As a result, it was held that a duty was owed in relation to the granting of certified extensions to the work but not in relation to earlier work such as the preparation of drawings and survey work.

Whether a fuller or more rigorous analysis of the relationship would produce the same answer as that given in *Salliss* must be doubtful, especially in the light of the ever greater obstacles that have been placed in the way of claims for economic loss.

Helpful as *Salliss*, in particular, is to a contractor pressing such a claim against an architect, it is more likely that an architect could now defend such a claim by pointing either generally to the very tight constraints on economic loss claims or more specifically to the general idea that contractual arrangements may reduce or limit the duty of care if the presence of a tortious duty contradicts the apparent intention of the parties. This was raised on similar facts in *Pacific Associates* v. *Baxter* [1990] QB 993 (see p. 99 above) which stated that no duty arose against the background of the particular set of contracts in the case and said of the *Salliss* (at p. 1020) that it 'apparently overlooks the contractual structure against which any reliance placed by the victim on the assumption of liability demonstrated by the proposed tortfeasor depends'.

So *Sallies* is doubted explicitly by Purchas LJ and implicitly by his colleagues, and any attempt to argue on its basis will need to be carried out very carefully, to see whether the duty of care proposed is compatible with the surrounding web of contractual provisions in the case in question.

Supervision

The remaining stage at which the builder's and the architect's separate functions interact lies of course in the responsibility of the architect to supervise the actual construction work while it is in progress. In the course of this, naturally the architect may discover that the contractor is executing his work badly and the question therefore arises as to whether the architect owes any duty in tort towards the contractor, as well as his obvious duty to his employer. Equally, it may also be the case that during the course of the construction work, the builder becomes aware that all is not well with the design or materials recommended by the architect and therefore the question arises here too of a possible role for tort.

The first point to make concerns the basic supervisory role of the architect. His contractual obligation to his employer, which in turn shapes the area of responsibility in which any tort duty may lie, is not an onerous one. The House of Lords emphasized this in *East Ham Corporation* v. *Bernard Sunley & Sons* [1966] AC 406. It was pointed out forcefully that the architect's duty is to the employer and not to the contractor (see Lord Pearson at p. 449) and that continual supervision is of course impractical (see Lord Upjohn at p. 443). In any event the duty is not an absolute one; the terms and other circumstances of the contract will affect the standard of care (see, for example, *Cotton* v. *Wallis* [1955] 1 WLR 1160). Thus the standard of care will be variable and may be quite low, even if a duty of care exists in tort between architect and builder.

CLAYTON v. WOODMAN & SON (BUILDERS) LTD

[1962] 1 WLR 585   Court of Appeal

The plaintiff was a bricklayer, working for contractors, who was injured when a wall collapsed. This occurred after the plaintiff had questioned whether it should not have been demolished, but the architect, Berry, had examined the site and insisted on adherence to the original plan. It was held that the builders were liable, but that the architect was not.

PEARSON LJ (at p. 593):

> In this appeal the two effective parties are the builder and the architect, but for the purpose of determining the result of the appeal one has to consider whether there was a liability of the architect to the plaintiff, the injured workman. It is quite plain, in my view, both as a general proposition and under the particular contract in this case, that the builder, as employer, has the responsibility at common law to provide a safe system of work, and he also has imposed on him under the Building (Safety, Health and Welfare) Regulations 1948, the responsibility of seeing that those regulations are complied with, so that everything is as safe for the workman as it reasonably can be. It is important that that responsibility of the builder should not be overlaid or confused by any doubt as to where his province begins or some other person's province ends in that respect. The architect, on the other hand, is engaged as the agent of the owner for whom the building is being erected, and his function is to make sure that in the end, when the work has been completed, the owner will have a building properly constructed in accordance with the contract, plans, specification and drawings and any supplementary instructions which the architect may have given. The architect does not undertake (as I understand the position) to advise the builder as to what safety precautions should be taken or, in particular, as to how he should carry out his building operations. It is the function and the right of the builder to carry out his own building operations as he thinks fit, and, of course, in doing so, to comply with his obligations to the workman.
>
> It is against that background that one has to see this case. There are in the contract with which we are concerned several provisions which are entirely consistent with and which support what I have said as to the position of the builder.

(and at p. 594):

> What else is it said that the architect should have done or how else is it said he should have acted? If it is said that he was wrong in omitting to alter the specification and thereby impliedly requiring that it should be carried out, the obstacle to suggesting that there was anything improper in his conduct is the fact that in so doing he was carrying out his duty to the building owner: it was his duty to direct the incorporation of the wall if (as he thought, and it would

seem to be right), that was in the best interests of the building owner. But it cannot be right, in my view, to impose on the architect two conflicting duties in this situation: his duty to the owner to insist on the performance of the contract, and some other duty supposed to be owed to the builder or to the builder's workman to make a variation in the specification in the circumstances of the case.

Secondly, it might be suggested that the fault of the architect was in not advising the builder, through his existing representative on the site, the plaintiff, as to how the work required by the specification should be executed. If he had done so, the architect would have been stepping out of his own province and into the province of the builder. It is not right to require anyone to do that, and it is not in the interests of the builder's workpeople that there should be a confusion of functions as between the builder on the one hand and the architect on the other. I would hold that it was plainly not the architect's duty to do that. It will be observed that he had at any rate no pre-existing duty to do that. He was not asked to give any such advice and he did not profess to give any such advice, and I cannot see that it can be regarded as fault on his part that he did not step out of his province and advise the builder in what manner the builder should carry out his own building operations.

Thirdly, it might be suggested that the architect should have given a warning to the builder's workman, the plaintiff, as to how the work should be done or that there was some risk involved in doing it in a particular way. But there, also, it seems to me that that would have been stepping out of his own province and entering that of the builder. He was entitled to assume that the work would be carried out properly, that the builder knew his own business and would properly perform his own operations.

Fourthly, it might be suggested that the architect should have stopped the progress of the work. There again there was no need to do that unless he could assume that the builder did not know his own business and was not going to do his own work in the right way.

Having regard to those considerations I am unable to accept the judge's view, which is very closely related to the particular facts of the particular case and does not involve any different view from that which I have expressed as to the division of functions in a normal case between the builder on the one side and the architect on the other. The judge took the view that Berry had taken the responsibility of giving an order to the workman to do something which in the circumstances was dangerous. If I had thought there was evidence that Berry directed the workman to do something which he knew or ought to have known would be done in such a manner that it would be dangerous, then it might well be that some duty would have been imposed on him. But the distinguishing feature here from that hypothetical case is that all Berry in fact did was to decide not to vary the contract, with the effect that work would be done which could be performed in a safe manner. It is not shown by the evidence that the manner of work was necessarily unsafe, and in my view it would be wrong to impose on the architect in this case any extraneous duty which would involve a completely additional task for him and would involve interference by him in the sphere of the builder. That might result in confusion of functions and in the function of the builder not being performed by him.

The case clearly asserts a firm divide between the separate areas of responsibility of builder and architect. However, it was not long before a different view appeared to be adopted.

CLAY v. A.J. CRUMP & SONS LTD

[1964] 1 QB 533   Court of Appeal

The plaintiff was an employee of contractors. He was injured when an old wall collapsed. This wall was originally planned for demolition but its retention was approved by the architect after a telephone conversation with the demolition contractors, but without inspection of the wall which was in a palpably dangerous condition.

ORMEROD LJ (at p. 555):

> It was contended on behalf of the architect that he was employed under a contract with the owners and in consequence was answerable to them alone if by any act or omission he was in breach of that contract. It may be that there was a time when this view of the law would have prevailed. Decisions in recent years, however, have broadened the basis upon which persons may be found liable if they are in default in the performance of their contractual duties and in considering whether the architect in this case owed a duty to the plaintiff other questions have to be taken into account than the contractual liabilities of the architect to the building owner.

(and at p. 556):

> Is this a case in which it can be said that the plaintiff was so closely and directly affected by the acts of the architect as to have been reasonably in his contemplation when he was directing his mind to the acts or omissions which are called in question? In my judgment, there must be an affirmative answer to that question. The architect, by reason of his contractual arrangement with the building owner, was charged with the duty of preparing the necessary plans and making arrangements for the manner in which the work should be done. This involved taking precautions or giving instructions for them to be taken so that the work could be done with safety. It must have been in the contemplation of the architect that builders would go on the site as the whole object of the work was to erect buildings there. It would seem impossible to contend that the plaintiff would not be affected by the decisions and plans drawn up by the architect. But [counsel for A.J. Crump & Sons Ltd] has argued with some force that the plaintiff cannot be regarded as being in the class of persons to whom a duty was owed by the architect as there was ample opportunity for inspection of the wall in question by the architect and by other

people, including the plaintiff himself, before the wall actually collapsed. This, of course, is true enough, but is not, in my judgment, an end of the matter.

(and at p. 558):

The facts in this case are that the architect, instead of going himself to look at the wall to decide whether it was safe to be left, spoke to the demolition contractor, who in turn took the opinion of the foreman, and the architect acted on that opinion. Having come to a decision that the wall could be left, he appears not to have taken any further step to satisfy himself of its safety, although it is abundantly clear that there were opportunities for him to examine it. There can be no doubt on these facts that the architect was negligent. No one has suggested that the wall, left as it was, was safe. The contrary view has been generally accepted and, as I said earlier, that is borne out by the fact that the wall fell down without anything further being done with regard to it. There is no evidence of the kind of weather there was during the weeks when the site was left before the builders came on to it, but equally there is no evidence of the effect such weather might have had in changing the stability of the wall. The architect chose to rely upon the opinion of the demolition contractor. This must have been wrong if the evidence called before the judge is anything to go by. And, for my part, I can see no reason why it should be said that because an architect, instead of making sure for himself, accepts the opinion of another man whose opinion is given either negligently or certainly without sufficient examination, the architect is free from liability. He has done nothing more, as I see it, than appoint an agent to act for him to give a decision which it was his duty to give himself. In those circumstances, I fail to see how it can be said that the injury to the plaintiff was not caused by the negligence of the architect. It was urged that even if the architect and the demolition contractor were each in turn negligent, that negligence was not the cause of the injury to the plaintiff as there was negligence on the part of the builder, who did not ensure that the premises were reasonably safe before allowing his workmen to go on to them, and in those circumstances was guilty of negligence which was the cause of the accident to the plaintiff. It was urged on the part of the builders that they were entitled to rely on the demolition contractors as having left the premises in a safe condition and that nothing other than a cursory examination was required of them. That may, of course, be so, but the evidence appears to have been that even a cursory examination would have disclosed the state of the wall and to an experienced eye the fact that it was dangerous. It may be that there was negligence in some degree on the part both of the demolition contractors and the builiders. If there was such negligence, it may be that it was a contributory cause of the accident. It cannot, however, in my judgment, absolve the architect from a share in the blame. To hold otherwise would be to hold that an architect, or indeed anyone in a similar position, could behave negligently by delegating to others duties he was under an obligation to perform and escape liability by the plea that the injuries caused were caused by the negligence of that other person and not of himself. I do not accept that as being the true position in law and I am satisfied that it is not the law as laid down by Lord Atkin in *Donoghue* v. *Stevenson* or by the judgments in any of the other cases to which we have been referred.

The key factual distinction which justifies the different approach in *Clay* is that the architect carelessly departed from his original plan, whereas in *Clayton* he reaffirmed it, and the separate conclusions reached may be acceptable on that basis. However, the judgment of Ormerod LJ is notably short on authority, in particular, *Clayton* itself. In fairness, Davies LJ (at p. 570) did consider *Clayton*, and distinguished it by pointing to passages which seemed to support that the architect did not do anything negligent in allowing work to proceed in accordance with the plans. Discussion of this issue has subsequently continued.

OLDSCHOOL v. GLEESON (CONSTRUCTION) LTD

(1976) 4 BLR 105 High Court (Official Referee's Business)

The plaintiff sued Gleeson for damage caused by defective work. Gleeson sought a contribution from the second defendant (Taylor, Whalley & Spyra), consulting engineers to the project. It was held that they were not liable.

JUDGE STABB QC (at p. 123):

> I do not think that the consulting engineer has any duty to tell the contractors how to do their work. He can and no doubt will offer advice to contractors as to various aspects of the work, but the ultimate responsibility for achieving the consulting engineer's design remains with the contractors. To take the present case as an example, I have no doubt that it was the contractor's duty to set whatever shoring might have been necessary. It was also for them to decide upon the sequence of excavation that was to be adopted and how such excavation was to be temporarily supported if required. Both the district surveyor and Mr Gabriel [of the second defendants] asked the contractors to provide a sequence of operations, but neither obtained a satisfactory reply. If the contractors had said, for example, that they planned to excavate first down to footing level along the whole length of the party wall and thereafter to excavate the rest of No. 31, the consulting engineer might well have pointed out the undesirability or even the danger of adopting that course: but I do not think that he was under a duty to direct the contractors, for instance, to excavate in strips up to the party wall. It was the responsibility of the contractors to decide upon the method and sequence of excavation so as to achieve the consulting engineer's design; but if, for example, they planned to excavate the hoist pit without any temporary support, and so informed the consulting engineer, then as a matter of common sense the consulting engineer

would intervene to prevent that which was described as 'an act of incredible folly'.

From the evidence which I have heard and from the contemporaneous documents I am satisfied that the second defendants adequately fulfilled their duty of supervision. Mr Gabriel persistently drew Mr Craven's [of Gleesons] attention to the inadequacy of the shoring, although he was not, in my view, duty bound to do so. He warned him of the risk that he was running. He emphasised the necessity to blind the excavated ground by the party wall at the earliest opportunity. He told Mr Craven on 11 March, before any excavation of the hoist pit had started, to put in the sheeting or strutting, in the form of precast concrete planks, which were on the site, and not to excavate further until this was done. In the circumstances he had no reason to foresee that further excavation would be carried out until this was done. He personally visited the site seven times in the six weeks period from 29 January to the date of collapse, and in spite of Mr Craven's evidence to the contrary, I believe that Mr Gabriel advised and warned Mr Craven in the manner which he described, but that Mr Craven for some reason best known to himself ignored the advice and warning, and noticeably did not seek the assistance of Mr Gabriel at any time, more particularly at the critical time when cracks started to appear on 13 March, and when the danger signals should have been apparent. [Counsel] agreed that, if the contractor had any doubts about how the excavation should be done or how the temporary support should be set or how the underpinning should be carried out, then they should have asked the consulting engineer, but this they never seem to have done.

As I have already indicated, I do not consider that the consulting engineer's duty of supervision extends to instructing the contractors as to the manner in which they are to execute the work, and I think that that is probably accepted by the first defendants to a large extent. What is said, however, is that when the consulting engineer knows or ought to know that the contractors are heading into danger whereby damage to property is likely to result, then he owes the contractors a duty of care to prevent such damage occurring. If he sees the contractors not taking special precautions without which a risk of damage to property is likely to arise, then he the consulting engineer cannot sit back and do nothing. I am not sure that the consulting engineer's duty extends quite that far but, even if it does, I do not believe that he is under a duty to do more than warn the contractors to take the precautions necessary, and in so far as those precautions consisted here of shoring and providing temporary support and immediate blinding in excavations in the vicinity of the party wall, I am satisfied that Mr Gabriel gave Mr Craven ample warning.

Accordingly, it follows that I am not satisfied that the second defendants failed adequately to perform their duty of supervision.

He went on to discuss whether any liability may arise to the builder in respect of negligently executed design work (at p. 131):

It seems abundantly plain that the duty of care of an architect or of a consulting engineer in no way extends into the area of how the work is carried out. Not only has he no duty to instruct the builder how to do the work or what safety precautions to take but he has no right to do so, nor is he under any duty to the builder to detect faults during the progress of the work. The architect, in that respect, may be in breach of his duty to his client, the building owner, but this does not excuse the builder for faulty work.

I take the view that the duty of care which an architect or a consulting engineer owes to a third party is limited by the assumption that the contractor who executes the work acts at all times as a competent contractor. The contractor cannot seek to pass the blame for incompetent work onto the consulting engineer on the grounds that he failed to intervene to prevent it. I do not consider that this view conflicts in any way with the decision in *Clay* v. *A.J. Crump & Sons Ltd* [1964] 1 QB 533, in which an architect was held partially responsible for the personal injuries suffered by a contractor's workman caused in part by the architect having allowed a dangerously defective wall to remain standing on the site when the contractors came in. The responsibility of the consulting engineer is for the design of the engineering components of the works and his supervisory responsibility is to his client to ensure that the works are carried out in accordance with that design. But if, as was suggested here, the design was so faulty that a competent contractor in the course of executing the works could not have avoided the resulting damage, then on principle it seems to me that the consulting engineer responsible for that design should bear the loss.

Judge Stabb therefore exonerates the consulting engineers in respect of both their supervisory function and their design work. Supervision does not detract from the primary responsibility of the builder for the proper execution of the work, and faulty design can only lead the architect to be responsible if its faults are such that the work could not be done without causing damage. This type of situation can of course potentially lead to losses to the contractor too. Could he claim?

VICTORIA UNIVERSITY OF MANCHESTER v. HUGH WILSON AND LEWIS MOMERSLEY

(1984) 1 Const LJ 162   High Court (Official Referee's Business)

Tile cladding on a recently constructed building occurred. The plaintiff took action against those responsible. Their claim against the architects was settled, but the contractors were held fully liable. The question then arose as to whether the architects could be liable to the contractors. After reviewing *Clayton*, *Clay* and *Oldschool*, JUDGE NEWEY QC went on (at p. 168):

In *Arenson* v. *Arenson* [1977] AC 405, a case concerned with a valuer, Lord Salmon said at p. 438 when referred to *Sutcliffe*:

'The architect owed a duty to his client, the building owner, arising out of the Contract between them to use reasonable care in issuing his Certificate. He also, however, owed a similar duty of care to the Contractor arising out of their proximity.'

In *Junior Books Ltd* v. *Veitchi Co. Ltd* [1982] 3 All ER 201, a case which was not concerned with an architect and contractors, but with an employer and sub-contractors, the House of Lords held that there was sufficient proximity between them for it to be possible for the sub-contractors to have owed a duty of care to the employer when laying a floor not to create economic loss.

I think that in view of the House of Lords cases Judge Stabb's words in *Oldschool* must now be regarded as too widely expressed. Because of proximity an architect may sometimes owe a duty of care to contractors even in relation to how they carry out their work. If, for example, an architect knew that on a site with which they were concerned contractors or sub-contractors were making a major mistake which would involve the contractors in expense, I think that the architect would probably owe a duty to the contractors to warn them. In those circumstances the architect would not be instructing the contractors in how to do their work, but merely warning them of the probable consequences of persistence in the particular method which they had adopted.

In this case, however, the architects clearly did not discover that the sub-contractors were doing or had done their work badly. They had no knowledge which they could communicate. They knew that the contractors were themselves under an express duty to inspect the sub-contractors' work and had no reason to believe that the contractors would not inspect properly. In all the circumstances I conclude that the architects did not owe the contractors a duty to exercise when inspecting. It follows that the architects were not in breach of duty to the contractors.

A duty, but no breach finding, was a convenient way of holding that no liability should exist even at the height of the post-*Anns* spread of tortious duties of care. Given that that was the background of this case, it can hardly be regarded as the strongest authority today. Nevertheless, it remains as a potentially valuable precedent for contractors to employ against architects.

The final issue which then arises is whether there is any correlative duty on the contractor to warn the architect that the design is giving problems and may be defective.

EQUITABLE DEBENTURE ASSETS CORPORATION LTD v. WILLIAM MOSS

(1984) 1 Const LJ 131   High Court (Official Referee's Business)

Sub-contractors, Alpine, who had gone into liquidation supplied defective curtain walling for the plaintiffs' [EDAC] new offices. Action was therefore taken against Moss, the contractor, and Morgan, the architect.

JUDGE NEWEY QC (at p. 132):

> Morgan's duty to supervise and inspect must also have continued until practical completion.
>
> Apart from obligations in contract and in the absence of any agreement to the contrary, Morgan plainly owed to EDAC a duty of care in negligence, which was, I think, precisely the same as their duty in contract. Morgan's duty in negligence obviously extended to design and to supervision and inspection.

No surprise there, as Judge Newey establishes that the architect would be liable to his employer. But Judge Newey goes on (at p. 134):

> I think that if on examining the drawings or as a result of experience on site Moss formed the opinion that in some respect the design would not work, or would not work satisfactorily, it would have been absurd for them to have carried on implementing it just the same. In my view if the directors of EDAC and of Moss had been asked at the time when the contract was made what Moss should do in those circumstances, they would have agreed at once that Moss should communicate their opinion to Morgan. I think, therefore, that in order to give efficacy to the contract the term requiring Moss to warn of design defects as soon as they came to believe that they existed was to be implied in the contract.
>
> I think that as part of Moss's duty of care in negligence they owed EDAC and Morgan a duty to inform the latter of design defects known to them.
>
> Since Moss did not call any witnesses of fact and there was no direct evidence from any other source as to what their employees did or did not do, whether Moss came to know of facts which must have led them to believe that the design was defective has necessarily to be a matter of inference.
>
> Morgan sent to Moss Alpine's drawings as contract drawings. It can be readily inferred that an employee or employees of Moss looked at the drawings. Whether Moss had had much experience of curtain walling, I do not know. Since, however, Mr Rae, an architect with experience of curtain walling, spent a day and a half examining the drawings without detecting defects, I cannot possibly conclude that they were so obvious that Moss must necessarily have come to know of them.

Again this needs to be judged against the background of the time. Since the decision in this case, the House of Lords in *Peabody* (see pp. 12 above and pp. 228 below) has put great emphasis on policy considerations in general and on the respective abilities and responsibilities of all those involved in the construction process. It is not certain that a court today would hold that the builder owed a duty to the architects given their likely level of respective qualification though it should be noted that Judge Newey in the instant case did emphasize the experienced character of the builders.

The uncertainty of this matter is added to by the decision in the

Southern Water case. The question arose here too as to whether sub-contractors owed a duty to the supervisors of the project, here consulting engineers.

SOUTHERN WATER AUTHORITY v. CAREY

[1985] 2 All ER 1077   High Court (Official Referee's Business)

JUDGE SMOUT QC (at p. 1092):

> Counsel for the first defendants, however, puts the argument in yet another way. He contends that there was a duty on the sub-contractors in tort to warn the first defendants of defects which they knew or ought to have known. He relies on *Equitable Debenture and Assets Corp Ltd* v. *William Moss Group Ltd* (1984) 1 Const LJ 131, a decision of his Honour Judge John Newey QC, and indirectly on the decision of the Canadian Supreme Court in *Brunswick Construction Ltd* v. *Nowlan* (1974) 49 DLR (3rd) 93 and the decision of his Honour Judge Hawser QC in *Twins Transport Ltd* v. *Patrick & Brocklehurst* (1983) 25 Build LR 65. I remind myself that we are considering whether there be a duty in tort and not in contract; in that respect, the only passages in point in the *Equitable Debenture and Assets Corp. Ltd* judgment are those where Judge Newey concludes there was a duty between Alpine, as sub-contractor involved in design, and Morgan, the architect. However, as Judge Newey stated, Alpine had admitted liability and had taken no part in the hearing and it appears that there was no legal argument there developed that can assist me in the instant case. The *Brunswick Construction* and the *Twins Transport* decisions do not in this respect lay down principles that bear on the facts before me.
>
> I can well understand that there may be a duty on a sub-contractor to warn an architect or consulting engineer in particular circumstances, as, for example, if the sub-contractor is a builder who knows of some fault in the architect's plans or knows of a failure of the consulting engineer to provide for any necessary expansion joints. In such a case, the sub-contractor who proceeds to build regardless of his knowledge may well be in breach of a duty in tort, and it may be that such a duty is owed not only to the building owner but also to the architect or consulting engineer. However, those circumstances are very different from the instant case. I am not satisfied that a sub-contractor has any separate liability to his supervising officer arising out of any failure of the sub-contractor to warn the supervising officer of any defect that there may be in the sub-contractor's work. Such a stern proposition would, in my opinion, extend the nature of duty of care in tort well beyond the present horizon.
>
> Accordingly, I have come to the conclusion that, howsoever the matter is argued, no duty of care is owed by the fourth or the fifth defendants to the first defendants in relation to the quality of fitness of the work in the circumstances pleaded.

Again it seems that a contractor possessed with expertise is the type of contractor who may be vulnerable to this novel duty; the issue remains to be decided. However, again it should be remembered that, as the contractual provisions are silent as to the existence of a duty, this may now be a factor that will make it more difficult to establish a duty in tort, at least where there is some form of direct contractual link between builder and architect.

Architects – summary

The risk of architects and others such being sued remains high, but is mitigated by several factors. Restrictive duties of care, especially in cases of economic loss, will limit the range of claimants. Realistic attitudes to the standard of design prevail, after the fright of *Eckersley*, and the standard required in supervision is not great. Likewise, there is a growing understanding in the courts of the respective role of supervisor and supervised.

Chapter 10

Local Authority Liability

As we have seen in Chapter 7, the roles of the builder and of the local authority in defective premises litigation have been intertwined ever since this area of law began to develop. In particular, from *Dutton* v. *Bognor Regis Urban District Council* [1972] 1 QB 375 to *Murphy* v. *Brentwood District Council* [1990] 2 WLR 414, [1991] 1 AC 398, the issue of whether a duty of care is owed, and in respect of what, has been one where there has been much common case-law involving judicial consideration of both parties' duties of care.

Equally, there are several points at issue only in the realm of local authority liability. Therefore this chapter will begin by highlighting the position of the local authority in relation to duty of care before going on to look at specific issues in local authority liability.

Local Authorities' duty of care

(a) Duty to avoid threats to health and safety

This was the duty created by *Anns* v. *London Borough of Merton* [1978] AC 728 out of the statutory context of the Public Health legislation and the Building Regulations. This framework was there to protect the health and safety of owners and occupiers and the powers given to local authorities were there to avoid threats to their health and safety. However, *Murphy* now asserts that such threats disappear once work is done to prevent them materializing and that therefore the cost of such work counts as purely economic loss in most cases (see pp. 34 above). As discussed in relation to the builder (see p. 153 above), possible exceptions exist where one part of a building is harmed by another separate part, probably installed by another contractor and also where work is done to avert a threat to persons outside the premises, such as passers-by on the highway. In each case, there are grounds, taken from *Murphy*, for arguing that these are cases of physical damage, where a duty of care framed in terms of mainstream negligence remains at least arguable. This applies equally to local authorities and builders.

(b) Duty to avoid causing economic loss

In cases other than the exceptions just noted, the House of Lords in *Murphy* categorize preventative repair work as economic loss. However, once again, this does not, as such, bar claims in respect of such losses against local authorities, but rather simply means that claims will only succeed if they can be forced within the highly restrictive criteria for such a duty laid down by *Junior Books* v. *Veitchi* [1983] AC 520 [see pp. 22 above].

If we remind ourselves of the special factors which gave rise to liability in that case, it can be seen that key elements were the special skills of Veitchi, the reliance placed thereon by Junior Books, and Veitchi's knowledge both of that reliance and of the likelihood of economic loss arising given their actual knowledge of the proposed uses of the building. All this should arise in a relationship akin to a contractual one.

By comparison, the local authority, although not without some expertise and, of course, local knowledge, does not seem to have the level of special skill as a well-known contractor such as Veitchi, and if the owner of a building is relying on anyone, it is (or should be) his own architects, engineers and surveyors, rather than the local authority. Similarly, the level of involvement in the project of the local authority is far less than that of an expert nominated sub-contractor such as Veitchi. The whole relationship between building owner and local authority does not arise out of contract or anything akin to it. Given all these factors, as well as the need to examine the purpose of the statutory powers in question (see pp. 228, below), it is very difficult to envisage any realistic situation where a claim for economic loss will be likely to succeed against a local authority in respect of the operation of their powers. The only possible type of action where, conceivably, a local authority may face liability for economic losses is in relation to any careless misstatement they may make. It is likely that a building owner may well rely on certain statements, especially those based on local knowledge, and if made carelessly such statements could perhaps lead to liability under *Hedley Byrne* v. *Heller* [1964] AC 465 and *Caparo Industries plc* v. *Dickman* [1990] 2 AC 605.

(c) Duty not to cause injury to person or property

In *Murphy*, no injury was caused (happily) to the occupiers or their personal property. As a result, the case is concerned only with the broader duty, to avoid future threats to safety, and does not tell us what the position would be if the house had collapsed suddenly, and without warning, causing injury or damage.

If this relatively unusual scenario were to occur, general principle might suggest that a duty of care would arise. If a local authority carelessly inspects the foundations of a building, harm to its future users and their property is clearly foreseeable. Further, that finite group of users can be reasonably said to be in a relationship of proximity with the local authority, at least as far as their personal injuries are concerned. It would be harsh to involve policy considerations against the physically injured and indeed to allow their claims, for personal injury at least, would be entirely in accordance with, not contrary to, the purposes of the legislation.

However, the House of Lords appear to have their doubts.

MURPHY v. BRENTWOOD DISTRICT COUNCIL

[1991] 1 AC 398   House of Lords

For facts, see pp. 31.

LORD MACKAY (at p. 457):

> I should make it clear that I express no opinion on the question whether, if personal injury were suffered by an occupier of defective premises as a result of a latent defect in those premises, liability in respect of that personal injury would attach to a local authority which had been charged with the public law duty of supervising compliance with the relevant building byelaws or regulations in respect of a failure properly to carry out such duty.

Lord Keith and Lord Jauncey expressed themselves in similar terms as having open minds on this issue. Lord Oliver is more ambivalent, noting that because the claim in question was one for economic loss it fell outside the purposes of the legislation. It could be deduced from this alone that a claim for personal injury (but less likely one for damage to personal property) may fall within the purposes of the legislation and thus be permitted. However, he also comments that it was not clear that there was any justification for assuming that there was any right of tortious action from the statute for breach of statutory duty, but this is a different, and more restrictive, tort from negligence.

The obvious areas where an argument can be put against the imposition of a duty are twofold. First, in the area of proximity, the precise phrase used since *Yuen Kun-Yeu* v. *Attorney-General of Hong Kong*

[1988] AC 175 (see p. 16 above) is that there needs to be 'close and direct proximity'. It is the directness that poses the problem here. Ultimately, the plaintiff's house falls down because the builder has built it badly; his is surely a position of direct proximity to the plaintiff. In contrast, the local authority is, at the very least, in a less proximate position.

Secondly, careful thought needs to be given to the policy considerations. One of the constant themes to emerge from the case-law of recent years is that policy will be used to deny a duty if there is a fear that the defendant may adopt unduly defensive procedures to avoid the threat of liability.

ROWLING v. TAKARO PROPERTIES LTD

[1988] AC 473   House of Lords

Takaro were involved in a New Zealand tourism project. The New Zealand Government refused consent to Takaro to issue shares to an interested Japanese investor. Takaro argued unsuccessfully, inter alia, that the relevant Minister, Rowling, had been negligent in not giving his consent to Takaro's application.

LORD KEITH (at p. 502) gave reasons to defeat Takaro's claim:

> The third is the danger of overkill. It is to be hoped that, as a general rule, imposition of liability in negligence will lead to a higher standard of care in the performance of the relevant type of act; but sometimes not only may this not be so, but the imposition of liability may even lead to harmful consequences. In other words, the cure may be worse than the disease. There are reasons for believing that this may be so in cases where liability is imposed upon local authorities whose building inspectors have been negligent in relation to the inspection of foundations, as in *Anns* v. *Merton London Borough Council* [1978] AC 728 itself; because there is a danger that the building inspectors of some local authorities may react to that decision by simply increasing, unnecessarily, the requisite depth of foundations, thereby imposing a very substantial and unnecessary financial burden upon members of the community. A comparable danger may exist in cases such as the present, because, once it became known that liability in negligence may be imposed on the ground that a minister has misconstrued a statute and so acted ultra vires, the cautious civil servant may go to extreme lengths in ensuring that legal advice, or even the opinion of the court, is obtained before decisions are taken, thereby leading to unnecessary delay in a considerable number of cases.

Lord Keith asserts that *Anns* and its successor have created a risk of overkill in the performance of the powers of local authorities. It would be rash to predict that a court will fly against this firm view in the near future, even given the lack of authority or evidence cited to support Lord Keith's view. Indeed, Lord Keith reiterated his view in *Murphy* (at p. 923).

Whether a court will follow general principle, and conclude that there is a duty on the local authority to avoid personal injury to occupiers or will follow the hints of Lord Keith and deny that such a duty exists on policy grounds must be regarded as an open question at present. However, it is suggested that the role of policy factors should be of lesser importance in a case of personal injury and that there a duty should prevail; claims for property damage alone are probably not within the purposes of the legislation and seem to be more likely to become a casualty of this particular dispute. It is therefore assumed that, at the very least, it is highly likely that the local authority will be under a duty to avoid causing personal injury. It is accordingly necessary to consider the incidents of that duty of care.

Powers to act

It is important to distinguish carefully between the two separate types of statutory obligation under which local authorities may find themselves. A duty is a mandatory obligation – the authority *shall* do something. Whether an action in tort arises from a statutory duty is a matter independent of the tort of negligence. Some statutes state specifically that an action will arise; others remain silent and the courts have to try and deduce the intentions of Parliament on a matter on which Parliament has remained silent – an intriguing guessing game. However, such actions fall outside the scope of this work.

Powers, on the other hand, are purely permissive in character: an authority *may* do something or not, as it wishes. The legislation simply gives authorization for action if the authority wishes to act. This then means that the first question will be whether the authority has to act or can be found negligent in not acting.

ANNS v. LONDON BOROUGH OF MERTON

[1978] AC 728   House of Lords

For facts, see pp. 9 above.

LORD WILBERFORCE (at p. 753):

> What then is the extent of the local authority's duty towards these persons?
> Although, as I have suggested, a situation of 'proximity' existed between the

council and owners and occupiers of the houses, I do not think that a description of the council's duty can be based upon the 'neighbourhood' principle alone or upon merely any such factual relationship as 'control' as suggested by the Court of Appeal. So to base it would be to neglect an essential factor which is that the local authority is a public body discharging functions under statute: its powers and duties are definable in terms of public not private law. The problem which this type of action creates, is to define the circumstances in which the law should impose, over and above, or perhaps alongside, these public law powers and duties, a duty in private law towards individuals such that they may sue for damages in a civil court. It is in this context that the distinction sought to be drawn between duties and mere powers has to be examined.

Most, indeed probably all, statutes relating to public authorities or public bodies, contain in them a large area of policy. The courts call this 'discretion' meaning that the decision is one for the authority or body to make, and not for the courts. Many statutes also prescribe or at least presuppose the practical execution of policy decisions: a convenient description of this is to say that in addition to the area of policy or discretion, there is an operational area. Although this distinction between the policy area and the operational area is convenient, and illuminating, it is probably a distinction of degree; many 'operational' powers or duties have in them some element of 'discretion'. It can safely be said that the more 'operational' a power or duty may be, the easier it is to superimpose upon it a common law duty of care.

I do not think that it is right to limit this to a duty to avoid causing extra or additional damage beyond what must be expected to arise from the exercise of the power or duty. That may be correct when the act done under the statute *inherently* must adversely *affect* the interest of individuals. But many other acts can be done without causing any harm to anyone – indeed may be directed to preventing harm from occurring. In these cases the duty is the normal one of taking care to avoid harm to those likely to be affected.

Let us examine the Public Health Act 1936 in the light of this. Undoubtedly it lays out a wide area of policy. It is for the local authority, a public and elected body, to decide upon the scale of resources which it can make available in order to carry out its functions under Part II of the Act – how many inspectors, with what expert qualifications, it should recruit, how often inspections are to be made, what tests are to be carried out, must be for its decision. It is no accident that the Act is drafted in terms of functions and powers rather than in terms of positive duty. As was well said, public authorities have to strike a balance between the claims of efficiency and thrift (du Parcq LJ in *Kent* v. *East Suffolk Rivers Catchment Board* [1940] 1 KB 319, 338): whether they get the balance right can only be decided through the ballot box, not in the courts. It is said – there are reflections of this in the judgments in *Dutton* v. *Bognor Regis Urban District Council* [1972] 1 QB 373 – that the local authority is under no duty to inspect, and this is used as the foundation for an argument, also found in some of the cases, that if it need not inspect at all, it cannot be liable for negligent inspection: if it were to be held so liable, so it is said, councils would simply decide against inspection. I think that this is too crude an argument. It overlooks the fact that local authorities are public bodies operating under statute with a clear responsibility for public health in their area. They must, and in fact do, make their discretionary decisions responsibly and for reasons which accord with the statutory purpose; see *Ayr Harbour Trustees* v. *Oswald* (1883) 8 App Cas 623, 639, *per* Lord Watson:

'the powers which [section 10] confers are discretionary . . . But it is the plain import of the clause that the harbour trustees . . . shall be vested with, and shall avail themselves of, these discretionary powers, whenever and as often as they may be of opinion that the public interest will be promoted by their exercise.'

If they do not exercise their discretion in this way they can be challenged in the courts. Thus, to say that councils are under no duty to inspect, is not a sufficient statement of the position. They are under a duty to give proper consideration to the question whether they should inspect or not. Their immunity from attack, in the event of failure to inspect, in other words, though great is not absolute. And because it is not absolute, the necessary premise for the proposition 'if no duty to inspect, then no duty to take care in inspection' vanishes.

Passing then to the duty as regards inspection, if made. On principle there must surely be a duty to exercise reasonable care. The standard of care must be related to the duty to be performed – namely to ensure compliance with the byelaws. It must be related to the fact that the person responsible for construction in accordance with the byelaws is the builder, and that the inspector's function is supervisory. It must be related to the fact that once the inspector has passed the foundations they will be covered up, with no subsequent opportunity for inspection. But this duty, heavily operational though it may be, is still a duty arising under the statute. There may be a discretionary element in its exercise – discretionary as to the time and manner of inspection, and the techniques to be used. A plaintiff complaining of negligence must prove, the burden being on him, that action taken was not within the limits of a discretion bona fide exercised, before he can begin to rely upon a common law duty of care. But if he can do this, he should, in principle, be able to sue.

Is there, then authority against the existence of any such duty or any reason to restrict it? It is said that there is an absolute distinction in the law between statutory duty and statutory power – the former giving rise to possible liability, the latter not, or at least not doing so unless the exercise of the power involves some positive act creating some fresh or additional damage.

My Lords, I do not believe that any such absolute rule exists: or perhaps, more accurately, that such rules as exist in relation to powers and duties existing under particular statutes, provide sufficient definition of the rights of individuals affected by their exercise, or indeed their non-exercise, unless they take account of the possibility that, parallel with public law duties there may coexist those duties which persons – private or public – are under at common law to avoid causing damage to others in sufficient proximity to them. This is, I think, the key to understanding of the main authority relied upon by the appellants – *East Suffolk Rivers Catchment Board* v. *Kent* [1941] AC 74.

LORD SALMON (at p. 767):

I am convinced that if an inspection of the foundations did take place, the council, through its building inspectors, owed a duty to the future tenants and occupiers of the maisonettes to exercise reasonable care and skill in carrying out that examination. The failure to exercise such care and skill may be shown to have caused the damage which the plaintiffs have suffered. The fact that the inspection was being carried out under a statutory power does not exclude the common law duty of those carrying it out to use reasonable care and skill – for it cannot in any way diminish the obvious proximity between the inspectors and the prospective tenants and their assignees.

It has, however, been argued on the council's behalf that, since it was under no obligation to inspect the foundations, had it failed to do so, it could not be liable for the damage caused by the inadequacy of the foundations. Accordingly, so the argument runs, if the council decided to inspect the foundations in the exercise of its statutory powers, it owed the prospective tenants and their assignees no duty to inspect carefully because, even if the inspection was carried out negligently, the prospective tenants and their assignees would be no worse off than if there had been no inspection. I reject this argument and confess that I cannot detect that it has even any superficial attraction. The council is given these statutory powers to inspect the foundations and furnished with public funds to enable the powers to be used for the protection of (amongst others) prospective purchasers of the buildings which are to be built upon them. If, when the council exercises these powers, it does so negligently, it must be obvious that those members of the public in the position of the present plaintiffs are likely to suffer serious damage. The exercise of power without responsibility is not encouraged by the law. I recognise that it may not be practical to inspect the foundations of every new building. This, however, is no excuse for a negligent inspection of such foundations as are inspected. When a council negligently exercises its powers of inspection, it should be and I believe is responsible in law to those who suffer damage as a result of that negligence.

I do not think that there is any danger that the responsibility which, in my view, lies upon the council is likely to lead to any flood of litigation. It is not a common occurrence for foundations to give way, nor for their inspection to be negligently carried out. If the foundations do give way, there is no warranty by the council which has inspected them that they are sound. The council is responsible only if it has exercised its powers to inspect; and the defects in the foundations, would have been detected by an inspection carried out with reasonable care and skill. It seems to me to be manifestly fair that any damage caused by negligence should be borne by those responsible for the negligence rather than by the innocents who suffer from it.

These two passages from *Anns* highlight the nature of the power and the nature of the associated liability in negligence. A decision *bona fide* to carry out no inspections or only a very limited number is perfectly acceptable to the law of tort (administrative law may intervene if such decisions are reached for the wrong reasons). These are matters of discretion and policy. But once it is decided to operate the statutory powers, it is a different matter. The authority is then under an obligation to use reasonable care in that operation.

Any citation from *Anns* is these days entitled to be greeted with a raised eyebrow or two. However, all that the recent attack on *Anns* has done is to overrule its application in building cases because of the confusion between physical damage and economic loss. The passages cited above are simply illustrative of general principle and still apply today insofar as negligence actions are available against local authorities.

Statutory powers in context

Another way in which restrictions have arisen on negligence claims against local authorities is by careful examination of the purpose of the statute and then limiting negligence claims to within that ambit.

GOVERNORS OF THE PEABODY DONATION FUND v. SIR LINDSAY PARKINSON & CO.

[1985] AC 210   House of Lords

For facts, see p. 12 above.

LORD KEITH (at p. 241):

> The purpose for which the powers contained in paragraph 15 of Part III of Schedule 9 have been conferred on Lambeth is not to safeguard building developers against economic loss resulting from their failure to comply with approved plans. It is in my opinion to safeguard the occupiers of houses built in the local authority's area, and also members of the public generally, against dangers to their health which may arise from defective drainage installations. The provisions are public health measures.

(and at p. 242):

> It is sufficient to hold that Lambeth owed no duty to Peabody to activate their paragraph 15 powers, notwithstanding that they might reasonably have foreseen that failure to do so would result in economic loss to Peabody, because the purpose of avoiding such loss was not one of the purposes for which these powers were vested in them. I find myself in respectful agreement with the following passage in the judgment of Slade LJ in the court below *ante*, p. 227E–H.

'Can it have been the intention of the legislature, in conferring on a borough council power to enforce against a defaulting site owner requirements made by it in accordance with paragraph 13 of Part III of Schedule 9, to protect such owner against damage which he himself might suffer through his own failure to comply with such requirements? In my opinion, this question can only be answered in the negative. This particular power exists for the protection of other persons – not for that of the person in default. I say nothing about the case where a local authority have failed to make known their requirements or where they have made requirements of an inadequate or defective nature. However, I can see no justification for extending the law of negligence by imposing on a local authority, over and above their public law powers and duties under paragraphs 13 and 15, a duty to exercise their powers of enforcement under paragraph 15(2), owed in private law towards a site owner, who, whether with or without personal negligence, disregards the proper requirements of the local authority, duly made under paragraph 13 and duly communicated to him or persons authorised to receive them on his behalf.'

Thus the number of prospective plaintiffs is reduced at a stroke. These statutory obligations were in place to protect the health and safety of occupiers. If the defects in the drains had led to an epidemic of disease amongst the residents, an action by them would have been allowed. However, the claim by the developer was, quite apart from being notably lacking in merit, way outside the purposes of the statute and thus impermissible. The implication of this ruling for commercial plaintiffs soon became clear.

INVESTORS IN INDUSTRY COMMERCIAL PROPERTIES LTD v. SOUTH BEDFORDSHIRE DISTRICT COUNCIL

[1986] QB 1034   Court of Appeal

For facts, see p. 188 above.

SLADE LJ (at p. 1061):

We think there can be no doubt that, whether or not he is the intended occupier, the building owner who, with personal carelessness on his part, permits the foundations of his building to be constructed in breach of the relevant building regulations, is owed no duty of care by the local authority under *Anns* principles. But what of the building owner who does so with no personal carelessness, because he has employed apparently competent independent experts?

We recognise that, on one reading of the *Peabody* case [1985] AC 210 a building owner guilty of no personal carelessness could be regarded as falling within the relevant duty of care, even if he is in breach of the relevant building regulations. Some support for this view is in particular to be derived from Lord Keith's reference to *Dennis's* case [1983] QB 409 without disapproval. Nevertheless, in deciding whether or not the alleged duty of care exists in the present case, we are, on the authority of the *Peabody* case, entitled and bound to take into consideration whether it is just and reasonable that it should exist. We feel no doubt that it is not just and reasonable. Mutatis mutandis, the passage which we have cited from Lord Keith's speech, at p. 241, seems to us fully to apply to the facts of the present case. Observance of the provisions of the Building Regulations 1972 was incumbent on, amongst others, the plaintiffs who put in train the building project. We are prepared to assume in favour of the plaintiffs that, despite their previous experience as building developers, they had no proper understanding of the risks which they were running in proceeding with the building operation on this unfilled site without a proper soil investigation and without piled foundations. They no doubt relied on the advice of their architects, consulting engineers and contractors and in the event were badly let down, at least by their consulting engineers. In our opinion, however, it would be neither reasonable nor just to impose on the local authority a liability to indemnify the plaintiffs against loss resulting from such disastrous reliance.

We can see no difference in principle, and it has not been submitted that there is any material difference, between the powers to secure or enforce compliance with building regulations which have been conferred on a local authority by the Public Health Act 1936 and those conferred on it by paragraph 15 of Part III of Schedule 9 to the Act of 1963. We will refer to all these powers collectively as 'supervisory powers' and venture to think that the following five propositions of law can be derived from the *Anns* case [1978] AC 728 and the *Peabody* case when considered together, in the light of the overruling of the *Acrecrest* case [1983] QB 260:

(1) The purpose for which the legislature has conferred the supervisory powers over building operations on local authorities is to protect the occupiers of buildings built in the local authority's area and also members of the public generally against dangers to health or personal safety. It is not to safeguard the building developer himself against economic loss incurred in the course of a building project, or indeed anyone else against purely economic loss.

(2) In view of the statutory purpose of these supervisory powers (though this point does not arise for decision in the present case) it may well be that on the basis of the *Anns* case a local authority, in exercising their statutory powers, will be held to owe a duty to a subsequent occupier other than the original building owner to take reasonable care to ensure that the building is erected in accordance with the building regulations, so as not to cause danger to health or personal safety of the occupier.

(3) Where this duty of care has been broken and, as a result, the condition of the property gives rise to danger to the health or safety of persons present on the premises, an occupier to whom the duty is owed may be at liberty to restore the property to a condition in which such danger is

eliminated, and having done so, to recover the amount of any such necessary expenditure from the local authority. Whether or not he will have such right of recovery must depend on the particular facts of the case.

(4)   On the basis of the *Peabody* case, however, a local authority, in exercising these supervisory powers, will normally owe no duty to an original building owner, because it is normally incumbent on the building owner himself to ensure that the building is erected in accordance with the relevant building regulations, and it cannot have been the intention of the legislature that, save perhaps in exceptional circumstances, a local authority could owe a duty to a person who is in such breach.

(5)   A fortiori, again on the basis of the *Peabody* case, a local authority in exercising these supervisory powers, will normally owe no duty to an original building owner who has had the benefit of the advice of architects, engineers and contractors and has relied on it. In such circumstances it will normally be neither reasonable nor just to impose upon the local authority a liability to indemnify the building owner against liability resulting from such reliance.

(and at p. 1063):

We are firmly of the view that the legislature, in imposing on local authorities for the general protection of the public the relevant statutory obligations under section 64 of the Public Health Act, cannot have been intending to protect a building developer such as the plaintiffs against damage which they themselves may suffer through their failure to comply with the relevant building regulations – or to entitle them to an indemnity from their fellow ratepayers against the consequences of any such failure.

We can see no sufficient reason for regarding the facts of the present case as exceptional in any relevant aspect. For the reasons already stated, we consider that the plaintiffs at all material times had themselves been in breach of the building regulations. Though we think the judge was right to conclude that the local authority were under a duty and consequent liability to the plaintiffs on the basis of the law as it appeared from the *Acrecrest* decision [1983] QB 260 to be, we think that numbers (4) and (5) of the five propositions listed above, which we derive from the *Peabody* case [1985] AC 210, fairly and squarely apply to the present case. In our judgment, therefore, the local authority, in considering whether or not to approve the plans and in their subsequent inspections of the site, owed no duty of care to the plaintiffs, whatever duty they may have owed to other persons. We therefore propose to allow the local authority's appeal, so far as it relates to the plaintiffs.

Needless to say, the subsequent changes in the law in this area mean that it is actual personal injury, not merely its threat, that will be needed for the plaintiff to have any chance of success. Equally, these restrictions placed on claims by the building owner were not intended to deny and did not have the effect of denying, claims by the ordinary subsequent purchaser.

JONES v. STROUD DISTRICT COUNCIL

[1986] 1 WLR 1141   Court of Appeal

Mr and Mrs Jones purchased their house in 1975. It had been built in 1964. Subsidence became evident after the drought of 1976. The local authority, it was agreed, had been negligent in the exercise of their statutory powers, and were held to owe a duty.

NEILL LJ (at p. 1149) reviewed the five points made by Slade LJ in the *Investors* case, and continued:

> To these five propositions it is necessary to add a sixth, based on Lord Fraser's speech in the *Pirelli* case [1983] 2 AC 1, 18, to the effect that any duty owed by the local authority is a duty owed to the owners – I would add: and occupiers – of the property as a class, and that if time runs against one owner or occupier it also runs against all his successors in title. No owner or occupier in the chain can have a better claim than his predecessors in title.
>
> It seems to me to follow from these propositions, and in particular from the third proposition set out in the judgment of Slade LJ, that until the condition of the property gives rise to danger to the health or safety of persons present on the premises no breach of the duty of care has taken place and accordingly no cause of action has arisen.
>
> It was argued on behalf of the council in the present appeal that it would be unjust if the cause of action against the council were to arise in a particular case at a later date than the cause of action against the builder because this circumstance might affect the right of the council to recover against the person primarily responsible for the damage.
>
> I see the force of this argument but the answer to it is that the obligations of the local authority and of the builder are not co-terminous. In many cases an action for negligent work may lie against the builder where there is no claim whatever against a local authority. I have, therefore, come to the conclusion that, unless the doctrine of 'doomed from the start' can apply, no cause of action arose against the council in the present case until some time after the end of the drought in 1976.

Of course, since *Murphy*, the Jones' claim would be unsuccessful as being for pure economic loss only but it is suggested that the case would still be applicable to a claim in respect of personal injury, assuming such claims still lie against the local authority.

Not all building owners are major financial operators such as in *Investors*; similarly not all subsequent purchasers are ordinary folk such as the Jones. The logic of Slade LJ's fifth proposition in *Investors* is that a commercial subsequent purchaser, who will only purchase after reliance on his own surveyors, etc, will not have a cause of action against the local authority, and in any event is now likely only to suffer economic loss. This aspect was the key element in the unsuccessful claim by a subsequent commercial purchaser in *Hambro Life Assurance plc* v. *White Young & Partners* (1987) 38 BLR 17. The ignorant and/or unadvised building owner is in a more difficult position, however. It was suggested that a duty might be owed by a council to such a plaintiff in the unreported case of *Dean* v. *Arnold* (1987).

By looking at the purpose of the statutory powers in the field of building control, the courts can identify both the type of damage and the category of plaintiff envisaged by the legislature and both these will restrict claims substantially.

Local authorities are under many other statutory obligations. Within the building control area, it seems reasonable to assume that all the powers and duties are there so as to promote health and safety, but other powers in relation to premises exist too, for example to take over and demolish dangerous property or to inspect work carried out under improvement grants. Principle suggests that the same analysis of the purpose of the powers should be carried out here too.

CURRAN v. NORTHERN IRELAND CO-OWNERSHIP HOUSING ASSOCIATION LTD

[1987] AC 718   House of Lords

For facts, see p. 14 above.

LORD BRIDGE (at p. 727):

> First, the statutory power which the authority is alleged to have negligently failed to exercise or to have exercised in a negligent way must be specifically directed to safeguarding the public, or some section of the public of which the plaintiff asserting the duty of care is a member, from the part' :ular danger which has resulted; in the *Anns* case the danger of buildings being erected on inadequate foundations. Secondly, the power must have been such that its due exercise could have avoided the danger; if the defect in the foundations had been discovered before they were covered up, the authority had power under the statute to require that the defect be made good. Thirdly, the non-exercise or negligent exercise of the power must have created a hidden defect which cannot subsequently be discovered and remedied before damage results.
>
> My Lords, the Order which must be considered in the instant case not only

confers on the Executive no powers of control of building operations analogous to those on which the decision in the *Anns* case depended, it confers no powers of such control at all. Once approval has been given to an application for an improvement grant, the Executive have no powers under the Order to control the building owner, still less the builder whom he chooses to employ, in the execution of the works. Their only power is to withhold payment of the grant, or of an instalment of the grant, if the works or the relevant part of the works have not been executed to their satisfaction. With every respect to the Court of Appeal, to hold that the Executive owe a duty of care to a building owner which they can only perform by refusing to pay him the grant they have approved seems to me an almost bizarre conclusion. In so far as article 60(5) of the Order imposes any duty on the Executive to satisfy themselves that the grant-aided works have been properly executed, as opposed to conferring a power to withhold payment if not so satisfied, it seems to me clear that the purpose of imposing any such duty is for the protection of the public revenue, not of the recipients of the grant or their successors in title. I can conceive of no reason why the legislature should be thought in this single provision to have intended to duplicate the kind of control of building operations which is entrusted in Northern Ireland, as in England, to local authorities so that persons interested in dwellings improved with the aid of a grant should enjoy a double protection.

In addition, any attempt to consider the imposition of civil liability in respect of powers as yet unconsidered by the courts must now be considered against the background of the general antipathy of the courts towards local authority liability in recent years. Indeed, it is clear that local authorities will owe a duty of care only in highly restrictive circumstances:

(1) in respect of actual personal injury only (and that in no way for certain); and
(2) to uninformed purchasers, possibly second or subsequent purchasers only.

Such is all that remains of the spectre of liability that has haunted local authorities since *Dutton* and *Anns* first came onto the scene no more than two decades ago.

Standard of care

The poor plaintiff who has struggled this far to claim against the local authority has of course yet to start to prove that the authority was in fact negligent i.e. fell below the standard of care of the reasonable authority. It is clear that the standard to be expected is not a high one.

STEWART v. EAST CAMBRIDGESHIRE DISTRICT COUNCIL

(1979) 252 EG 1105   High Court

The plaintiffs owned a house which had been built on the site of an old gravel pit. Severe settlement subsequently occurred. The local authority employed engineers to investigate the site and subsequently its building inspector, Hazelwood, inspected the foundation trenches and approved the work. The authority were found not to have been at fault.

SIR DOUGLAS FRANK QC (at p. 1107):

The question then is whether Mr Hazelwood, knowing the site to be on made ground, carried out his inspection with reasonable care or, on the other hand, whether he was negligent in allowing the house to be built without seeing that the foundations were adequate. It was finally conceded by counsel for the plaintiffs that Mr Hazelwood could only satisfy the duty which it is said was incumbent upon him by having boreholes sunk to the bottom of plot 5, that is to a depth of 20 ft, for otherwise he could not be sure of there being firm ground.

The case of *Anns* v. *Merton London Borough Council* [1978] AC 728 established or confirmed the principle that there is a duty on a building inspector to exercise reasonable care to ensure that the regulations applicable to foundations are complied with. I bear in mind that a building inspector has limited qualifications; ordinarily he is not a civil engineer, architect or chartered surveyor and cannot be expected to be vested with the expertise of those professions. Mr Hazelwood had to rely on the advice of those qualified to give it, such advice being based on a soil exploration and analysis. He carried out a routine inspection and complied with the advice of the soil engineers in that the trial pits, namely, the trenches, were carefully inspected and found unexceptional. [Counsel for Stewart] says that that was not enough. At first he said that Mr Hazelwood should have used an auger, but when it was pointed out that an auger would not reach the bottom of the filled land, he said that the duty of care was to sink boreholes to the bottom of the site and that was the essence of the plaintiffs' case. It seems to me that what is enshrined in that submission is the proposition that it is the function of the local authority to carry out for a developer the functions of a civil engineer in ascertaining the suitability of a site for building and advising on the design of that building. It puts the building inspector in the position of consulting engineer, insurer and warrantor.

The plaintiffs' contention might not only involve design of the foundations, for the size and type of the foundations might depend on the design and layout of the superstructure. The burden on the local authority and the cost to the ratepayer would be immense and I cannot conceive that either Parliament or the House of Lords in *Anns* case ever envisaged that such a burden would be placed on the building inspector and the local authority. In my judgment the building inspector, armed as he was with the technical advice and following it with reasonable care, was under a duty to ensure that the foundations were in accordance with the requirements made in his letter, but no more. Indeed, it is not established that even if further boreholes had been taken, the results would not have been the same as those found by Foundation Engineering Ltd.

I have great sympathy for the plaintiffs who bought a house which will cost a large sum to put into a fit state, but really they have only themselves to blame for not having the house surveyed, for a survey almost certainly would have revealed the defects. Be that as it may, it follows from the conclusion which I have reached that the action against the defendants must fail.

There are clear overtones here to the policy considerations that have shaped the duty of care issued in respect of local authorities. The authority plays only a 'walk-on' part as a secondary or junior party with only limited expertise and resources. This becomes especially important against the background emphasized in *Peabody* of the high standard to be expected from the professional architects and engineers involved in major construction projects.

Apportionment

The same factors that can act to limit the duty, and to make the breach finding difficult will still act in favour of the local authority even if the plaintiff is successfully against it. For them, as we have seen in Chapter 6, the court will have to decide how to apportion liability between builder and local authority. The convention has developed here that the local authority will only bear a limited share of the liability, normally 25 per cent.

WORLOCK v. S.A.W.S.

(1982) 265 EG 774   Court of Appeal

For facts, see pp. 158 above.

ROBERT GOFF LJ (at p. 780):

I turn finally to the local authority's appeal on the question of contribution. The learned judge dealt with this matter as follows:

There remains the question of contribution between the first and second defendants. I was referred to a case decided by Sir Douglas Frank, sitting as a deputy High Court judge, in which he took the view that the proper approach was to attribute three-quarters of the blame to an architect and 25 per cent of the blame to a local authority in a case which has got similarities to the case here. I consider that in any case in which one is looking at the failure to provide proper foundations, the primary responsibility must remain with the builder. However, in this particular case, the supervisory role performed by the authority was, as they must have been aware, of particular importance. There was no architect and Mr Hicks [the builder], as I have already made clear, was very much in need of assistance. Having considered all the circumstances, it seems to me that the proper apportionment between the authority and [the builder] is to attribute 60 per cent of the blame to [the builder] and 40 per cent to the local authority and I order a contribution between the two defendants on that basis.'

[Counsel for Worlock] accepted that there was a general trend in cases of this kind, subject always to special circumstances, to order contribution on the basis of 75 per cent responsibility resting on the builder and 25 per cent on the local authority. However, his complaint in the present case was that the learned judge, in departing from that apportionment, took into account matters which he ought not to have taken into account. Even assuming that the absence of an architect, and the fact that [the builder] was very much in need of assistance, were facts that could legitimately be taken into account by the judge, this could only be so if those facts were known to the local authority and there was no evidence to support the judge's conclusion that the local authority must have been aware of this. Having been shown the relevant passages of the evidence, I am satisfied that this submission is well founded. It follows that this court is entitled to interfere with the judge's exercise of his discretion, and to substitute its own view for his. In my judgment, the primary responsibility resting on the builder, a proper apportionment in the present case would have been to accept the conventional approach of attributing 75 per cent of the blame to the builder and 25 per cent to the local authority. I for my part would therefore allow the local authority's appeal on contribution to that extent.

I think it only fair to the local authority to place it on record that they are appealing on the issue of contribution as a matter of principle, being anxious about the effect of the learned judge's decision on other cases; and that they intend, even if successful in their appeal as to contribution, to consider whether they should require [the builder] to bear any greater proportion of the plaintiff's claim than the learned judge ordered. Speaking for myself, I respect, even sympathise with, this attitude. Moreover, I consider that it would be only right, having heard argument on the point, to express an opinion on the question whether, if the evidence had shown that the local authority was aware that [the builder] was very much in need of assistance, and that no architect was retained to supervise the work, these would have been proper matters for the judge to have taken into account. In my judgment, they would not. However inexperienced the builder, the fact remains that the primary responsibility rests upon him. The duties of the local authority remain the same, so far as their duties in respect of the regulations are concerned. No doubt most local authorities, if aware of the builder's inexperience, would go out of their way to afford him assistance; but in the event of there being a breach of duty by both, as in the present case, I cannot see that the local authority's awareness of the builder's inexperience should enhance their relative share of the responsibility. Indeed, there must be something inherently wrong in the proposition that the greater the incompetence of the builder, the less his relative share of responsibility. So, even if the local authority in the present case had, on the evidence, been aware of [the builder's] inexperience, and of the fact that there was no architect to supervise his work, I would still have held that the learned judge had erred in taking those matters into account, and I would still have ordered contribution on the basis of 75 per cent to the builder, 25 per cent to the local authority. I cannot help suspecting that the learned judge, in making the apportionment that he did, may have been moved to some extent by his very understandable sympathy for [the builder].

In the result, in my judgment the builder's and the local authority's appeals on liability should be dismissed; but the local authority's appeal on contribution should be allowed to the extent I have indicated.

Of course this assumes the other parties remain to be sued.

In any event, a less generous view has subsequently been taken by the Court of Appeal.

RICHARDSON v. WEST LINDSEY DISTRICT COUNCIL

[1990] 1 WLR 522   Court of Appeal

The plaintiff carried out home extension works by converting his roof space into additional rooms. Plans were seen by the building inspector of the defendant but never formally approved. In fact the plans were not detailed enough to show whether the building regulations were complied with or not. The work was subsequently executed by the plaintiff and other locals. No action was found to exist against the authority.

SLADE LJ (at p. 539):

> However, to my mind, common sense suggests that a local authority can reasonably expect the owner of property who is embarking on a building project to obtain such advice and assistance from competent persons as he may require to enable him to comply with the relevant building regulations and do the job properly. Save in exceptional circumstances, justice and reasonableness do not, in my view, entitle a building owner to expect the local authority itself to assume the role which would and should be played by competent advisers. Nor can justice and reasonableness ordinarily require a local authority to ascertain precisely what advice and assistance the building owner is receiving. If a building owner relies on his own supposed competence but the work is in fact too difficult for him, or if he relies on other persons

whose advice and assistance prove inadequate, he may find himself in difficulties. In my judgment, however, it would generally be neither just nor reasonable to impose upon the local authority a liability to indemnify him against loss resulting from such reliance. It is irrelevant that on the same facts justice and reasonableness may require the imposition on the authority of a duty owed to subsequent occupiers (as in *Anns* itself).

Slade LJ then went on to consider whether this was an exceptional case which was how Lord Keith, in *Peabody*, had justified the earlier decision in *Dennis* v. *Charnwood District Council* [1983] QB 409:

Exceptional cases may still arise where, on the particular facts it either is or ought to be clear to the local authority that the building owner is specifically relying on approval of the relevant plans (or on a favourable inspection) by the local authority as amounting to an assurance that he can safely proceed to build in accordance with them. That, as I read *Peabody*, is the basis upon which their Lordships in that case considered that on its particular facts the decision in *Dennis* was to be justified. To repeat a few of Lord Keith's words [1985] AC 210, 244–245.

'The plaintiffs were in breach of certain material provisions of the relevant bye-laws dealing with the adequacy of foundations, but the fact remains that plans showing the intended foundations had been submitted with their authority and had been approved. *This approval might reasonably be taken as an indication that the foundations were satisfactory . . .*' [Emphasis added.]

Mr Young submitted in effect that, in expressing the view that the *Dennis* decision was to be justified on the basis stated by Lord Keith, their Lordships were adopting and approving a general principle that, save where justice and reasonableness require a different decision on particular facts, the common law duty of care referred to earlier in this judgment is owed by the local authority to any original building owner/occupier of property who is liable to suffer injury to his health or safety as a result of any breach of the relevant duty on the part of the local authority, provided only that his personal negligence was not the sole cause of his subsequent loss. For my part, I am unable to read the reference to *Dennis* in *Peabody* as adopting any such general principle.

Was this, then, an 'exceptional case'?

SLADE LJ (at p. 540):

The plaintiff may well be regarded as unfortunate, because he may well have been let down by one or more of his advisers in this building operation. However, I accept the submission made by [counsel] on behalf of the council that the circumstances of the present case are not exceptional so as to make it just and reasonable to impose a duty of care on the council towards the plaintiff building owner. Reliance is placed by the plaintiff on the judge's finding that the plaintiff had no technical knowledge. However, as the judge found, the council 'were not to know that'; 'he acted as though he had technical knowledge.' In *Peabody* the fact that the plaintiffs 'had no personal knowledge or understanding of what was going on' did not avail them: see [1985] AC 210, 241E, *per* Lord Keith. It was also urged on behalf of the plaintiff that he did not have the benefit of professional advisers. I am not satisfied on the evidence that the council were aware of this fact. Nor do I think it was incumbent on it to make inquiries as to the nature of the advice available to him. [Counsel] on behalf of the council pointed out that one important cause of the plaintiff's troubles was that he did not employ a builder who was then responsible for carrying out all aspects of the building works. In effect he acted as his own builder. He rejected the builder (Mr Lovelock) who was a specialist in this type of work. he did not want an architect, but used Mr Bennett, the draftsman who had been introduced by Mr Lovelock. He employed, on a labour only basis, Mr Staples, who was not a fully qualified tradesman and who had been working in a fertiliser factory. He employed other tradesmen direct and bought his own materials. On his own evidence, he acted in this way so that he would have better control of what was going on and because he thought it would save him money. His desire to achieve economies in this manner is understandable, but cannot, in my judgment, have operated to impose a higher duty on the local authority, or to make it just and reasonable for him to seek compensation, of which the burden would ultimately fall on his fellow ratepayers.

The facts of the present case have certain features in common with those of *Dennis* [1983] QB 409 in that in that case also the plaintiff and his family were known to be intended occupiers of the extension and the council, in approving the plans or in being estopped from denying their approval of them, could reasonably be understood as having indicated their opinion that they involved no risk to the health or safety of the occupier of the extension. The present case, however, is in my view plainly distinguishable from *Dennis* on its facts, if only for two reasons. First, there is no evidence that the council ever gave any specific approval (written or otherwise) in respect of the final plan 266/82 for building regulation purposes. In due course the plaintiff no doubt became entitled to assume that they had been approved because the council had never rejected them. In my judgment, however, he could not reasonably have understood the council's silence as constituting any kind of assurance by the council addressed to him personally that the projected building operations involved no risk to his health or safety as intended occupier; and indeed he did not suggest that he did so regard them. Secondly, the defect in the plans in *Dennis* (as in *Anns*) related to a defect in the foundations of the property, a defect which would probably become completely hidden as soon as they were covered up. In *Curran* v. *Northern Ireland Housing Association* [1987] AC 718, 728, Lord Bridge of Harwich (with whose speech all their Lordships concurred) stated the following as one of the three elements fundamental to the ratio of *Anns*:

'Thirdly, the non-exercise or negligent exercise of the power must have created a hidden defect which cannot subsequently be discovered and remedied before damage results.'

[Counsel for Richardson] submitted that in the present case the defects in the roof were hidden upon completion of the building works in a similar way to foundations. However, the concealment of the defects cannot have been nearly so final and complete as that which could reasonably have been foreseen as likely to occur in *Dennis* as soon as faulty foundations were covered up.

More broadly, I cannot accept *Dennis* as authority for the general proposition (or as intended to be authority for the general proposition) that when a local authority is negligent in passing plans which do not comply with the relevant building regulations (which in any given case may be several in number) the building owner will have a remedy in damages against it, provided only that it is reasonably foreseeable that resulting defects in the proposed building will give rise to risks of injury to his health or safety as occupier. Such a proposition would, in my judgment, impose on local authorities a duty (and on its ratepayers a potential financial burden) which went far beyond what was fair and reasonable. For the protection of the public generally, Parliament, in enacting the Public Health Acts and the regulations made thereunder, has placed a number of obligations on building owners and has in effect imposed duties on local authorities to see that those obligations are complied with. In my judgment, it cannot have contemplated that, save in exceptional circumstances, a failure to police the activities of the building owner with sufficient care would entitle the building owner himself to a personal remedy in damages against the policing authority. As I have indicated, I understand this to have been a major feature of the approach of the House of Lords to the question before it in *Peabody*, even though on the facts of that case there was no question of risk of injury to the health or safety of the building owner.

As Bingham LJ points out in his judgment, this is a harsh decision potentially denying a remedy to quite a large group of plaintiffs. The typical home extension does not involve an architect and often the work, like that of Richardson, is carried out to cope with the demands of a growing family, so the temptation to minimize expense is understandable. However, the decision in the very different circumstances of *Investors* appears to preclude their claim.

Local authority – summary

It is important to emphasize that many of the cases between *Anns* and *Murphy* are misleading insofar as they concentrate on the threat to health and safety that was then, but is no longer, at the heart of the liability of the local authority. Now, claims can, at best, be confined to cases of actual physical injury, possibly damage to property and, conceivably, economic loss caused by some negligent misstatements. Even within these confined categories, the courts have learnt that the role of local authority is peripheral to the roles of the other main players and have adjusted their perceptions of duty, breach and apportionment accordingly.

A further principal area of restriction is the habit, since *Peabody*, of examining with care the purpose of the statute and the identity of its proposed beneficiaries and using these elements to limit, often severely, the scope of the duty owed in negligence. The fear that a local authority may be left with 100 per cent of the damages bill because of the non-availability of co-defendants, noted in Chapter 6, still remains in theory, but the number of cases where a local authority will be liable for anything at all will be few and far between, thus minimizing the threat.

Chapter 11

Liability of Surveyors for Valuation and Survey

Although in fact the role of the surveyor or valuer is far removed from that of the architects, engineers and quantity surveyors who form part of the building team, nevertheless in the law there is much common ground between these two categories of potential defendants. Both, obviously, are professional and therefore have to reach the standard of care of their peers, and not just that of the reasonable man. More significantly, both of the groups will find that much of their liability revolves around statements, rather than acts; in particular, any claim made against a valuer will be in respect of his statements as to the value of the property, albeit one made in the light of an act of inspection, and this will lead to the claim being framed under the principles laid down in *Hedley Byrne* v. *Heller* [1964] AC 465.

However, in at least one major way, the legal position of the surveyors is very different. He does not play a role, primary or secondary in the actual creation of the building. Rather, he is simply on hand, often at a much later stage in its history, to testify as to its value and/or condition. Obviously this will reflect any defects that are present, but, to the employer, the only loss that can arise from negligence in the course of valuation is an economic loss.

Duty of care – to whom?

The surveyor will of course be employed by contract. He will therefore be in a very close relationship with his employer, who will often be the intending purchaser, but who may be, in an important class of recent cases, the building society, employing him to ensure that their mortgage loan is based upon adequate security. The employer can sue in contract or, if he so elects, in the tort of negligence, subject, as ever, to the *Tai Hing* ruling (p. 70, above) having only a limited effect, as suggested previously.

The question of whether third parties can also sue the surveyor in negligence has become important in recent years, especially but not exclusively in cases where, as is common practice, the surveyor's report to the building society after what is usually a fairly cursory inspection is released to the intending purchaser.

YIANNI v. EDWIN EVANS & SONS

[1982] QB 438   High Court

The plaintiffs wished to buy a house with mortgage finance. The defendants were engaged by a building society and reported that the property was worth £15,000, the asking price. A loan of £12,000 was accordingly advanced. Their survey work was negligent and major defects subsequently became apparent. The plaintiffs successfully claimed against the surveyors.

PARK J (at p. 454):

> I conclude that, in this case, the duty of care would arise if, on the evidence, I am satisfied that the defendants knew that their valuation of 1, Seymour Road, in so far as it stated that the property provided adequate security for an advance of £12,000, would be passed on to the plaintiffs, who, notwithstanding the building society's literature and the service of the notice under section 30 of the Building Societies Act 1962, in the defendants' reasonable contemplation would place reliance upon its correctness in making their decision to buy the house and mortgage it to the building society. What therefore does the evidence establish?
>
> These defendants are surveyors and valuers. It is their profession and occupation to survey and make valuations of houses and other property. They make reports about the condition of property they have surveyed. Their duty is not merely to use care in their reports, they have also a duty to use care in their work which results in their reports. On the instructions of the building society, the defendants sent a representative to 1, Seymour Road to make a survey and valuation of that property. He knew that the object of the survey was to enable the defendants, his employers, to submit a report to the building society for the use of the directors in discharging their duty under section 25 of the Act of 1962. The report therefore had to be directed to the value of the property and to any matter likely to affect its value. The defendants knew therefore that the director or other officer in the building society who considered their report would use it for the purpose of assessing the adequacy of 1, Seymour Road as security for an advance. There is no evidence that the society had access to any other reports or information for this purpose or that the defendants believed or assumed that the building society would have any information beyond that contained in their report. Accordingly, the defendants knew that the director or other officer of the society who dealt with the

plaintiffs' application would rely upon the correctness of this report in making on behalf of the society the offer of a loan upon the security of 1, Seymour Road. The defendants therefore knew that the plaintiffs would receive from the building society an offer to lend £12,000 which sum, as the defendants also knew, the plaintiffs desired to borrow. It was argued that, as the information contained in the defendants' report was confidential to the directors, the defendants could not have foreseen that the contents of their report would be passed on to the plaintiffs. But the contents of the report never were passed on. This case is not about the contents of the entire report; it is about that part of the report which said that 1, Seymour Road was suitable as security for a loan of £12,000. The defendants knew that that part would have to be passed on to the plaintiffs, since the reason for the plaintiffs' application was to obtain a loan of £12,000. Accordingly, the building society's offer of £12,000, when passed on to the plaintiffs, confirmed to them that 1, Seymour Road was sufficiently valuable to cause the building society to advance on its security 80 per cent of the purchase price. Since that was also the building society's view the plaintiffs' belief was not unreasonable.

It was argued that there was no reasonable likelihood that the plaintiffs would rely upon the fact that the defendants had made a valuation report to the building society, or alternatively that the defendants could not reasonably have foreseen or contemplated first, that the plaintiffs would rely upon the valuation in the report, or second, they would act unreasonably in failing to obtain an independent surveyor's report for their own guidance. These submissions were founded upon the fact that the defendants would know that the plaintiffs would have been provided with the building society's literature and that the building society, for its own protection would serve with their offer the statutory notice pursuant to section 30 of the Act. Now these defendants plainly are in a substantial way of business in London as surveyors and valuers. The documents show that they have an address at Down Street, Mayfair, and another in Lavender Hill, SW11. They must have on their staff some members of the Royal Institute of Chartered Surveyors. The terms of the building society's request to them to value 1, Seymour Road indicated that they had regularly carried out valuations for the Halifax, and no doubt for other building societies. Mr Hunter's [an expert witness] evidence is that for some six years over 90 per cent of applicants for a building society mortgage have relied upon the building society's valuation, as represented by the society's offer of an advance, as a statement that the house in question is worth at least that sum. These applicants, and in particular applicants seeking to buy houses at the lower end of the property market, do not read building society literature, or if they do they ignore the advice to have an independent survey and also the terms of the statutory notice. Mr Hunter's evidence was unchallenged. No witness was called to suggest that he had in any way misrepresented the beliefs, conduct and practice of the typical applicant. I think that Mr Hunter was telling me what was common knowledge in the professional world of building societies and of surveyors and valuers employed or instructed by them. I am satisfied that the defendants were fully aware of all these matters.

The defendants' representative who surveyed and valued 1, Seymour Road noted the type of dwellinghouse it was; its age, its price and the locality in which it was situated. It was plainly a house at the lower end of the property market. The applicant for a loan would therefore almost certainly be a person of modest means who, for one reason or another, would not be expected to obtain an independent valuation, and who would be certain to rely, as the plaintiffs in fact did, on the defendants' valuation as communicated to him in the building society's offer. I am sure that the defendants knew that their valuation would be passed on to the plaintiffs and that the defendants knew that the plaintiffs would rely upon it when they decided to accept the society's offer.

For these reasons I have come to the conclusion that the defendants owed a duty of care to the plaintiffs because, to use the words of Lord Wilberforce in *Anns* v. *Merton London Borough Council* [1978] AC 728, 751H, there was a sufficient relationship of proximity such that, in the reasonable contemplation of the defendants, carelessness on their part might be likely to cause damage to the plaintiffs.

I turn now to consider whether there are any considerations which ought to negative or to reduce or limit the scope of the duty or the class of person to whom it is owed. [Counsel for Edwin Evans & Sons] submitted that for a number of reasons of policy the plaintiffs should have no remedy against the defendants. First, he said a decision in favour of the plaintiffs would encourage applicants for a mortgage to have no independent survey of the house they wished to buy. I can see nothing objectionable in a practice which would result in a house being surveyed once by one surveyor. In my view, the Abbey National, since September 1980, have adopted a sensible procedure for dealing with the survey problem, if it is a problem. Mr Hunter said that as a matter of courtesy the Abbey National now disclose their valuation report to the applicants. He also said: 'We felt that we had information which had been obtained by qualified and experienced people and it was of benefit to give that information to the applicant.' In addition, the Abbey National are about to introduce a report on condition and valuation so that, as Mr Hunter put it, the applicant has the choice of either the standard building society mortgage valuation report or the report on condition and valuation which covers the popular conception of structural survey market valuation and mortgage valuation.

[Counsel for Edwin Evans & Sons] also submitted that if the defendants were held liable to the plaintiffs no professional man would be able to limit his liability to a third party, even if he could do so to his own client. He would be at the mercy of a client who might pass on his report to a third party and, as defects in the property he had surveyed might not manifest themselves for many years, he would be likely to remain under a liability for those defects he ought to have detected for a very long period, and at the end of the period, for an unlimited amount by way of damages. In my view, the only person to whom the surveyor is liable is the party named in the building society's 'Instructions to Valuer' addressed to him. That party, as well as the building society, has to be regarded as his client. That does not seem to me to be unreasonable, since, to his knowledge, his fee for the valuation is paid by that party to the building society which hands it over to him. On this submission, it can also be said that the surveyor's report is concerned with the valuation of a dwellinghouse, the condition of which is important only in so far as it affects its value at that time. It is common knowledge that in the ordinary way, the market value of a dwellinghouse is not static. Consequently, a valuation made at one time for the purpose of assessing its suitability as security for a loan would be of limited use.

This case represented a major development. Park J appeared to be using the *Anns* decision to justify an extension of liability for negligent misstatement to a third party who had no direct dealings with the surveyor. Indeed, it was even employed in a case involving commercial property, in *Stevenson* v. *Nationwide Building Society* (1984) 272 EG 663. However, the facts that the plaintiffs ultimately paid for the survey, the knowledge by the surveyor of their involvement and the increasing likelihood that the report will actually go to the purchaser all contribute to a justification of the decision. However, its controversial nature inevitably meant that it needed further consideration.

SMITH v. ERIC S. BUSH: HARRIS v. WYRE FOREST DISTRICT COUNCIL

[1990] 1 AC 831   House of Lords

These two appeals, heard together, were on similar facts. Each plaintiff applied for a mortgage (Smith to a building society, Harris to the defendant local authority). A standard valuation survey was carried out by, respectively, an independent surveyor and a surveyor employed by the local authority. As a result, mortgage facilities were extended to the plaintiffs, but defects which the surveys should have picked up subsequently materialized.

LORD TEMPLEMAN (at p. 843)

> The common law imposes on a person who contracts to carry out an operation an obligation to exercise reasonable skill and care. A plumber who mends a burst pipe is liable for his incompetence or negligence whether or not he has been expressly required to be careful. The law implies a term in the contract which requires the plumber to exercise reasonable skill and care in his calling. The common law also imposes on a person who carries out an operation an obligation to exercise reasonable skill and care where there is no contract. Where the relationship between the operator and a person who suffers injury or damage is sufficiently proximate and where the operator should have foreseen that carelessness on his part might cause harm to the injured person, the operator is liable in the tort of negligence.

Manufacturers and providers of services and others seek to protect themselves against liability for negligence by imposing terms in contracts or by giving notice that they will not accept liability in contract or in tort. Consumers who have need of manufactured articles and services are not in a position to bargain. The Unfair Contract Terms Act 1977 prohibits any person excluding or restricting liability for death or personal injury resulting from negligence. The Act also contains a prohibition against the exclusion or restriction of liability for negligence which results in loss or damage unless the terms of exclusion or the notice of exclusion satisfies the requirements of reasonableness.

These two appeals are based on allegations of negligence in circumstances which are akin to contract. Mr and Mrs Harris paid £22 to the council for a valuation. The council employed, and therefore paid, Mr Lee, for whose services as a valuer the council are vicariously liable. Mrs Smith paid £36.89 to the Abbey National for a report and valuation and the Abbey National paid the appellants for the report and valuation. In each case the valuer knew or ought to have known that the purchaser would only contract to purchase the house if the valuation was satisfactory and that the purchaser might suffer injury or damage or both if the valuer did not exercise reasonable skill and care. In these circumstances I would expect the law to impose on the valuer a duty owed to the purchaser to exercise reasonable skill and care in carrying out the valuation.

In *Cann* v. *Wilson* (1888) 39 ChD 39, approved by this House in *Hedley Byrne & Co. Ltd* v. *Heller & Partners Ltd* [1964] AC 465, a valuer instructed by a mortgagor sent his report to the mortgagee who made an advance in reliance on the valuation. The valuer was held liable in the tort of negligence to the mortgagee for failing to carry out the valuation with reasonable care and skill.

A valuer who values property as a security for a mortgage is liable either in contract or in tort to the mortgagee for any failure on the part of the valuer to exercise reasonable skill and care in the valuation. The valuer is liable in contract if he receives instructions from and is paid by the mortgagee. The valuer is liable in tort if he receives instructions from and is paid by the mortgagor but knows that the valuation is for the purpose of a mortgage and will be relied upon by the mortgagee.

LORD GRIFFITHS (at p. 864):

I have come to the conclusion that *Yianni* [1982] QB 438 was correctly decided. I have already given my view that the voluntary assumption of responsibility is unlikely to be a helpful or realistic test in most cases. I therefore return to the question in what circumstances should the law deem those who give advice to have assumed responsibility to the person who acts upon the advice or, in other words, in what circumstances should a duty of care be owed by the adviser to those who act upon his advice? I would answer – only if it is foreseeable that if the advice is negligent the recipient is likely to suffer damage, that there is a sufficiently proximate relationship between the parties and that it is just and reasonable to impose the liability. In the case of a surveyor valuing a small house for a building society or local authority, the application of these three criteria leads to the conclusion that he owes a duty of care to the purchaser. If the valuation is negligent and is relied upon damage in the form of economic loss to the purchaser is obviously foreseeable. The necessary proximity arises from the surveyor's knowledge that the overwhelming probability is that the purchaser will rely upon his valuation, the evidence was that surveyors knew that approximately 90 per cent of purchasers did so, and the fact that the surveyor only obtains the work because the purchaser is willing to pay his fee. It is just and reasonable that the duty should be imposed for the advice is given in a professional as opposed to a social context and liability for breach of the duty will be limited both as to its extent and amount. The extent of the liability is limited to the purchaser of the house – I would not extend it to subsequent purchasers. The amount of the liability cannot be very great because it relates to a modest house. There is no question here of creating a liability of indeterminate amount to an indeterminate class. I would certainly wish to stress that in cases where the advice has not been given for the specific purpose of the recipient acting upon it, it should only be in cases when the adviser knows that there is a high degree of probability that some other identifiable person will act upon the advice that a duty of care should be imposed. It would impose an intolerable burden upon those who give advice in a professional or commercial context if they were to owe a duty not only to those to whom they give the advice but to any other person who might choose to act upon it.

LORD JAUNCEY (at p. 871):

I prefer to approach the matter by asking whether the facts disclose that the appellants in inspecting and reporting must, but for the disclaimers, by reason of the proximate relationship between them, be deemed to have assumed responsibility towards Mrs Smith as well as to the building society who instructed them.

There can be only an affirmative answer to this question. The four critical facts are that the appellants knew from the outset:

(1)   that the report would be shown to Mrs Smith;
(2)   that Mrs Smith would probably rely on the valuation contained therein in deciding whether to buy the house without obtaining an independent valuation;
(3)   that if, in these circumstances, the valuation was, having regard to the actual condition of the house, excessive, Mrs Smith would be likely to suffer loss; and
(4)   that she had paid to the building society a sum to defray the appellants' fee.

In the light of this knowledge the appellants could have declined to act for the building society, but they chose to proceed. In these circumstances they must be taken not only to have assumed contractual obligations towards the building society but delictual obligations towards Mrs Smith, whereby they became under a duty towards her to carry out their work with reasonable care and skill. It is critical to this conclusion that the appellants knew that Mrs Smith would be likely to rely on the valuation without obtaining independent advice. In both *Candler* v. *Crane, Christmas & Co.* [1951] 2 KB 164 and *Hedley Byrne & Co. Ltd* v. *Heller and Partners Ltd* [1964] AC 465, the provider of the information was the obvious and most easily available, if not the only available source of that information. It would not be difficult therefore to conclude that the person who sought such information was likely to rely upon it. In the case of an intending mortgagor the position is very different since, financial considerations apart, there is likely to be available to him a wide choice of sources of information, to wit, independent valuers to whom he can resort, in addition to the valuer acting for the mortgagee. I would not therefore conclude that the mere fact that a mortgagee's valuer knows that his valuation will be shown to an intending mortgagor of itself imposes upon him a duty of care to the mortgagor. Knowledge, actual or implied, of the mortgagor's likely reliance upon the valuation must be brought home to him. Such knowledge may be fairly readily implied in relation to a potential mortgagor seeking to enter the lower end of the housing market but non constat that such ready implication would arise in the case of a purchase of an expensive property whether residential or commercial. [Counsel] for the appellants conceded that if there had been no disclaimer they must fail. For the reasons which I have just given I consider that this concession was rightly made.

An unusual case for the late 1980s indeed, with the House of Lords not seeking to restrict liability in negligence, but rather supporting a hitherto controversial decision. True, the speeches are carefully drawn and it is evident that everything hinges on the clear actual knowledge of the surveyors that the plaintiff would be relying on their report in proceeding with the purchase and associated mortgage transaction. This will, however, usually be the case and *Smith* represents a useful weapon for the purchaser of a defective house, and the more useful a weapon in the light of the growing restriction placed on claims against those involved in the actual construction process.

However, as we have already seen, (pp. 27, above) within a few months of their decision in *Smith*, the House of Lords returned to view the whole areas of *Hedley Byrne* liability. How does the more restrictive approach now taken fit in with *Smith*?

CAPARO INDUSTRIES PLC v. DICKMAN

[1990] 2 AC 605   House of Lords

For facts see p. 27 above.

LORD OLIVER (at p. 638):

The most recent authority on negligent misstatement in this House – the two appeals in *Smith* v. *Eric S. Bush* and *Harris* v. *Wyre Forest District Council* [1990] 1 AC 831 which were heard together – does not, I think, justify any broader proposition than that already set out, save that they make it clear that the absence of a positive intention that the advice shall be acted upon by anyone other than the immediate recipient – indeed an expressed intention that it shall not be acted upon by anyone else – cannot prevail against actual or presumed knowledge that it is in fact likely to be relied upon in a particular transaction without independent verification. Both appeals were concerned with surveyors' certificates issued to mortgagees in connection with the proposed purchases for which the mortgagees were contemplating making advances. In each case there was an express disclaimer of responsibility, but in each case it was known to the surveyor that the substance of the report (in the sense of what was important to a purchaser) – that is to say whether or not any repairs to the property were considered essential – would be made known by the mortgagee to the purchaser, the plaintiff in the action, and would be likely to be acted upon by him in entering into a contract to purchase the property. In so far as the case was concerned with the effects of the disclaimer, it does not require consideration in the present context, but there are important passages in the speeches in this House bearing upon the questions which arise on this appeal and indicative of the features which, in that case, led their Lordships to conclude that the necessary relationship of proximity existed between the surveyors and the purchasers of the respective properties. Lord Templeman deduced the relationship from a combination of factors. He said, at pp. 847–8:

'I agree that by obtaining and disclosing a valuation, a mortgagee does not assume responsibility to the purchaser for that valuation. But in my opinion the valuer assumes responsibility to both mortgagee and purchaser by agreeing to carry out a valuation for mortgage purposes knowing that the valuation fee has been paid by the purchaser and knowing that the valuation will probably be relied upon by the purchaser in order to decide whether or not to enter into a contract to purchase the house.'

After referring, inter alia, to the speeches quoted (at p. 247–250, above) Lord Oliver went on (at p. 641):

Thus *Smith* v. *Eric S. Bush* [1990] 1 AC 831, although establishing beyond doubt that the law may attribute an assumption of responsibility quite regardless of the expressed intentions of the adviser, provides no support for the proposition that the relationship of proximity is to be extended beyond circumstances in which advice is tendered for the purpose of the particular transaction or type of transaction and the adviser knows or ought to know that it will be relied upon by a particular person or class of persons in connection with that transaction. The judgment of Millett J. in the recent case of *Al Saudi Banque* v. *Clarke Pixley* [1990] Ch 313 (decided after the decision of the Court of appeal in the instant case) contains an analysis of the decision of this House in *Smith* v. *Eric S. Bush* and concludes – and I agree – that it established a more stringent test of the requirements for proximity than that which had been applied by the Court of Appeal in the instant case. At p. 335–6 of his judgment, Millett J gives what I find a helpful analysis of that case and of the features which distinguished it from the *Hedley Byrne* case and from the instant case:

'In each of the cases considered by the House of Lords, therefore, there was a tripartite transaction in which the valuation could realistically be regarded as provided by the valuer to the purchaser. In each of the cases the valuation was given to the mortgagee with the intention of being acted on by him in a specific transaction known to the valuer, viz the making of a mortgage offer in connection with a specific transaction of house purchase, and in the knowledge that the valuation or the gist of the valuation would be communicated to the purchaser and would in all probability be relied upon by him in deciding whether to go ahead with the very transaction for which the mortgage offer was sought. This was a much more restricted context in which to found a duty of care than was present in the *Caparo* case, for there was in contemplation not only a particular and identified recipient of the information to whom the defendant knew that it would be communicated, but a particular and known purpose for which he could foresee that it would be relied on.

In *Hedley Byrne* [1964] AC 465 and the cases which followed it, the statement was made directly to the plaintiff with the intention that the plaintiff should act upon it. The *JEB Fasteners* case [1983] 1 All ER 583 can be supported only on the basis that the statement was impliedly confirmed directly to the plaintiff without any such intention, but with a particular transaction in contemplation, and it was foreseeable that the plaintiff would rely upon it in that transaction. In *Caparo's* case [1989] QB 653 it was made to the plaintiff without any such intention and without any particular transaction in contemplation, but it was foreseeable that the plaintiff might rely upon it in some unknown future transaction. In *Smith* v. *Eric S. Bush* [1990] 1 AC 831 it was made to a third party with the intention that he should act upon it in a known and contemplated transaction, but in the knowledge that it would be communicated to the plaintiff and would almost certainly be relied upon by him in connection with a transaction without which the transaction of the third party could not proceed.

My Lords, no decision of this House has gone further than *Smith* v. *Eric S. Bush*, but your Lordships are asked by the respondents to widen the area of responsibility even beyond the limits to which it was extended by the Court of Appeal in this case and to find a relationship of proximity between the adviser and third parties to whose attention the advice may come in circumstances in which the reliance said to have given rise to the loss is strictly unrelated either to the intended recipient or to the purpose for which the advice was required. My Lords, I discern no pressing reason of policy which would require such an extension and there seems to me to be powerful reasons against it. As Lord Reid observed in the course of his speech in *Hedley Byrne* [1964] AC 465, 483, words can be broadcast with or without the consent of foresight of the speaker or writer; and in his speech in the same case, Lord Pearce drew attention to the necessity for the imposition of some discernible limits to liability in such cases. he said, at p. 534:

'The reason for some divergence between the law of negligence in word and that of negligence in act is clear. Negligence in word creates problems different from those of negligence in act. Words are more volatile than deeds. They travel fast and far afield. They are used without being expended and take effect in combination with innumerable facts and other words. Yet they are dangerous and can cause vast financial damage. how far they are relied on unchecked ... must in many cases be a matter of doubt and difficulty. If the mere hearing or reading of words were held to create proximity, there might be no limit to the persons to whom the speaker or writer could be liable.

As I have already mentioned, it is almost always foreseeable that someone, somewhere and in some circumstances, may choose to alter his position upon the faith of the accuracy of a statement or report which comes to his attention and it is always foreseeable that a report – even a confidential report – may come to be communicated to persons other than the original or intended recipient. To apply as a test of liability only the foreseeability of possible damage without some further control would be to create a liability wholly indefinite in area, duration and amount and would open up a limitless vista of uninsurable risk for the professional man.

Thus, the *Smith* approach survives the restrictions placed in *Caparo* on claims for negligent misstatement. It looks increasingly anomalous, as being a rare exception where someone other than the immediate recipient of a statement is held to be entitled to rely upon it and take action in negligence if it subsequently turns out to have been made incorrectly, but equally is now evidently well-entrenched in the law. It is also clear that it will be difficult to interpret in practice.

BEAUMONT v. HUMBERTS

[1990] 49 EG 46   Court of Appeal

The plaintiffs bought a country house for £110,000 in 1984. They arranged for a survey by Mr Thomas a partner of the defendants. The plaintiffs then sought a mortgage of £30,000 and their bank approached the surveyor who confirmed a reinstatement value figure of £175,000 to the property. The plaintiffs insured for this figure. In 1985, a fire destroyed the property and actual reinstatement costs were found to be £300,000. It was held that no claim could be allowed, with the Court of Appeal finding by 2–1 that there was no duty owed (and a different 2–1 that there was no breach).

STAUGHTON LJ (at p. 48):

The liability to a house purchaser of a surveyor or valuer engaged by a mortgagee has recently been considered by the House of Lords in *Smith's* case. It is undesirable and unnecessary to add in this case to the volume of law on negligent advice and economic loss. The parties are agreed, on the basis of *Smith's* case, that for a duty of care to be established

(i)   there must be sufficient proximity between the task of the surveyor or valuer and the affairs of the mortgagor,

(ii)   it must be foreseeable that the mortgagor is likely to suffer damage if the surveyor or valuer is negligent, and possibly

(iii) it must be just and reasonable to impose liability.

It is proximity which is the crucial element in the present case. The word has replaced neighbour, the concept of Lord Atkin in *Donoghue* v. *Stevenson* [1932] AC 562, where it was no doubt derived from the 20th chapter of Exodus. It describes not physical closeness but a degree of connection between the conduct of one person and that of another. In *Smith's* case (at p. 816) Lord Griffiths found proximity from the surveyor's knowledge of the overwhelming probability that the purchaser would rely on his valuation and from the fact that the surveyor obtained the work only because the purchaser was willing to pay the fee. Here there is a further element of proximity, in that Mr Thomas had previously acted for Mr Beaumont in this very transaction under a contract between them. But in the course of the argument it became clear tht knowledge of likely reliance was the factor in dispute between the parties. We were referred to the case of *Caparo Industries Plc* v. *Dickman* [1990] 2 WLR 358 at p. 405, where Lord Jauncey of Tullichettle required more than a possibility of reliance – it should be probable, if not highly probable.

There is no direct evidence that Mr Thomas in fact knew that Mr Beaumont was likely to rely on his reinstatement value; Mr Thomas was never asked. The question then is whether such knowledge can be inferred from other evidence or (which is in practice much the same thing) whether he ought to have known.

The answer is by no means easy. It is clear enough that if Mr Thomas had given a reinstatement value of, for example, £300,000 instead of £175,000, Mr Beaumont would have insured for that amount. But that is because he would have been required to do so by the bank, if he was to obtain the loan. It is another question whether Mr Beaumont, if he had believed Mr Thomas to be unreliable or if Mr Thomas had never been asked by the bank for a reinstatement value, would have obtained one himself from another source. In cross-examination Mr Beaumont said that, on completion, he would have had a proper reinstatement valuation if he had felt that one was required. The judge did not accept that evidence.

The judge's conclusion is criticised, on the ground that he may have based it on his finding that there was:

'absolutely nothing in the correspondence to suggest that Mr Beaumont had it in mind to have an independent valuation made for his own purposes at any time.'

Whether that was a sufficient ground for rejecting Mr Beaumont's evidence, there was certainly other evidence which would justify the judge's conclusion. First, there is the fact that Mr Beaumont did not ask Mr Thomas for any valuation when he gave instructions for the structural survey. Second, his letter of 8 February 1984 showed reluctance to incur another fee for further advice. Third, he was content to make his own estimate of £170,000 for reinstatement value on 30 September 1983. Fourth, he wrote on 27 February 1984 that Mr Thomas might be receiving 'additional minor queries' from the bank. Fifth, I cannot see a great deal of sense in postponing a definitive solution of the insurance figure until completion. So far as I recall, Mr Beaumont is not said to have known that Mr Thomas would be asked about it until the bank's letter of 9th April 1984, by which time he had already filled up a proposal form. In my judgment, there was ample evidence to support the judge's conclusion that Mr Beaumont would not have obtained an independent valuation; and we would not be justified in departing from it. It may be that Mr Beaumont was relying on Mr Thomas to give careful advice to the bank in some respects – for example, not to derogate from the sale value of the property or its conditions. But I do not see that he was relying on Mr Thomas to advise on an appropriate figure for reinstatement value.

I am not sure whether it is logically possible for Mr Thomas to have known that Mr Beaumont was going to rely on his valuation, or that Mr Thomas ought to have known that, if in fact Mr Beaumont was not going to rely on it. In case there is such a possibility, I turn to consider the position of Mr Thomas. Not all the facts which I have mentioned in the previous paragraph were known to him. He did, however, know that Mr Beaumont had not asked him for any valuation; he knew that Mr Beaumont regarded the bank as likely to raise 'additional minor queries'; he may well have regarded his advice to the bank as essentially ancillary to the advice he had already given to Mr Beaumont; and he knew that Mr Beaumont was a company director, a man of substance, capable of taking decisions for himself. Against that, he had been warned by the bank that a copy of his report might be forwarded to Mr Beaumont. I agree with the judge's conclusion when he said:

'I do not consider that Mr Thomas contemplated or should have contemplated, that Mr Beaumont was relying on him to provide Mr Beaumont with an accurate reinstatement figure.'

In the circumstances, I hold that Mr Thomas did not owe Mr Beaumont a duty of care when giving advice to the bank.

TAYLOR LJ (at p. 52):

The present case does not concern valuation at the lower end of the market. Nor, however, was it in the 'very expensive' category alongside blocks of flats and industrial properties to which Lord Griffiths referred. It is important, therefore, to look at all the specific circumstances, the probabilities and the expectations of the parties.

Mr Thomas became involved because Mr Beaumont had instructed him to do a structural survey, but not a valuation, of the property in September 1983. He had in fact surveyed the property 11 years earlier for a different client. When Mr Beaumont applied for a mortgage, the bank needed to have a valuation. For that, Mr Beaumont was going to have to pay, as is normal. He was reluctant to incur the expense of another full survey, so he suggested that the bank go to Mr Thomas, in whom at that time he had confidence. He wrote personally to Mr Thomas indicating that he (Mr Thomas) 'may be receiving additional minor queries' from the bank and saying he would be grateful if Mr Thomas would assist in that regard. When Mr Thomas was instructed by the bank, he was told it was assumed he would have no objection to disclosure of his report to the clients, i.e. Mr Beaumont, to whom the account for his fee should be sent.

Thus, Mr Thomas knew he had been brought in because of his previous professional relationship with Mr Beaumont. He knew Mr Beaumont would become aware of his valuation because:

(a) he was having to pay for it;
(b) the bank had indicated it was likely the report would be disclosed to Mr Beaumont; and
(c) Mr Beaumont would clearly be told the valuation because he would be required by the bank to insure for not less than that sum.

In view of the terms in which Mr Beaumont had written to Mr Thomas, it must have seemed unlikely that he would have obtained a valuation from another surveyor. Why should he, when he was paying for one from the surveyor upon whom he had recently placed reliance? It is true that before Mr Thomas put in his report Mr Beaumont had, off his own bat, arranged insurance in the sum of £170,000, but in reliance on the report the cover was increased to £175,000. If the non-negligent figure should have been higher and Mr Thomas had advised it, no doubt Mr Beaumont would have insured, and been required to insure by the bank, for the higher figure.

In my judgment, the test of proximity here is whether when Mr Thomas gave his valuation it was clear that, because of the close relationship stemming from the history I have recited, Mr Beaumont was likely to rely and act upon that valuation. In my view, although thrift may have made Mr Beaumont content to back his own hunch that £170,000 was a sound insurance figure rather than voluntarily to seek a valuation, once he knew of and had to pay for the bank's valuation from his own surveyor it was (in Lord Griffiths' words) 'an overwhelming probability' that he would rely upon it. These considerations go to foreseeability but they also show the close proximity between mortgagor and surveyor in this particular case. There was no question here of the bank's getting a valuation from a surveyor unknown to Mr Beaumont who might or might not, therefore, have placed reliance upon it. Here there had been a contractual relationship between Mr Thomas and Mr Beaumont over the structural survey and in a sense the valuation had come as an addendum to that survey, paid for by Mr Beaumont, albeit solicited by the bank and reaching Mr Beaumont via the bank.

DILLON LJ (at p. 53):

Mr Beaumont had not indeed at any time himself asked Mr Thomas for an insurance valuation of Heastige House, on a reinstatement or any other basis, but in the circumstances of this case there was, in my judgment, nothing whatsoever to suggest to Mr Thomas that Mr Beaumont, having been given the figure of Mr Thomas' insurance reinstatement valuation, would go off to some other valuer to get a rival quotation for himself as a cross-check on Mr Thomas' valuation.

On these facts, I have no doubt that the criteria in *Smith* v. *Eric S. Bush* [1989] 2 WLR 790 are satisfied and Mr Thomas owed a duty of care to Mr Beaumont as well as to the bank.

It was submitted that all that was said in *Smith* v. *Eric S. Bush*, and the case decided at the same time of *Harris* v. *Wyre Forest District Council*, only applies to purchases, financed by mortgage, of properties at the lower end of the housing market by 'ordinary' people in humble circumstances. A mortgagee's surveyor would have to assume that such a purchaser would rely on his valuation for the mortgagee and would not get another valuation from another valuer; but it was submitted that the mortgagee's surveyor was not required to make any such assumption if the purchaser was a more sophisticated person buying a considerably more expensive house. But I see no need to explore those distinctions in the present case, where Mr Thomas was first instructed by Mr Beaumont rather than by the bank. The obvious reason why Mr Beaumont got the bank to instruct his own surveyor, Mr Thomas, rather than instructing some other surveyor of the bank's choice, was that Mr Beaumont did not want to pay the bills of two surveyors. The position of proximity which Mr Thomas accepted when he agreed to accept the bank's instructions is thus essentially the same as the position in *Smith* v. *Eric S. Bush*, and the consequence is that Mr Thomas owed a duty of care to Mr Beaumont as well as to the bank in making his valuation.

The judge held that Mr Thomas owed no duty of care to Mr Beaumont in making the valuation because the course of business between Mr Beaumont and Mr Thomas was that when the former wanted the latter to do something for him, he asked him, and he did not ask him to do a valuation (although he had asked him to assist the bank with their enquiries). The judge also found as a fact – and I would not interfere with his finding – that Mr Beaumont would not have commissioned his own insurance valuation had one not been forthcoming via the bank. But these points are in no way inconsistent with Mr

Beaumont's relying on Mr Thomas' valuation, when the amount of it, as a reinstatement valuation, was sent to him by the bank, and thus in no way inconsistent with Mr Thomas' duty to appreciate that Mr Beaumont was likely to rely on his valuation.

Truly, then, a case finely balanced, but ultimately holding that the surveyor's factual knowledge of the purchaser and likely reliance on his valuation is the key element in upholding the existence of a duty. This is so even though the plaintiff was not obviously strapped for cash, but simply was prudent enough not to waste money on two surveys.

Beaumont shows that the vague policy reasons favouring the plaintiff in *Smith* are going to be difficult to apply in different contexts; in this respect the split in the Court of Appeal is probably as significant as the result.

Other cases have arisen where surveyors have been held to be liable to third parties. It is now necessary to examine them, and consider whether and to what extent they are still applicable today.

SINGER AND FRIEDLANDER LTD v. JOHN D. WOOD & CO.

(1977) 243 EG 212   High Court

The defendants carried out a valuation of farm land owned by Lyon, with whom they had a contract. The plaintiffs, a merchant bank, lent a substantial sum to Lyon in the light of the defendants' valuation of Lyon's land. Lyon subsequently went into liquidation and the plaintiffs lost their money. They claimed successfully against the valuers.

WATKINS J (at p. 212):

> It was, apparently, the practice of some, if not all, finance houses when asked to loan money to land developers to rely upon a valuation of the land provided by valuers engaged for that purpose by the developers. In doing so, the finance house informed the valuer direct that it would be placing reliance upon his valuation. Accordingly, the valuer was made to understand that he owed a duty not only to his clients, the developers, but also to the finance house to use reasonable skill and care in carrying out the valuation. In the instant case it is admitted by counsel on behalf of the defendants, that in circumstances which I shall have to describe in some detail the defendants came to owe such a duty to the plaintiffs. Moreover, [counsel for] the defendants, who actually did the valuation on their behalf, agrees that he knew when providing the valuation to the plaintiffs that they were relying upon it for the purpose of considering what if anything to lend to Lyon Homes Ltd.
>
> The duty owed was of the nature described in the case of *Hedley Byrne & Co. Ltd* v. *Heller & Partners* [1964] AC 465.

He went on to hold that the valuer had fallen below the appropriate level of care, in particular by not making proper checks with the local planning authority. This seems to be a case of where there would still be a claim in negligence today. The defendants had actual knowledge of the fact that the finance house would be relying on their valuation of Lyon's land.

BOURNE v. MCEVOY TIMBER PRESERVATION LTD

(1975) 237 EG 496   High Court

Bourne was buying a property from a company (LC & W). A building society survey discovered rot. LC & W arranged for the defendants to do an examination and carry out the necessary works, but subsequently Bourne paid the bill for the work and was named as the client on a guarantee. Subsequently, further untreated rot was discovered and Bourne sought to sue the defendants in negligence.

BRISTOW J (at p. 497):

> In my judgment I have to test whether there was sufficient proximity between the plaintiff and the defendants to give rise to a duty situation by asking myself this question: at the time the defendants made their inspection and reported to LC & W, their principals, did they know, or ought they to have known, that the purchaser of the house might well be affected in the decisions which he took by the contents of their report? In my judgment the answer to that question on the evidence in this case must be 'yes'. The defendants knew the house was being 'tarted up' for sale. The defendants knew their report might go to the mortgagees. The fact that it might go to mortgagees meant that their findings must affect the value people would put on the house. What was the right value to put upon the house must affect LC & W, the sellers, the mortgagees (if any), and most probably the third person concerned in the sale transaction, the buyer. The defendants regarded the buyer as the beneficiary of their work, if they got the job. LC & W would fill in the buyer's name on the 20-year guarantee of their work, and the defendants would honour the guarantee in the hands of the occupiers from time to time of the house.

However, he went on to find that there was no breach of duty by the defendants. It seems clear that if Bourne's identity and involvement were known to the defendants when they were carrying out their inspection, then the finding that a duty was owed to Bourne would still be justifiable today. On the facts, however, it would seem that at the time of the work they were only aware of the existence of potential purchasers as a class, and their actual knowledge of Bourne himself only transpired later on.

This would not seem to satisfy the *Caparo* test, such potential purchasers being a possibly large class and difficult to determine accurately, and this case would now be decided against the plaintiff at the first stage of duty of care.

SHANKIE-WILLIAMS v. HEAVEY

(1986) 279 EG 316   Court of Appeal

The defendant was employed to report on the condition of the timbers of a ground floor flat by its owner Whittaker. The property was part of a house divided into three flats, one on each floor. Mrs and Mrs Shankie-Williams (the first and second plaintiffs) were endeavouring to purchase the ground floor flat and the defendant's work was in response to their building society surveyor's report; the purchase subsequently went ahead. The following month, a Mr Daulby (the third plaintiff) purchased the first floor flat having seen the defendant's guarantee of his work. Both flats were subsequently found to be suffering from rot. All their claims were unsuccessful.

MAY LJ (at p. 17):

As to the first and second plaintiffs, the defendant clearly knew that Mr Whittaker wanted a report or guarantee from him which he could show to prospective purchasers of the property that he was asked to inspect and report upon. [Counsel for Shankie-Williams] contends that in those circumstances in any event the appellant owed the first and second plaintiffs a duty of care in and about the work that he was instructed by Mr Whittaker to do: there was that degree of proximity, that adequate nexus, between him and the work he was asked to do and the two plaintiffs to found an appropriate duty of care. Putting it in another way, he ought clearly to have had them in contemplation when carrying out the work which he had been asked to do.

We have been referred to the well-known case of *Candler v. Crane, Christmas & Co.* [1951] 2 KB 164, and in particular to the dissenting judgment of Denning LJ (as he then was) which the House of Lords subsequently upheld in the equally well-known case of *Hedley Byrne & Co. Ltd* v. *Heller & Partners Ltd* [1964] AC 465. We have also been referred to the well-known case of *Anns* v. *Merton London Borough Council* [1978] AC 728. For my part I think it unnecessary to refer in detail to those cases. Suffice it to say that I am satisfied, on the facts which I have outlined, that the defendant in this case did owe the prospective purchaser of the ground-floor flat, which he was inspecting and reporting upon, an appropriate duty of care.

(and at p. 318):

Did that breach of duty cause the first and second plaintiffs subsequent damage when the dry rot showed itself two years later? The answer to that question will be in the affirmative only if the defendant's report was something which induced them to go ahead with the purchase, in the course of the conveyancing of which they had paused, pending the eradication of the dry rot which had been found to be there. Was the negligent report by the defendant a cause of the first and second plaintiffs' subsequent damage? The learned judge found on this point that it was 'the certificate and its associated document that decided them [that is to say, the first two respondents] to purchase'. On the material before us, what the learned judge intended to mean by his reference to an associated document to the certificate is, if I may respectfully say, ambiguous. It could have been the estimate to which I have referred, which contained no reference to no evidence of dry rot; or it could have been the report which contained the reference to 'no evidence of . . . dry rot present at the time of our inspection'. The only evidence before the learned judge on this point was that it could have been the estimate. The evidence in any event was somewhat vague, and if by 'associated document' the learned judge intended to refer to that estimate, then the *prima facie* negligence contained in the report cannot be said to have been causative of the parties' decision to proceed with their purchase of the lease and subsequent damage. If, on the other hand, when the learned judge spoke of the 'associated document' he is to be taken as having referred to the report, then, with respect, having considered the notes of evidence contained in the papers, there was no evidence to support that finding.

In those circumstances I am driven to the conclusion, in so far as the occupants of the ground floor, the first and second plaintiffs, are concerned, that the learned judge erred in holding that there was sufficient causation to prove a link between any breach of duty on the part of the defendant and the ultimate damage sustained by the first and second plaintiffs. Consequently, in my view, the learned judge erred in his conclusion that the defendant was liable in damages in respect of the remedial works in so far as the first and second plaintiffs were concerned.

As to the third plaintiff, the occupier of the first-floor flat, in my judgment the relevant inquiry in law must be whether in any event the defendant owed him any duty of care in and about his inspection and report in October 1981. In other words, as I posed the question in relation to the first two plaintiffs, ought the defendant to have had in mind a potential purchaser of the first-floor flat when on the instructions of the absent Mr Whittaker he was inspecting and reporting upon the ground-floor flat? As I have said, the defendant's instructions were to inspect only the ground floor, particularly the floor boards and the timbers beneath. He asked Mr Whittaker how he was to describe that particular part of the property. He was told that it was thereafter to be known as no 80 Colney Hatch Lane, and he so described it.

In my judgment, on the facts as I have outlined them, there was an insufficient nexus between the appellant and the intending purchaser of a lease of the first-floor flat.

I ask myself, in the circumstances of this case, where is one to stop in this inquiry? If the appellant owed to the third plaintiff a duty of care, when reporting on the ground-floor flat, to be occupied by the first and second plaintiffs, did he owe a similar duty to any intending purchaser of the second-floor flat? on the material before us, it seems that the second-floor flat, at any rate at that time, was going to be occupied by Mr Whittaker himself; but if not, would anybody who had taken a lease of that flat (as did the third plaintiff of the first-floor flat) have been able to contend that the defendant owed to him a duty of care in relation to his inspection of the ground floor? I think not.

If, contrary to that view, the answer has to be in the affirmative, then I pose the question: 'What about an adjoining occupier?' Supposing that this was a company converting not one house but a row of five houses in the street. Did an appellant in those circumstances, asked to inspect and report only upon the ground-floor flat at no 80, owe any duty to intending lessees of flats in the yet to be converted adjoining premises at, let us say, 78 Colney Hatch Lane? Again, I think not.

In my judgment the appellant owed to the third plaintiff no duty of care in respect of his inspection and report of the ground floor.

Clearly the unsuccessful Daulby claim would remain so today and in the absence of any proof of reliance, the claim by Shankie-Williams would also fail. But again it does appear that the Court of Appeal would, but for this fact, have allowed their claim, even though again the identity of the particular plaintiffs does not appear to have been known to the defendant. Now, after *Caparo* and *Smith*, the plaintiffs would have been unsuccessful at this stage too.

Claims by third parties against surveyors will be easy to formulate in the context of valuations by building society surveyors, thanks to *Smith*. However, outside of that context, such claims will only be viable today in cases of clear actual knowledge by the surveyors of the presence and identity of the third party, on facts such as those of *Singer and Friedlander*.

Standard of care

The usual general principles that define the standards of care of a professional man apply here. That said, two distinct features arise in the course of the valuation process. These are the especially subjective nature of the activity, and also the range of different types of survey, from the most detailed to the relatively scanty, that the valuer may be carrying out.

SINGER AND FRIEDLANDER LTD v. JOHN D. WOOD & CO

(1977) 243 EG 212   High Court

For facts, see p. 258 above.

WATKINS J (at p. 213):

The valuation of land by trained, competent and careful professional men is a task which rarely, if ever, admits of precise conclusion. Often beyond certain well-founded facts so many imponderables confront the valuer that he is obliged to proceed on the basis of assumptions. Therefore, he cannot be faulted for achieving a result which does not admit of some degree of error. Thus, two able and experienced men, each confronted with the same task, might come to different conclusions without any one being justified in saying that either of them has lacked competence and reasonable care, still less integrity, in doing his work. The permissible margin of error is said by Mr Dean [expert witness], and agreed by Mr Ross [expert witness], to be generally 10 per cent either side of a figure which can be said to be the right figure, i.e. so I am informed, not a figure which later, with hindsight, proves to be right but which at the time of valuation is the figure which a competent, careful and experienced valuer arrives at after making all the necessary inquiries and paying proper regard to the then state of the market. In exceptional circumstances the permissible margin, they say, could be extended to about 15 per cent, or a little more, either way. Any valuation falling outside what I shall call the 'bracket' brings into question the competence of the valuer and the sort of care he gave to the task of valuation. So, for example, if the right figure in December 1972 for the value of 131 acres of Manor Farm was £1,500,000, the sum of £2,000,000 would fall well outside the bracket, given that exceptional circumstances governed the valuation. With these views those who advise the plaintiffs agree, or at least do not dissent from.

The way in which a valuer should conduct himself so as to fulfil his duty to a merchant bank, or any other body or person, varies according to the complexity or otherwise of the task which confronts him. In some instances the necessary inquiries and other investigations preceding a valuation need only be on a modest scale. In others a study of the problem needs to be in greater depth, involving much detailed and painstaking inquiries at many sources of information. In every case the valuer, having gathered all the vital information, is expected to be sufficiently skilful so as to enable himself to interpret the facts, to make indispensable assumptions and to employ a well-practised professional method of reaching a conclusion; and it may be to check that conclusion with another reached as the result of the use of a second well-practised method. In every case the valuer must not only be versed in the value of land throughout the country in a general way, but he must inform himself adequately of the market trends and be very sensitive to them with particular regard for the locality in which the land he values lies. Whatever conclusion is reached, it must be without consideration for the purpose for which it is required. By this I mean that a valuation must reflect the honest opinion of the valuer of the true market value of the land at the relevant time, no matter why or by whom it is required, be it by merchant bank, land developer or prospective builder. So the expression, for example, 'for loan purposes' used in a letter setting out a valuation should be descriptive only of the reason why the valuation is required and not as an indication that were the valuation required for some other purpose a different value would be provided by the valuer to he who seeks the valuation. It might, however, be an indication that the valuer, knowing the borrowing of money was behind the request for valuation, acted with even more care than usual to try to be as accurate as possible.

If a valuation is sought at times when the property market is plainly showing signs of deep depression or of unusual buoyancy or volatility, the valuer's task is made much more difficult than usual. But it is not in such unusual circumstances an impossible one. As Mr Ross said, valuation is an art, not a science. Pinpoint accuracy in the result is not, therefore, to be expected by he who requests the valuation. There is, as I have said, a permissible margin of error, the 'bracket' as I have called it. What can properly be expected from a competent valuer using reasonable skill and care is that his valuation falls within this bracket. The unusual circumstances of his task impose upon him a greater test of his skill and bid him to exercise stricter disciplines in the making of assumptions without which he is unable to perform his task; and I think he must beware of lapsing into carelessness or over-confidence when the market is riding high. The more unusual be the nature of the problem, for no matter what reason, the greater the need for circumspection.

The factors and activities which a competent valuer will consider and undertake in valuing land cannot be composed in such a way as to indicate an unvarying approach to every problem which confronts him. But a collection of them from which he will choose at will in a given circumstance must include the following:

(a)   The kind of development of the land to be undertaken.
(b)   The existence, if any, of planning permission in outline or in detail as to a part or the whole of the land. And if permission be for the building of houses, the situation and acreage of part of the land excluded from planning permission because, for example, of a tree preservation order, the need for schools and lay-out of roads and other things. Furthermore, the number of houses permitted or likely to be permitted to be built by the planning authority is a relevant, indeed a vital factor.
(c)   The history of the land, including its use, changes in ownership and the most recent buying prices, planning applications and permissions, the implementation or otherwise of existing planning permissions, the reason for the failure, if it be the fact that a planning permission has not been implemented.
(d)   The position of the land in relation to surrounding countryside, villages and towns and places of employment; the quality of access to it, the attractiveness or otherwise of its situation.
(e)   The situation obtaining about the provision of services, for example, gas, electricity, sewage and other drainage and water.
(f)   The presence, if it be so, of any unusual difficulties confronting development which will tend to increase the cost of it to an extent which affects the value of the land. A visit to the site must surely find a prominent place among physical activities to be undertaken.
(g)   The demand in the immediate localities for houses of the kind likely to

be built, with special regard to the situation of places of employment and increases to be expected in demand for labour. This will involve, inevitably, acquiring knowledge of other building developments recently finished or still in progress, especially having regard to rate of disposal, density and sale price of the houses disposed of. In this way the existence, if any, of local comparables, a valuable factor, can be discovered.

(h) Consultation with senior officers of the local planning authority is almost always regarded as an indispensable aid, likewise a knowledge of the approval planning policy for the local area; a study of the approved town or county map may prove rewarding.

(i) Whether ascertaining from the client if there have been other previous valuations of the land, and to what effect, should be undertaken is probably questionable because a valuer's mind should not be exposed to the possibilities of affectation by the opinion of others.

(j) If he is a man whose usual professional activities do not bring him regularly into the locality, or what is more important has never done so, he will obviously need to be especially careful in collecting as much relevant local knowledge as he can, possibly by consulting valuers who work regularly in the area.

(k) The availability of a labour force which can carry out the prospective development.

When this harvest of knowledge has been gathered in by the valuer with, on occasions, as must be accepted, help from a competent member or members of his staff or the firm for whom he works, he must assess the worth of the details of it. Some of it he may set aside either as being unreliable or for some other good reason. Other details may impress him as reliable as facts which speak for themselves or upon which he can, using his best judgment, rest assumptions without the making of which his task is rendered impossible. With these he has to try to penetrate the mists of the future and withal bring to bear upon him his training, skill and experience in order to produce a carefully-achieved conclusion in terms of monetary value.

Given the range of imponderables that have to be guessed, it is evident that it may be difficult for a plaintiff to establish that reasonable care has not been exercised in the course of valuation, unless the figure is wildly out of line, at which stage collective professional opinion is likely to turn strongly in favour of the plaintiff. On the other hand, the range of different fields embraced in the course of the valuation process can lead to surveyors being found liable in respect of matters some way from their core training.

DAISLEY v. B.S. HALL & CO.

(1973) 225 EG 1553 High Court

The plaintiffs purchased a house, High Trees, after having it surveyed by the defendants. The survey report noted some minor cracks and shrinkage, but did not connect these with the presence of nearby poplar trees. The defendants were held to be negligent in their conduct of the survey.

BRISTOW J (at p. 1555):

In April, 1970, Mr Daisley issued a writ against Halls complaining that they failed to carry out their survey with proper care and professional skill in that they failed to report upon the hazards of the house caused by the clay soil and the presence of the poplars, which might cause cracking of the external brickwork and internal plasterwork and which had caused or might cause settlement in the ground-floor floor slabs. The duty of a practitioner of any professional skill which he undertakes to perform by accepting instructions from his client is to see the things that the average skilled professional in that field would see, draw from what he sees the conclusions that the average skilled professional would draw, and take the action that the average skilled professional would take. It is clear on the evidence of all the experts called in this case that the risk to buildings on shrinkable clay subsoil adjacent to poplar trees is notorious and is one of the subjects which is included in a surveyor's education. In my judgment it follows that any surveyor who finds a house under 10 years old, built as close to a row of poplars as High Trees was, where two of those trees had been felled for no apparent reason other than to reduce the risk of root damage, is under a duty to ascertain by effective means the nature of the subsoil whether or not he finds evidence of settlement of the house which may be due to subsoil shrinkage. If he finds that the subsoil is shrinkable clay, then it is clearly his duty to consider whether damage which may be due to this cause has already taken place, because this must affect the seriousness of the risk of damage of which it is his duty to warn his client. If circumstances are such that the risk is high, clearly he should warn his client not to go on with the purchase. If circumstances are such that the risk is very small it must, in my judgment, be his duty to say so. Between the two extremes, the price which he advises his client it is sensible to pay must take account of the extent of the risk as he sees it and the possible cost of putting things right if damage does occur.

I am satisfied on the evidence in this case that a reasonably skilled surveyor, alive to the fact that this house was built on shrinkable clay subsoil, would have suspected that crack 3.5.11. might have been caused by subsoil shrinkage. Mr Ross Davis [a surveyor witness for Daisley] took this view. In my judgment any reasonable surveyor must then take the view that the shrinkage of the subsoil might be caused by the adjacent poplar trees which, applying the Building Research Station rule of thumb, were well in range. He would then take into account the age of the house, the size and number of the trees within range and the likelihood of damage from swelling of the clay if the trees were cut down, and advise his client on the extent of this risk accordingly. Mr Gascoigne-Pees [the surveyor] was alive to the importance of the subsoil, but in my opinion simply to ask his principal was not an effective way to find out what it was when the presence of the poplar trees made it important to know. Unhappily he did not recognise the trees as poplars, thinking, at that time, that all poplars are Lombardy poplars. In my judgment the average skilled surveyor would appreciate that the presence of poplars may be hazardous and that it really is essential to make sure of your tree recognition, because poplars are not the only trees of which there is more than one species. Not realising that the trees were poplars and that the subsoil was shrinkable clay, it is perhaps not surprising that Mr Gascoigne-Pees did not suspect that crack 3.5.11. might have a sinister origin and did not think that it would be wise to look for roots in the drains, which were only 35 ft from the nearest tree. His field experience was then very limited. Now that he has more, he himself says, very candidly, that though he does not think there were any signs of existing damage yet he would still have warned of the possibility of future damage even though he would have rated the risk as very small. In this case, in seeing whether the defendants are in breach of contract, it is not necessary to consider the extent of the risk. I am satisfied that any surveyor of average skill faced with this house, these trees and this subsoil would have given some warning, whatever his view of the origin of crack 3.5.11. Mr Gascoigne-Pees gave none, and so the defendants are in breach.

It may seem harsh at first sight to hold a surveyor liable for not being able to tell one type of poplar tree from another but it should be said in fairness that this was a matter clearly affecting the ground conditions very close to the house and this property was within the scope of his survey.

Clearly, a major factor in assessing the appropriate standard of care will be the nature of the survey required. Is it a full scale structural survey that is required or is it a mere valuation, usually for mortgage purposes, that is required? Different types of surveys involve greater or lesser detail and examination, and yet there will often be a factual dispute between the parties as to what was intended. The courts have to resolve this and do so by looking at the conduct and intention of the parties. Examples of this inherently factual process re *Buckland* v. *Mackesy & Watts* (1968) 208 EG 969 and *Sutcliffe* v. *Sayer* (1987) 281 EG 1452. In *Buckland*, the surveyor claimed that he had been asked merely to make some measurements, but the fact that he billed the work as survey work and, in his report, noted some minor defects (and that the plaintiff was in receipt of no other advice) meant that he was held liable for negligently failing to draw the attention of his client, the purchaser, to *inter alia*, wet rot, woodworm and cracked foundations. He had held himself out as doing more than just measurement work and was therefore liable for not fulfilling these greater responsibilities.

On the other hand, in *Sutcliffe* a valuation, carried out by an estate agent who was not qualified to carry out a full structural survey, was held to be just that, and not a fuller survey. As a result, the defendant was held not to have been negligent in his failure to draw the purchaser's attention to the likely difficulty of resale of that particular property; this was a matter for a full survey report, which had not been requested and was not provided. This decision may seem harsh on the (impoverished) purchasers of the house but in legal terms makes clear sense – it is very difficult to see how the plaintiffs could place any reliance or at least any reasonable reliance on the defendant in respect of matters on which he was not offering a view, nor had the expertise so to have done.

The question then arises within the agreed obligation as to the extent of the investigation in which the surveyor must engage. In particular the question arises as to whether the surveyor must become an investigator following up his suspicions and exploring matters in detail; although *Sutcliffe* does not expect this to happen in a mere valuation, several recent cases do seem to suggest that the surveyor is under this hand of positive duty.

MORGAN v. PERRY

(1974) 229 EG 1737   High Court

Morgan purchased a hillside property after a structural survey by the defendant. The report was entirely favourable. In fact earth movement plagued the areas and the house became valueless. It was held that the defendant was negligent.

JUDGE KENNETH JONES (at p. 1739):

What form, then, should an examination of the house have taken? Mr Davis [an expert witness] said that he would have looked for trouble, especially movement in the building. He would first have taken account of the fact that the house was built on a hillside. He would have considered the presence of repointing in a house only four years old as an indication that something was wrong. All these were matters he would have reported to his client. He would not have stopped there. He would have observed the roadway to look for repairs due to causes other than traffic. He would have asked himself, 'Why two gullies so close together?' Further examination would have revealed that one was twice the normal depth. He would then have got in touch with the highway authority. If he could not have found the final answer himself, he would have said so and advised calling in further experts. Left on his own, he would have advised his client not to bid.

The defendant failed to notice the repointing. He (the judge) did not accept his evidence that that could not be seen even today. He failed to notice the shape and extent of the crack over the kitchen door. He failed to notice in-filling in the cornice in the south-east bedroom or to observe that the cupboard doors were not running freely. He failed to notice the over-papering. He said that even if he had noticed it, it would not have surprised him. It must be well known that, human nature being what it was, from time to time vendors would take steps to cover up defects. That was at least part of the reason why the purchaser employed a surveyor. The defendant missed that, as he missed so much else. The totality of what was there to be seen, considered against the known potential dangers of a steeply-sloping site to which the defendant appeared to have given no real thought, ought undoubtedly to have aroused his suspicions and led him to a close examination of the house. This would have revealed the slope in the floors; he would have been led also to look at and think about the highway outside. He did not look at the road. An examination and consideration of the road would have further aroused his suspicions and led him to the highway authority. In his (Judge Jones's) judgment, the defendant failed to exercise the reasonable care and skill of a prudent surveyor which he had impliedly promised to exercise when he received instructions to survey the house. He committed a breach of his contract with the plaintiff, and provided him ultimately with a report which was so misleading as to be valueless.

This decision, on the facts of the particular site and given that it was a full survey, is not too surprising. The defendant should have been put on guard by the nature of the site and therefore should have gone further than might be usual in his investigations.

HOOBERMAN v. SALTER REX

(1985) 274 EG 151 High Court (Official Referee's Business)

The plaintiff purchased a maisonette with a flat roof terrace after having commissioned a structural survey from the defendants. In due course, leaks from the terrace area and associated problems of rot developed.

JUDGE SMOUT QC (at p. 155):

Accordingly I have to determine whether the defendants fell short of the standard of ordinary skill and care owed by competent surveyors in their failure to recognise and warn the plaintiff in March 1977 of the dangers inherent in the faulty felt upstands of the terrace, and in the lack of ventilation in the roof area between the terrace and the bedroom ceiling. Each defect was a substantial cause of the eventual damage. Both were readily apparent. Should their significance have been appreciated at the time of the inspection by Mr Heasman in March 1977?

Mr John Taphouse was the expert witness called by the plaintiffs. He is a chartered building surveyor and a Fellow of the Royal Institution of Chartered Surveyors, and of 30 years' professional experience. He drew attention to the British Standards Code of Practice No. 144 published in 1970.

Paragraph 2.4.2 of the code warns that roof decks may 'fail structurally from persistent moisture ingress arising from site exposure, condensation or leakage. Designers wishing to use these decks should ensure that every step is taken to minimise such risks'. Figure 1 at p. 32 of the code shows the recommended manner of attaching felt with a metal flashing against a vertical wall. It will be seen that the felt would thus be free to move up and down and is not bonded to the wall. Para. 3.8 of the code states:

> Where roofs having timber joists, or timber board, resin-bonded wood shipboard or plywood decks are constructed so that there is a space between ceiling and deck, this space must be ventilated to the open air to avoid fungal growth.

He expressed the view that flat roofs had already a bad reputation by 1977 and had been known for very many years to be particularly vulnerable. He said in evidence that despite the absence of any evidence of damp in 1977, some six years after construction, a surveyor should none the less look at such a roof very carefully and if he saw defects of construction then he should 'sound a warning of possible failure'. In his expert report that formed part of his evidence he states at p. 15:

> It is quite clear that there were sufficient warning signs in the visible roof construction to have enabled a competent surveyor to predict with a fair degree of certainty that leaks would occur and dry rot would break out as a consequence.

And he concluded that the survey report of a reasonably careful surveyor:

> would have emphasised that the roof void is unventilated and he should have sounded the strongest possible warning that this lack of ventilation may have already caused rot to occur or would certainly induce dry rot in the future. A surveyor should have advised his client to provide adequate ventilation to the roof structure at the earliest possible time.

Mr Leo Lewis, the expert called by the defendants, took a different view. He likewise was a Fellow of the Royal Institution of Chartered Surveyors with considerable experience. He would not accept that by 1977 timber roofs had a poor record and had to be regarded with suspicion. Nor did he believe that he himself would have warned as to the lack of ventilation. He put his point in this way:

> You may say any roof may leak and that would lead to dry rot. You can raise doubts that are too great – you have to be selective in your recommendations. What my client seeks from me is a balanced judgment. There is no way in which a house of antiquity is going to be in good order. You would not get anywhere by saying that certain features do not accord with modern requirements.

He conceded, however, that:

> The absence of cover flashings were such that a reasonably competent surveyor would know was bad practice in 1977.

and further:

> A reasonably competent surveyor would seek to find out if there were any angle fillets.

Although Mr Lewis emphasised the likelihood of condensation from the erroneous position of the vapour barrier, and that the defective pointing was another very likely cause of moisture, he none the less had to accept that six pints of water coming through the ceiling indicated a leak in the roof.

Mr Heasman [the defendant's surveyor] was a frank witness. He was palpably honest. He described his role in this way:

> When I give a survey, my object is to advise my client as to any reason of a structural nature that should prevent him from purchasing the property, and also to advise him of any necessary repairs.

and later:

> I was there to advise the plaintiff as to whether he should buy the property: no evidence that he should not.

and again:

> The purpose of my survey was not to alarm a client who clearly wanted to buy the property . . . if a real risk of serious problems, of course, I would warn him. When carrying out a survey, I think a surveyor has to have regard to the salient points – to the points likely to be of concern to the client.

When pressed in cross-examination with regard to the lack of ventilation he said:

> I did not mention lack of ventilation between ceiling and roof joists. I was not aware of lack of ventilation because I was not looking for ventilation. At that time not mandatory for ventilation . . . for paragraph 3.8 is a code of practice (CP 144) and not a by-law.

When asked if he did not know that journals at that time were warning surveyors as to the lack of ventilation in regard to flat roofs, he replied:

> Ventilation was not required by statute. The journal is like a White Paper that might become an Act of Parliament. Until the code of practice is a by-law, it is not essential. If I saw something that did not accord with the recommended practice, I would not wish to alarm the purchaser.

And as to the felt upstands:

> The felt was laid directly into the parapet wall. That is contrary to what is recommended to be the best practice. It is not uncommon. I noted that the felt was put directly into the parapet wall. That did not concern me to the extent that I should refer to client as a matter of importance. No defect was visible. It had been in position for some years. I looked to see if felt upstands had suffered any tearing. If I had seen any evidence of stresses I would have noted them.

In fact he noted in his survey report – 'Flashings in felt OK . . .'.

In my view Mr Heasman was under a misconception as to his role. He was there to inspect the property so far as reasonably practicable so as to report candidly upon its condition. No doubt he has to be selective and determine what aspects are important and what are unimportant, but whether his conclusions are comfortable or uncomfortable to the client is immaterial. I feel that Mr Heasman was not exercising a sufficiently independent judgment to provide a reliable assessment of the condition of the property. Furthermore I conclude that in limiting his role to that of considering whether or not the client should buy the property and as to what repairs were necessary, he overlooked the need to consider what were the potential dangers that needed to be avoided. I bear in mind Mr Lewis' warning to the court against hindsight. None the less, I am satisfied that in the state of knowledge in the construction industry and among surveyors in 1977 it was generally appreciated that flat roofs were vulnerable, that their construction had to be viewed with caution, that the defects in respect of the upstands and lack of ventilation at these premises were apparent and were contrary to good practice, and that a competent surveyor should have appreciated at that time that they reflected a serious potential danger. I have to say that in this instance he fell below the standards of ordinary care and skill and that he failed to appreciate the significance of the defects and accordingly failed to warn the plaintiff of serious potential danger.

In the result I accept the conclusion of Mr Taphouse in his expert's report, that Mr Heasman 'should have advised the plaintiff that the roof was poorly designed and constructed and that it was likely to be troublesome'. Had the plaintiff been so advised, he could then have negotiated the purchase of the house in full knowledge of its true condition and could either have secured the house at a price he considered to be reasonable, or, if the vendor refused to negotiate, he could have taken the decision on whether to proceed with the purchase or not. Mr Taphouse added the comment that the direct result is that the plaintiff has suffered a great deal of expense, inconvenience and worry which could have been easily avoided.

Again this is a case of a full survey and so, again, it is not too harsh on the defendants to expect them in that context to go beyond the narrow question of whether the plaintiff should buy the property to the issue of what potential future problems there may be which should be explored.

Flat roofs were known to be a source of danger, and this must also be a relevant factor.

ROBERTS v. J. HAMPSON & CO.

[1990] 1 WLR 94   High Court

The plaintiffs purchased a bungalow in North Wales. This followed a valuation by the defendants arranged by the building society. The plaintiffs noted that the valuation of the property was at the price they had agreed to pay, and went ahead. In fact the valuation report omitted to note major defects, the plaintiffs later discovering serious dry rot and rising damp. They were successful in their action against the defendants.

IAN KENNEDY J (at p. 101):

> The first question is: what is the extent of the service that a surveyor must provide in performing a building society valuation? The service is, in fact, described in the Halifax's brochure. It is a valuation and not a survey but any valuation is necessarily governed by conditions. The inspection is, of necessity, a limited one. Both the expert surveyors who gave evidence before me agreed that with a house of this size they would allow about half an hour for their inspection on site. That time does not admit of moving furniture, nor of lifting carpets, especially where they are nailed down. In my judgment, it must be accepted that where a surveyor undertakes a scheme 1 valuation it is understood that he is making a limited appraisal only. It is, however, an appraisal by a skilled professional man. It is inherent in any standard fee work that some cases will colloquially be 'winners' and others 'losers', from the professional man's point of view. The fact that in an individual case he may need to spend two or three times as long as he would have expected or as the fee structure would have contemplated, is something that he must accept. His duty to take reasonable care in providing a valuation remains the root of his obligation. In an extreme case, as Mr Caird [the plaintiff's expert witness] said, a surveyor might refuse to value on the agreed fee basis, though any surveyor who too often refused to take the rough with the smooth would not improve his reputation. If, in a particular case, the proper valuation of a £19,000 house needs two hours work, that is what the surveyor must devote to it.
>
> The second aspect of the problem concerns moving furniture and lifting carpets. Here again, as it seems to me, the position that the law adopts is simple. If a surveyor misses a defect because its signs are hidden, that is a risk that his client must accept. But if there is specific ground for suspicion and the trail of suspicion leads behind furniture or under carpets, the surveyor must take reasonable steps to follow the trail until he has all the information which it is reasonable for him to have before making his valuation. Thus, I think it

is entirely reasonable for Mr Jones [the surveyor] to say that he took his damp readings along the walls at intervals of three or four feet unless there was furniture in the way, but it does not follow from that that he was relieved from moving furniture if an evident defect extended behind it.

The next and central question is whether the defendants owed a duty to the plaintiffs in respect of the care with which they performed the valuation. [Counsel for J. Hampson & Co.] submits that the plaintiffs were outside the scope of the defendants' duty of care. She submits that the scope of a duty is not to be found in any broad concept of neighbourhood nor, now, in the test proposed by Park J in *Yianni* v. *Edwin Evans & Sons* [1982] QB 438. She submits that that case was wrongly decided and should not be followed. She submits that the correct test is a tripartite one: there must be proved the fact of reliance, the reasonableness of reliance and the foreseeability of reliance. I will dispose at once of the fact of reliance for it is not in question that the plaintiffs did rely on the valuation provided by the defendants.

(and at p. 104):

In my judgment there is clearly a sufficient proximity between the defendants and the plaintiffs in the present case. The defendants undoubtedly knew from the very fact that a scheme 1 survey was being undertaken that it was highly unlikely that the plaintiffs were relying upon some other professionally based information as to the property. They knew clearly of the terms of the Halifax's notes to applicants and therefore that there was no disclaimer of liability. Mr Jackson [a witness], in the course of his evidence, said that in 1984 there was a realisation among surveyors that valuations were read and relied upon by borrowers. I am fully satisfied that there has been here established a duty on the part of the defendant towards the plaintiffs to take reasonable care in the making of their valuation.

I turn to the question of breach. The only breach suggested is in relation to Mr Jones's assessment of the rot and its cause. His description is a laconic one: 'a certain amount of rot'. The likelihood is that the skirting board was badly infested in May 1984. Although the speed at which dry rot can spread is not accurately predictable, it can be as fast as 14 millimetres a week. There are no circumstances proved here which would suggest any re-awakening of dormant rot. There was no installation of central heating, no flood or dripping pipe, no carpeting of a previously uncarpeted area. The house had been superficially well cared for and this does not appear to have been the case of a house whose previous occupiers could not afford to heat it, where the arrival of a new owner with the inclination and means to maintain a higher temperature may trigger the spread of rot. In any event the months between the survey and the discovery of the outbreak were summer and autumn months when a change in the living pattern of the occupants would be of less significance.

Although I recognise that there were features suggesting residual dampness, particularly the absence of spalling and the fact that the new plaster which had been put on the lower part of this wall made it the more difficult to detect the extent to which dampness was rising up it, I consider that Mr Jones made inadequate investigations to determine whether the rot was indeed so localised as he, as I believe, assumed. Mr Jackson agreed, as it seems to me wholly correctly, that a surveyor must be conservative if there is something present in a house which might be serious. He said that if there was evidence of past repairs and yet evidence of rot in one room there should be a conservative approach, and that the surveyor should recommend further investigation. His own test was that if the repairs were likely to cost 10 per cent of the proposed purchase price he would certainly have recommended a retention. He said that if he was uncertain as to the position he would recommend a retention and further investigation. In that event, as he described it, the building society would require quotations to be obtained before proceeding with the matter. In my judgment Mr Jones was wanting in that he did not take any test of the wood blocks of the floor, a test which I am satisfied would have revealed some dampness to put him upon further inquiry, even if there were not traces of rot itself. I was also very surprised by his evidence that he did not trace the rot further because of the presence of a piece of furniture. In my judgment, he failed to appreciate the difference between his ordinary duty in relation to parts of the structure which were hidden by furniture or by carpets and his duty where there was, what I have called, a trail of suspicion. I am satisfied that there was in this case a lack of reasonable care, albeit restricted to this one matter.

This case goes much further than the previous two. Here this was not a full structural survey but only a valuation, yet the surveyor was still held to be liable. However, two essential features enable the decision to be justified. Firstly, the defendant surveyor had been put on guard as to the likely presence of damp and had not followed up the suspicions that he had developed by moving furniture and carpets; though this was not part of the normal valuation, he had become aware of facts which should have made him go beyond what was normal. In this respect the case is on all fours with *Morgan* and *Hooberman*, in both of which the circumstances were such as to put the surveyor on guard. This contrasts with *Whalley* v. *Roberts* [1990] 6 EG 104, where the defect was neither usual nor obvious, and the surveyor therefore not liable. Secondly, there are few problems in finding reliance here. In fact the plaintiffs clearly were relying on the surveyor; he was qualified so that reliance was reasonable (c.f. *Sutcliffe* v. *Sayer*) and, in the circumstances of a typical young couple buying a house at the limit of their financial resources, that reasonable reliance was known to the defendants.

The courts do seem intent on ensuring a high standard of professional conduct even in the simplest valuation report. *Cross* v. *David Martin and Mortimer* [1989] 10 EG 110 is another such case where the general conditions – soil, poplar trees and a sloping site – should have put the surveyor on guard against the risk of subsidence and thus led him to investigate further than might otherwise have been usual. However, it does seem that the surveyor may be able to fulfil his obligation by clear warnings about future risks in his report – *Eley* v. *King and Chasemore* [1989] 22 EG 109.

In general, the key feature, it is suggested, which distinguishes these cases is the presence of something suspicious. Once suspicion is aroused then, irrespective of the type of survey in question, the surveyor needs to explore further to see whether the suspicion is grounded or not. Where, however, there is nothing to put the reasonable surveyor on watch, it seems that his duties remain at a low level (e.g. *Whalley* v. *Roberts* (1990) 6 EG 104).

Disclaimers of liability

It has been seen that different surveys will serve different functions and be carried out at different levels, and that this alone will in some cases affect the standard of care. In many cases, however, the surveyor will wish to make clearer what the purpose of the survey has been, the degree of reliance to be placed upon it and the extent of the responsibility which the surveyor is prepared to accept by setting these matters as part of the survey report. In particular, this may take the form of a disclaimer, where the surveyor will expressly disclaim any responsibility for the consequences of relying, in whole or in part, on the report. This is a common feature of building society valuations, which will, typically, point out that they are not full structural surveys and that no warranty or representation is made to the purchaser in respect of the various statements made by the surveyor to the building society.

Two important related questions arise in relation to the use of such disclaimers by surveyors:

(i)   do such disclaimers act to affect their liability?
(ii) are such disclaimers caught by the statutory controls on exclusion clauses, considered in Chapter 5?

Effect of disclaimers

There is no doubt that appropriately drafted disclaimers can obviate liability in respect of statements by professional men. The classic example of this is none other than *Hedley Byrne* v. *Heller* [1964] AC 465 itself where the new liability for negligent misstatements created by the House of Lords was not available to the plaintiffs since the advice given by the defendants was accompanied by a disclaimer of liability which was held to be effective by removing the necessary element of reliance from the situation: the two simple words 'without responsibility' had this dramatic effect.

SMITH v. ERIC S. BUSH;
HARRIS v. WYRE FOREST DISTRICT COUNCIL

[1990] 1 AC 831 House of Lords

For facts, see p. 247 above.

LORD GRIFFITHS (at p. 856):

> At common law, whether the duty to exercise reasonable care and skill is founded in contract or tort, a party is as a general rule free, by the use of appropriate wording, to exclude liability for negligence in discharge of the duty. The disclaimer of liability in the present case is prominent and clearly worded and on the authority of *Hedley Byrne & Co. Ltd v. Heller & Partners Ltd* [1964] AC 465, in so far as the common law is concerned effective to exclude the surveyors' liability for negligence. The question then is whether the Unfair Contract Terms Act 1977 bites upon such a disclaimer.

It is now quite clear that, at common law, a disclaimer of liability is capable of negating what would, but for it, be a clear duty of care. This is, as in *Hedley Byrne* liability itself, irrespective of whether the parties are in a contractual or a non-contractual relationship. It is of course dependent upon the disclaimer being drafted in appropriate terminology but it is clear from *Smith* and other, earlier, cases that the standard form of disclaimer by surveyors in their reports to the building society are appropriate, and effective.

Disclaimers and statutory control

Does the Unfair Contract Terms Act 1977, with its controls on exclusion clauses in favour of consumers, apply to disclaimers? Two views have over the years been expressed, and these revolved around the definition of an exclusion clause.

Unfair Contract Terms Act 19777 s. 13 (1)

13(1) To the extent that this Part of this Act prevents the exclusion or restriction of any liability it also prevents—

(a) making the liability or its enforcement subject to restrictive or onerous conditions;

(b) excluding or restricting any right or remedy in respect of the liability, or subjecting a person to any prejudice in consequence of his pursuing any such right or remedy;

(c) excluding or restricting rules of evidence or procedure; and (to that extent) sections 2 and 5 to 7 also prevent excluding or restricting liability by reference to terms and notices which exclude or restrict the relevant obligation or duty.

In many ways this is a very broad definition; this enables one school of thought to argue, simply, that the presence of the disclaimer limits the remedies open to the plaintiff by cancelling out negligence liability. The other, more subtle, argument, is that disclaimers have a different effect from exclusion clauses. They do not exclude liability, but rather are one of the features that make up the definition of the liability itself. In other words, they prevent liability from ever arising, rather than provide an exclusion of it. The more subtle argument, however, has not prevailed.

SMITH v. ERIC S. BUSH;
HARRIS v. WYRE FOREST DISTRICT COUNCIL

[1990] 1 AC 831   House of Lords

For facts, see p. 247 above.

LORD TEMPLEMAN (at p. 848):

In *Harris* v. *Wyre Forest District Council* [1988] QB 835, the Court of Appeal (Kerr and Nourse LJJ and Caulfield J) accepted an argument that the Act of 1977 did not apply because the council by their express disclaimer refused to obtain a valuation save on terms that the valuer would not be under any obligation to Mr and Mrs Harris to take reasonable care or exercise reasonable skill. The council did not exclude liability for negligence but excluded negligence so that the valuer and the council never came under a duty of care to Mr and Mrs Harris and could not be guilty of negligence. This construction would not give effect to the manifest intention of the Act but would emasculate the Act. The construction would provide no control over standard form exclusion clauses which individual members of the public are obliged to accept. A party to a contract or a tortfeasor could opt out of the Act of 1977 by declining in the words of Nourse LJ, at p. 845, to recognise 'their own answerability to the plaintiff'. Caulfield J said, at p. 850, that the Act 'can only be relevant where there is on the facts a potential liability'. But no one intends to commit a tort and therefore any notice which excludes liability is a notice which excludes a potential liability. Kerr LJ, at p. 853, sought to confine the Act to 'situations where the existence of a duty of care is not open to doubt' or where there is 'an inescapable duty of care'. I can find nothing in the Act of 1977 or in the general law to identify or support this distinction. In the result the Court of Appeal held that the Act does not apply to 'negligent misstatements where a disclaimer has prevented a duty of care from coming into existence'; *per* Nourse LJ, at p. 848. My Lords this confuses the valuer's report with the work which the valuer carries out in order to make his report. The valuer owed a duty to exercise reasonable skill and care in his inspection and valuation. If he had been careful in his work, he would not have made a 'negligent misstatement' in his report.

Section 11(3) of the Act of 1977 provides that in considering whether it is fair and reasonable to allow reliance on a notice which excludes liability in tort, account must be taken of:

'all the circumstances obtaining when the liability arose or (but for the notice) would have arisen'.

Section 13(1) of the Act prevents the exclusion of any right or remedy and (to that extent) section 2 also prevents the exclusion of liability:

'by reference to ... notices which exclude ... the relevant obligation or duty'.

Nourse LJ dismissed section 11(3) as 'peripheral' and made no comment on section 13(1). In my opinion both these provisions support the view that the Act of 1977 requires that all exclusion notices which would in common law provide a defence to an action for negligence must satisfy the requirement of reasonableness.

LORD GRIFFITHS (at p. 857):

[The statutory provisions] indicate that the existence of the common law duty to take reasonable care, referred to in section 1(1)(b), is to be judged by considering whether it would exist 'but for' the notice excluding liability. The result of taking the notice into account when assessing the existence of a duty of care would result in removing all liability for negligent misstatements from the protection of the Act. It is permissible to have regard to the second report of the Law Commission on Exemption Clauses (1975) (Law Com. No. 69) which is the genesis of the Unfair Contract Terms Act 1977 as an aid to the construction of the Act. Paragraph 127 of that report reads:

'Our recommendations in this part of the report are intended to apply to exclusions of liability for negligence where the liability is incurred in the course of a person's business. We consider that they should apply even in cases where the person seeking to rely on the exemption clause was under no legal obligation (such as a contractual obligation) to carry out the activity. This means that, for example, conditions attached to a licence to enter on to land, and disclaimers of liability made where information or advice is given, should be subject to control . . . .'

I have no reason to think that Parliament did not intend to follow this advice and the wording of the Act is, in my opinion, apt to give effect to that intention.

LORD JAUNCEY (at p. 872):

The next question is whether the disclaimers by and on behalf of the appellants fall within the ambit of the Unfair Contracts Act 1977. In *Hedley Byrne & Co. Ltd* v. *Heller & Partners Ltd* [1964] AC 465, it was held that the disclaimer of responsibility made by the defendant bankers when giving the reference negatived any assumption by them of a duty of care towards the plaintiff. If the circumstances of this case had arisen before 1977 there can be no doubt that the disclaimers would have been effective to negative such an assumption of responsibility. Has the Act of 1977 altered the position? The relevant statutory provisions are sections 2(2), 11(3) and 13(1):

> '**2(2)**   In the case of other loss or damage, a person cannot so exclude or restrict his liability for negligence except in so far as the term or notice satisfies the requirement of reasonableness.
>
> **11(3)**   In relation to a notice (not being a notice having contractual effect), the requirement of reasonableness under this Act is that it should be fair and reasonable to allow reliance on it, having regard to all the circumstances obtaining when the liability arose or (but for the notice) would have arisen.
>
> **13(1)**   To the extent that this Part of this Act prevents the exclusion or restriction of any liability it also prevents—
>
> > (a) making the liability or its enforcement subject to restrictive or onerous conditions;
> > (b) excluding or restricting any right or remedy in respect of the liability, or subjecting a person to any prejudice in consequence of his pursuing any such right or remedy;
> > (c) excluding or restricting rules of evidence or procedure; and (to that extent) sections 2 and 5 to 7 also prevent excluding or restricting liability by reference to terms and notices which exclude or restrict the relevant obligation or duty.'

In the other appeal, *Harris* v. *Wyre Forest District Council* [1988] QB 835, the Court of Appeal held that the Act of 1977 did not apply. Nourse LJ, at p. 848, accepted the defendants' argument that a notice which prevented a duty of care from coming into existence was not one upon which section 2(2) bit. Kerr LJ said, at p. 854:

'For these reasons I agree with the judgments of Nourse LJ and Caulfield J that the effect of the Unfair Contract Terms Act 1977 on the disclaimer of responsibility and warning is of no relevance to the present case. One never reaches that issue, since it arises only if the existence of a duty of care and a breach of it have first been established.'

[Counsel for Wyre Forest District Council] in the *Harris* appeal supported the reasoning of the Court of Appeal and argued that the Act only applied to a disclaimer which operated after a breach of duty occurred. [Counsel for Bush] in this appeal adopted [the former's] argument.

My Lords, with all respect to the judges of the Court of Appeal, I think that they have overlooked the importance of section 13(1). The words 'liability for negligence' in section 2(2) must be read together with section 13(1) which states that the former section prevents the exclusion of liability by notices 'which exclude or restrict the relevant obligation or duty'. These words are unambiguous and are entirely appropriate to cover a disclaimer which prevents a duty coming into existence. It follows that the disclaimers here given are subject to the provisions of the Act and will therefore only be effective if they satisfy the requirement of reasonableness.

Indeed the House of Lords was unanimous in its view that a disclaimer fell within the scope of the 1977 Act. This is obviously a decision very favourable to the interests of house purchasers and, as such, may be justifiable on policy grounds. However, the decision, in its very broad interpretation of s.13, does make it difficult to see how any professional can define in advance the scope of his obligation if that definition in any way is, as is likely, going to make the enforcement of that liability more onerous without incurring the wrath of the 1977 Act.

As the speeches extracted above make clear, the effect of the decision in *Smith* that disclaimers are covered by the provisions of the 1977 Act means that they can only be upheld insofar as they are reasonable s.2(2); s.11(1); (see p. 94, above). How then does the general definition of reasonableness work out in this particular context?

SMITH v. ERIC S. BUSH;
HARRIS v. WYRE FOREST DISTRICT COUNCIL

[1990] 1 AC 831   House of Lords

For facts, see p. 247 above.

LORD TEMPLEMAN (at p. 849):

Both the present appeals involve typical house purchases. In considering whether the exclusion clause may be relied upon in each case, the general pattern of house purchases and the extent of the work and liability accepted by the valuer must be borne in mind.

Each year one million houses may be bought and sold. Apart from exceptional cases the procedure is always the same. The vendor and the purchaser agree a price but the purchaser cannot enter into a contract unless and until a mortgagee, typically a building society, offers to advance the whole or part of the purchase price. A mortgage of 80 per cent or more of the purchase price is not unusual. Thus, if the vendor and the purchaser agree a price of £50,000 and the purchaser can find £10,000, the purchaser then applies to a building society for a loan of £40,000. The purchaser pays the building society a valuation fee and the building society instructs a valuer who is paid by the building society. If the valuer reports to the building society that the house is good security for £40,000, the building society offers to advance £40,000 and the purchaser contracts to purchase the house for £50,000. The purchaser, who is offered £40,000 on the security of the house, rightly assumes that a qualified valuer has valued the house at not less than £40,000.

At the date when the purchaser pays the valuation fee, the date when the valuation is made and at the date when the purchaser is offered an advance, the sale may never take place. The amount offered by way of advance may not be enough, the purchaser may change his mind, or the vendor may increase his price and sell elsewhere. For many reasons a sale may go off, and in that case, the purchaser has paid his valuation fee without result and must pay a second valuation fee when he finds another house and goes through the same procedure. The building society which is anxious to attract borrowers and the purchaser who has no money to waste on valuation fees, do not encourage or pay for detailed surveys. Moreover, the vendor may not be willing to suffer the inconvenience of a detailed survey on behalf of a purchaser who has not contracted to purchase and may exploit minor items of disrepair disclosed by a detailed survey in order to obtain a reduction in the price.

The valuer is and, in my opinion, must be a professional person, typically a chartered surveyor in general practice, who, by training and experience and exercising reasonable skill and care, will recognise defects and be able to assess value. The valuer will value the house after taking into consideration major defects which are, or ought to be obvious to him, in the course of a visual inspection of so much of the exterior and interior of the house as may be accessible to him without undue difficulty. This appears to be the position as agreed between experts in the decided cases which have been discussed in the course of the present appeal.

(and at p. 851):

The valuer will not be liable merely because his valuation may prove to be in excess of the amount which the purchaser might realise on a sale of the house. The valuer will only be liable if other qualified valuers, who cannot be expected to be harsh on their fellow professionals, consider that, taking into consideration the nature of the work for which the valuer is paid and the object of that work, nevertheless he has been guilty of an error which an average valuer, in the same circumstances, would not have made and as a result of that error, the house was worth materially less than the amount of the valuation upon which the mortgagee and the purchaser both relied. The valuer accepts the liability to the building society which can insist on the valuer accepting liability. The building society seeks to exclude the liability of the valuer to the purchaser who is not in a position to insist on anything. The duty of care which the valuer owes to the building society is exactly the same as the duty of care which he owes to the purchaser. The valuer is more willing to accept the liability to the building society than to the purchaser because it is the purchaser who is vulnerable. If the valuation is worthless the building society can still insist that the purchaser shall repay the advance and interest. So, in practice, the damages which the valuer may be called upon to pay to the building society and the chances of the valuer being expected to pay, are less than the corresponding liability to the purchaser. But this does not make it more reasonable for the valuer to be able to rely on an exclusion clause which is an example of a standard form exemption clause operating in favour of the supplier of services and against the individual consumer.

(and at p. 852):

The valuer is a professional man who offers his services for reward. He is paid for those services. The valuer knows that 90 per cent of purchasers in fact rely on a mortgage valuation and do not commission their own survey. There is great pressure on a purchaser to rely on the mortgage valuation. Many purchasers cannot afford a second valuation. If a purchaser obtains a second valuation the sale may go off and then both valuation fees will be wasted. Moreover, he knows that mortgages, such as building societies and the council, in the present case, are trustworthy and that they appoint careful and competent valuers and he trusts the professional man so appointed. Finally, the valuer knows full well that failure on his part to exercise reasonable skill and care may be disastrous to the purchaser. If, in reliance on a valuation, the purchaser contracts to buy for £50,000 a house valued and mortgaged for £40,000 but, in fact worth nothing and needing thousands more to be spent on it, the purchaser stands to lose his home and to remain in debt to the building society for up to £40,000.

(and at p. 853):

It is open to Parliament to provide that members of all professions or members of one profession providing services in the normal course of the exercise of their profession for reward shall be entitled to exclude or limit their liability for failure to exercise reasonable skill and care. In the absence of any such provision valuers are not, in my opinion, entitled to rely on a general exclusion of the common law duty of care owed to purchasers of houses by valuers to exercise reasonable skill and care in valuing houses for mortgage purposes.

In the Green Paper 'Conveyancing by Authorised Practitioners' see Cm. 572, the Government propose to allow building societies, banks and other authorised practitioners to provide conveyancing services to the public by employed professional lawyers. The Green Paper includes the following relevant passages:

'3.10 There will inevitably be claims of financial loss arising out of the provision of conveyancing services. A bad mistake can result in a purchaser acquiring a property which is worth considerably less than he paid for it – because, for example, the conveyancer overlooked a restriction on use or the planning of a new motorway. The practitioner will be required to have adequate professional indemnity insurance or other appropriate arrangements to meet such claims.'

Annex paragraph 12:

'An authorised practitioner must not contractually limit its liability for damage suffered by the client as a result of negligence on its part.'

The Government thus recognises the need to preserve the duty of a professional lawyer to exercise reasonable skill and care so that the purchaser of a house may not be disastrously affected by a defect of title or an encumbrance. In the same way, it seems to me there is need to preserve the duty of a professional valuer to exercise reasonable skill and care so that a purchaser of a house may not be disastrously affected by a defect in the structure of the house.

The public are exhorted to purchase their homes and cannot find houses to rent. A typical London suburban house, constructed in the 1930s for less than £1,000 is now bought for more than £150,000 with money largely borrowed at high rates of interest and repayable over a period of a quarter of a century. In these circumstances, it is not fair and reasonable for building societies and valuers to agree together to impose on purchasers the risk of loss arising as a result of incompetence or carelessness on the part of valuers. I agree with the speech of my noble and learned friend, Lord Griffiths, and with his warning that different considerations may apply where homes are not concerned.

LORD GRIFFITHS (at p. 854):

It is clear, then, that the burden is upon the surveyor to establish that in all the circumstances it is fair and reasonable that he should be allowed to rely upon his disclaimer of liability.

I believe that it is impossible to draw up an exhaustive list of the factors that must be taken into account when a judge is faced with this very difficult decision. Nevertheless, the following matters should, in my view, always be considered.

(1) Were the parties of equal bargaining power. If the court is dealing with a one-off situation between parties of equal bargaining power the requirement of reasonableness would be more easily discharged than in a case such as the present where the disclaimer is imposed upon the purchaser who has no effective power to object.

(2) In the case of advice would it have been reasonably practicable to obtain the advice from an alternative source taking into account considerations of costs and time. In the present case it is urged on behalf of the surveyor that it would have been easy for the purchaser to have obtained his own report on the condition of the house, to which the purchaser replies, that he would then be required to pay twice for the same advice and that people buying at the bottom end of the market, many of whom will be young first-time buyers, are likely to be under considerable financial pressure without the money to go paying twice for the same service.

(3) How difficult is the task being undertaken for which liability is being excluded. When a very difficult or dangerous undertaking is involved there may be a high risk of failure which would certainly be a pointer towards the reasonableness of excluding liability as a condition of doing the work. A valuation, on the other hand, should present no difficulty if the work is undertaken with reasonable skill and care. It is only defects which are observable by a careful visual examination that have to be taken into account and I cannot see that it places any unreasonable burden on the valuer to require him to accept responsibility for the fairly elementary degree of skill and care involved in observing, following-up and reporting on such defects. Surely it is work at the lower end of the surveyor's field of professional expertise.

(4) What are the practical consequences of the decision on the question of reasonableness. This must involve the sums of money potentially at stake and the ability of the parties to bear the loss involved, which, in its turn, raises the question of insurance. There was once a time when it was considered improper even to mention the possible existence of insurance cover in a lawsuit. But those days are long past. Everyone knows that all prudent, professional men carry insurance, and the availability and cost of insurance must be a relevant factor when considering which of two parties should be required to bear the risk of a loss. We are dealing in this case with a loss which will be limited to the value of a modest house and against which it can be expected that the surveyor will be insured. Bearing the loss will be unlikely to cause significant hardship if it has to be borne by the surveyor but it is, on the other hand, quite possible that it will be a financial catastrophe for the purchaser who may be left with a valueless house and no money to buy another. If the law in these circumstances denies the surveyor the right to exclude his liability, it may result in a few more claims but I do not think so poorly of the surveyor's profession as to believe that the floodgates will be opened. There may be some increase in surveyors' insurance premiums which will be passed on to the public, but I cannot think that it will be anything approaching the figures involved in the difference between the Abbey National's offer of a valuation without liability and a valuation with liability discussed in the speech of my noble and learned friend, Lord Templeman. The result of denying a surveyor, in the circumstances of this case, the right to exclude liability, will result in distributing the risk of his negligence among all house purchasers through an increase in his fees to cover insurance, rather than allowing the whole of the risk to fall upon the one unfortunate purchaser.

I would not, however, wish it to be thought that I would consider it unreasonable for professional men in all circumstances to seek to exclude or limit their liability for negligence. Sometimes breathtaking sums of money may turn on professional advice against which it would be impossible for the adviser to obtain adequate insurance cover and which would ruin him if he were to be held personally liable. In these circumstances it may indeed be reasonable to give the advice upon a basis of no liability or possibly of liability limited to the extent of the adviser's insurance cover.

In addition to the foregoing four factors, which will always have to be considered, there is in this case the additional feature that the surveyor is only employed in the first place because the purchaser wishes to buy the house and the purchaser in fact provides or contributes to the surveyor's fees. No one has argued that if the purchaser had employed and paid the surveyor himself, it would have been reasonable for the surveyor to exclude liability for negligence, and the present situation is not far removed from that of a direct contract between the surveyor and the purchaser. The evaluation of the foregoing matters leads me to the clear conclusion that it would not be fair and reasonable for the surveyor to be permitted to exclude liability in the circumstances of this case. I would therefore dismiss this appeal.

It must, however, be remembered that this is a decision in respect of a dwelling house of modest value in which it is widely recognised by surveyors that purchasers are in fact relying on their care and skill. It will obviously be of general application in broadly similar circumstances. But I expressly reserve my position in respect of valuations of quite different types of property for mortgage purposes, such as industrial property, large blocks of flats or very expensive houses. In such cases it may well be that the general expectation of the behaviour of the purchaser is quite different. With very large sums of money at stake prudence would seem to demand that the purchaser obtain his own structural survey to guide him in his purchase and, in such circumstances with very much larger sums of money at stake, it may be reasonable for the surveyors valuing on behalf of those who are providing the finance either to exclude or limit their liability to the purchaser.

These passages show clearly the considerations of policy that helped to formulate the view taken by the House of Lords. The relative inequality as between the typical prospective house purchaser, faced with standard contract terms and (in all probability) short of cash, and the defendant surveyor, holding himself out as a professional man expert in his work is too great for such broadly drafted disclaimers to be acceptable.

It is evident that the decision will be applicable across a wide range of ordinary domestic property transactions. Equally, however, the speeches make clear that the unreasonableness of disclaimers by valuers will not be universal. The commercial purchaser may well be in a different position because of his superior bargaining position. The 'lack of resources' argument is difficult to employ in favour of the purchaser of a £12 million mansion. Whether a distinction will be drawn between purchasers at the limit of their financial resources and those who are not, therefore ultimately able to afford a survey of their own, remains to be seen; the logic of Lord Griffiths' remark suggests that they may well be seen as being in a different position but the decision in *Beaumont* v. *Humberts* (p. 253, above) may suggest that a more general application of *Smith* is likely.

A word is in order about the impact of *Smith* in general. At every stage it has to be seen as a decision favouring the interest of the consumer, albeit not invariably. A duty is owed to third parties; disclaimers are within the Unfair Contract Terms Act; it is *prima facie* unreasonable for a professional to exclude liability – all these propositions stand out from the many other cases that tend to have, in recent years, made it harder for plaintiffs to succeed in premises litigation. It cannot be that consumer purchasers are invariably in a better position than commercial purchasers as far as the law of tort is concerned: *Murphy* v. *Brentwood* alone destroys that particular proposition. What it may represent, however, is an acknowledgement that a purchaser who makes appropriate use of the services of a surveyor will have been acting responsibly to avoid the acquisition of defective premises and has gone out of his way to secure advice he thinks he can rely on. As such, the purchaser may well be entitled to a greater degree of protection from his surveyor's negligence than from that of others involved in the actual construction process.

Damages

The concentration of a surveyor's work on the value of the property, rather than an involvement in the creation of it, has major implications as far as the measure of damages is concerned. It is inevitable that, in most cases, the award of damages will focus on the value of the property rather than the usual measure of costs of repair. These two figures may of course be the same but equally market forces may mean that they will vary substantially.

PHILIPS v. WARD

[1956] 1 WLR 471   Court of Appeal

The plaintiff purchased a property after a surveyor's report valued it at £25,000. The survey failed to spot damage to timbers caused by beetle and worm. The true value of the property was found to be £21,000, but repairs would cost £7000. The plaintiff was held to be entitled only to the difference in value.

MORRIS LJ (at p. 475):

In my judgment, the damages to be assessed were such as could fairly and reasonably be considered as resulting naturally from the failure of the defendant to report as he should have done. I do not find any error of approach on the part of the official referee. It is said, however, that he was not warranted in proceeding on the basis which in fact constituted the first of the two alternative bases put forward in the statement of claim by the plaintiff, namely, the basis of the difference between the value of the property as it was described in the defendant's report and its value as it should have been described. In my view, however, that was the correct basis on the facts of this case.

If the plaintiff had discovered the omissions or errors in the defendant's report before he had contracted to buy, it may be that he would not have proceeded with his projected purchase. If he had wished to proceed and had received an estimate of £7000 as being the cost of doing work of restoration, it may be that he would have offered £18,000 and would have urged that that was an appropriate figure to pay, having regard to the fact that he would have to spend £7000. But the finding in this case shows that the property was, at the time of the sale, actually worth £21,000. Accordingly, the owner would not have been likely to sell for £18,000 property which was worth £21,000. It might at first seem surprising that the reduction in value below £25,000 should be less than the cost of executing the necessary work. But, apart from the fact that this was the finding of fact based on the general evidence given at the hearing in regard to valuations, it is to be remembered that property in which there is extensive new work by way of repair or restoration may have an increased life and may acquire an enhanced value. In the present case, if the repair work costing £7000 had been done in 1952, it seems very likely that the roof would then have been put into a condition so that it would last longer than it would have lasted had it been an old roof, though an old roof not impaired by the ravages of the death-watch beetle. Similar considerations might apply in regard to the replacement of the main timbers of the house. Furthermore, the effecting of the repair work costing £7000 would most probably have had the result that the annual cost of upkeep of the property would be diminished. But the plaintiff must not be placed in a better position by the award of damages than he would have been in had the defendant given a proper report. If the plaintiff had received £7000 damages in 1952 on the basis of what it would then cost him to do the repair work, and instead of doing the work had sold the property for £21,000, which was its value, he would have profited to the extent of £3000. The present case differs from *Baxter* v. *Gapp & Co.* [1939] 2 KB 271, where a negligently given valuation caused the plaintiff to advance money on mortgage to an unsatisfactory borrower, with the result that the plaintiff as mortgagee was put to expense and loss. The amount of such expense and loss was in that case the measure of the damage sustained by the plaintiff. In the present case, the plaintiff, being mindful to purchase a property and seeking advice as to its condition, paid £25,000 for it, which would have been its value if it had been in the condition described by the defendant, whereas its real value in its actual condition as it should have been described by the defendant was £21,000. In those circumstances, it seems to me that the figure of £4000 represented the proper measure of the damage sustained by the plaintiff, and I see no error of law in the approach of the official referee.

This remained the mainstream approach and *Philips* was widely used as authority for the measure of damages in survey cases.

PERRY v. SIDNEY PHILLIPS & SON

[1982] 1 WLR 1297   Court of Appeal

The plaintiff purchased a country property for £27,000 in 1976 after a surveyor's report. The report omitted several obvious defects, and had over-valued the property. In 1981, the plaintiff sold the house for £43,000. The plaintiff was held to be entitled to the difference in value in 1976.

LORD DENNING MR (at p. 1301):

> We now have to consider how the damages are to be assessed. The cases show up many differences. I need only draw attention to these:
> First, where there is a contract to build a wall or a house, or to do repairs to it, then if the contractor does not do the work or does it badly, the employer is entitled, by way of damages, to recover the reasonable cost of doing such work as is reasonable to make good the breach. The cost is to be assessed at the time when it would be reasonable for the employer to do it, having regard to all the circumstances of the case, including therein delay due to a denial of liability by the contractor or the financial situation of the employer. The work may not have been done even up to the date of trial. If the cost has increased in the meantime since the breach – owing to inflation – then the increased cost is recoverable, but no interest is to be allowed for the intervening period (see *Radford* v. *De Froberville* [1977] 1 WLR 1262 and *William Cory & Son Ltd* v. *Wingate Investments (London Colney) Ltd* (1980) 17 BLR 104), likewise if a wrongdoer damages his neighbour's house by nuisance or negligence, and the neighbour is put to expense in the repairing of it: see *Dodd Properties (Kent) Ltd* v. *Canterbury City Council* [1980] 1 WLR 433.
> Second, where there is a contract by a prospective buyer with a surveyor under which the surveyor agrees to survey a house and make a report on it – and he makes it negligently – and the client buys the house on the faith of the report, then the damages are to be assessed at the time of the breach, according to the difference in price which the buyer would have given if the report had been carefully made from that which he in fact gave owing to the negligence of the surveyor. The surveyor gives no warranty that there are no defects other than those in his report. There is no question of specific performance. The

contract has already been performed, albeit negligently. The buyer is not entitled to remedy the defects and charge the cost to the surveyor. He is only entitled to damages for the breach of contract or for negligence. It was so decided by this court in *Philips* v. *Ward* [1956] 1 WLR 471, followed in *Simple Simon Catering Ltd* v. *Binstock Miller & Co.* (1973) 117 SJ 529.

The former case was concerned with breach of contract by surveyors. It is their duty to use reasonable care and skill in making a proper report on the house. In our present case Messrs. Sidney Phillips & Son failed in that duty in 1976 when they made the negligent report. Mr Perry acted on the report in 1976 when he bought the house in July 1976. The general rule of law is that you assess the damages at the date of the breach: so as to put the plaintiff in the same position as he would have been in if the contract had been properly performed. Even if the claim be laid in tort against the surveyor, the damages should be on the same basis.

So you have to take the difference in valuation. You have to take the difference between what a man would pay for the house in the condition in which it was reported to be and what he would pay if the report had been properly made showing the defects as they were. In other words, how much more did he pay for the house by reason of the negligent report than he would have paid had it been a good report? That being the position, the difference in valuation should be taken at the date of the breach in 1976.

OLIVER LJ (at p. 1304):

The position as I see it is simply this, that the plaintiff has been misled by a negligent survey report into paying more for the property than that property was actually worth. The position, as I see it, is exactly the same as that which arose in *Philips* v. *Ward* [1956] 1 WLR 471, to which Lord Denning MR has already referred, and in the subsequent case of *Ford* v. *White & Co.* [1964] 1 WLR 885. It is said by [counsel for Perry] that this proposition is supported in some way by a more recent case. *Dodd Properties (Kent) Ltd* v. *Canterbury City Council* [1980] 1 WLR 433. That was a case in which the plaintiffs were claiming damages in tort against the defendants, they having removed a support to the plaintiffs' premises. It could not be suggested in that case that there was any other measure of damages than the cost of repair, the only question being the date at which the repairs ought to have been carried out; and the debate there was as to the date at which it was reasonable for the plaintiffs to having carried out the repairs. As I read the case, it merely exemplifies the general principle which is set out in the headnote to the case:

'. . . the fundamental principle as to damages was that the measure of damages was that sum of money that would put the injured party in the same position as that in which he would have been if he had not sustained the injury . . .'

and the question was what loss the plaintiff, acting reasonably, had actually suffered.

I see nothing in that case which justifies the proposition for which [counsel for Perry] contends that damages are to be assessed on the basis of some hypothetical value at the date of the trial because the plaintiff has chosen – as he did in this case – to retain the property and not to cut his loss by reselling it. I therefore am of the same view as Lord Denning MR that the right measure of damage is the measure suggested in both *Philips* v. *Ward* [1956] 1 WLR 471 and *Ford* v. *White & Co.* [1964] 1 WLR 885, which is simply the difference between what the plaintiff paid for the property and its value at the date when he obtained it.

Obviously the cost of repair and the diminution in value will often be connected. If my house needs £5000 of work to get it back into condition, its value is likely to fall by at least that amount, but other factors may depress the value further e.g. a buyer's general concern about lack of maintenance or, on the other hand, a buyer's acceptance of certain risks.

BOLTON v. PULEY

(1982) 267 EG 1160 High Court

The plaintiff here was suing his surveyor for his failure to have property reported on a boundary wall which subsequently collapsed. Repair work was calculated to cost £5900, but the plaintiff was held entitled to claim only £4425 under this head.

TALBOT J (at p. 1166):

It is clear, therefore, that the correct measure of the plaintiff's loss on these authorities is the difference between the price he paid and the price he would have paid if the defendant had given a proper report. I shall shortly be turning to the evidence on this point.

But before I come to the evidence on that point, it is necessary that I should consider the submission made by [counsel for Bolton] that I should not disregard the cost of repairs to the wall. The question was, how does one arrive at the diminution in value? He submitted that the way to do that was through a consideration of the cost of repairs. He drew my attention to a number of authorities. The first was *Wigsell* v. *The Corporation of the School for the Indigent Blind* (1882) 8 QBD 357. In that case the grantees of land had covenanted to build a wall. They did not do so and an action for damages was brought against them for breach of covenant. It was held by the Divisional Court that the true measure of damages was the pecuniary amount of the difference between the position of the plaintiff upon the breach of covenant and what it would have been if the contract had been performed. The court held also that the cost of building the wall was not the correct measure of damages. However, on similar facts, Oliver J, in *Radford* v. *De Froberville*

[1977] 1 WLR 1262, held that the correct measure of damages was the cost to the plaintiff of erecting a wall on his own land and not the amount by which the plaintiff's land as an investment property was diminished by the absence of the wall: he applied *Tito* v. *Waddell* [1977] Ch 106. In *Grove* v. *Jackman & Masters* (1950) 155 EG 182, in which the plaintiff claimed damages against the defendant surveyors for negligence in a survey of property which the plaintiff subsequently bought, Lord Goddard CJ held that the measure of damages was, *inter alia*, the cost of repairs. [Counsel for Bolton] cited another case on similar facts, *Leigh* v. *William Hill & Son* (1956) 168 EG 396, but I do not derive much assistance from that case. The Court of Appeal refused to interfere with a finding that the measure of damages for a surveyor's negligence was the cost of repairs, it would seem on the ground that no evidence had been given to the presiding judge at the Liverpool Court of Passage as to the difference in the value of the house with, or without, the defect complained of. It was in those circumstances that the Court of Appeal refused to allow the question of damages to be reviewed because of the decision of the Court of Appeal in *Philips* v. *Ward*.

The case which most supported contentions [by counsel for Bolton] that the cost of repairs must be considered is *Freeman* v. *Marshall & Co.* (1966) 200 EG 777. That was a case on similar facts, namely that a surveyor had been negligent in advising the plaintiff on the purchase of property. The report on the question of damages reads:

'So far as damages were concerned in the present class of case, all he (the judge) had to do was to read the headnote of a case in the Court of Appeal, *Philips* v. *Ward* (1956). That case said that the measure of damages was the difference between the fair value of the property if it had been described in the defendant's report and its value in its actual condition. On that basis it followed that the real value of the property was £13,500 less the cost of the repairs. The cost of the repairs was £550 and that was the sum he would award plus £147 for loss of rent.'

The facts, as reported, do not indicate that there was any particular reason that caused Lawton J to reach the real value of the property by deducting the cost of repairs.

(and at p. 1169):

I have formed the conclusion that a reasonable approach to a problem such as this is to accept the point of view that a purchaser, though he would seek a reduction in the price asked, would nevertheless be prepared not to insist upon a full reduction on the basis of the proposed cost of repairs, but would be prepared to negotiate about the matter. I do not think, therefore, that the proper approach is to reach the diminished value by merely deducting the cost of repairs as they would have been at the time of the purchase. I have, therefore, to try to relate that view to the facts of this case, having regard to the plaintiff and all that I know about him, particularly at the time of the purchase, and on the assumption (though this must be a matter of assumption because I have not heard the evidence of the vendors, a Mr and Mrs Pickles) that they, the vendors, would have been prepared to reduce the asking price in order to take some account of the expenditure that would be necessary on the boundary wall. Bearing in mind the price paid by the plaintiff and the cost of repairs which I have assessed and relating the one to the other, I would have thought that this is a case where the plaintiff might have consented to take the risk of 25 per cent of the cost of repairs and have then asked for, and obtained, a reduction of the purchase price by the remainder.

Therefore, taking the cost of £5900, making an allowance for the 25 per cent which I believe the plaintiff, in all probability, would have accepted, the reduced figure is £4425. I follow the principle (1) that in cases like this the measure of damages is the difference between the fair value of the property in the condition accepted by the plaintiff in reliance on the defendant's report and its value in its actual condition: I follow the principle (2) that you do not, in a straightforward case, take as the measure the cost of repairs found necessary to put the property in the represented condition. But, in order to arrive at the actual value in this case, i.e. with the wall in need of repair, I feel that I am bound to have regard to the cost of repair and then make an allowance of 25 per cent. It seems to me that to do so reflects the views expressed by Morris LJ and Romer LJ in *Philips* v. *Ward*. It makes an allowance for the fact that, on the assumption the plaintiff has the wall put in a sound condition, he is getting, as the result of repair, something better than he would have had if the wall had been an old wall in a stable condition. His liabilities for future repair would be less as the wall would have an 'increased life'. Put another way, if the plaintiff got as damages the full value of the repair work, he would be receiving from the defendant a betterment value for which the defendant should not be liable.

I therefore judge the defendant's liability to be in the sum of £4425.

Recently there has been a greater trend away from the traditional 'difference in value' approach as new fact situations have arisen which necessitate its modification.

HOWARD v. HORNE & SONS

[1989] 5 PN 136   High Court (Official Referee's Business)

The plaintiffs purchased a house for £200,250 after a survey report which indicated certain defects, which enabled the plaintiffs to bargain down the initial asking price, but which failed to disclose serious defects in the electrical wiring, which later cost £6500 to repair.

JUDGE BOWSHER QC (at p. 137):

[Counsel] on behalf of the defendants argued that the plaintiffs were entitled to no damages because they were not misled by the Report into buying above the market value. The factual basis for this submission was that it was agreed between experts that the defendants' valuation was about right for the property in the state which he described, disregarding fixtures and fittings and there was contained in the price paid by the plaintiffs an element attributable to fixtures and fittings valued by the first plaintiff at about £5000 to £7000. The price paid by the plaintiffs in the end was £200,250 inclusive of fixtures and fittings, and, submits counsel for the defendants, if the fixtures and fittings element is deducted, the price paid is less than valuation given in the negligent report, therefore the plaintiffs were not misled and suffered no damage. I agree with that submission as to the facts but not as to the conclusion.

For the plaintiffs, on the other hand, it was submitted that they did suffer damage in that they lost the opportunity to rely on the defective wiring in bargaining about the price.

The plaintiffs did use what was in the Report to bargain the price down from £202,500 to £200,250, a reduction of £2250. That reduction was achieved on the basis of the 12 recommendations in the report. Mr Howard admits that he knows nothing about property values, but I am sure that he used his skill as a trader to get the best reduction he could.

It is not clear what would have been the cost of complying with all the 12 recommendations but it would certainly have been substantially more than the reduction of £2250. Mr Howard had a budget of £10,000 for improvements to the property, but these included matters outside the 12 recommendations and on the other hand he did not intend to put all the 12 recommendations in hand at once.

If the report had made reference to the true state of the electric cabling, I have no doubt that Mr Howard would have used that information to secure a further reduction in the price. In the absence of evidence from the vendors, it is difficult to ascertain what reduction is likely to have been obtained.

It is a considerable tribute to Mr Howard's powers as a negotiator that in a strongly rising market he was able at the end of June, 1986 after receiving the defendants' Report to negotiate a reduction of a price offered in February, 1986. He demonstrated that ability again in July, 1987 when he sold the house privately at a price of £275,000 inclusive of curtains and carpets at a time when, according to the expert called for the defendants, the market value of the house was £235,000 exclusive of fixtures and fittings.

In the circumstances, I think it likely that Mr Howard would have negotiated a further reduction in the price he paid for the house if he had been told the true state of the wiring by the defendants.

I also take the view that the plaintiffs are entitled to damages to compensate them for loss of the opportunity to negotiate that reduction. I reject submissions [by counsel for Horne and Sons] as to the effect of *Perry* v. *Sidney Phillips and Son*. Oliver and Kerr LJJ stated the measure of damages where the plaintiff was persuaded by negligent advice to buy property at more than its true market value. They did not say that it was only in those circumstances that any damages at all should be payable to a purchaser who acted in reliance on a negligent surveyor's report.

Usually, where property has been bought in reliance on negligent advice from a surveyor, the damages will be calculated by reference to the diminution in value, but that is not an invariable rule. If it be proved that the plaintiff, if properly advised, would have been able to buy at a price below the market value (or in this case, even further below the market value) then he should be compensated for the loss of that opportunity.

(and at p. 138):

> It is now necessary to consider what is the likely reduction that the plaintiffs would have persuaded the vendors to make if the plaintiffs had been told of the true state of the electric wiring.
> The plaintiffs' expert, said that if the plaintiffs had been fully advised of the facts with regard to the wiring, he would expect that they could have negotiated a reduction in price of £2500 to £3000. On the other hand, the defendants' expert, who lives locally in Bourne End and has his office at High Wycombe and is clearly in touch with the local market, said that a prudent purchaser if he knew of the defects would have tried to negotiate a reduction in price, but a prudent vendor in those market conditions would have rejected the attempt. [The defendants' partner] himself said, that he would have expected a renegotiation to have achieved a reduction of £1500. I regard that statement as somewhat generous towards the plaintiffs, but since it was made by the partner of the defendants most concerned in this matter, I think that I should accept it.
> On that evidence, I find that on the balance of probabilities the plaintiffs, if properly advised, would have bought the property for £1500 less than they in fact paid and they are entitled to recover damages in that amount together with interest.

This comes close to giving damages for the loss of a chance. The plaintiffs lost the chance to renegotiate the price further; however, from the evidence given it was possible for the judge to assess precisely what loss the plaintiffs had in fact incurred.

STEWARD v. RAPLEY

[1989] 15 EG 198   Court of Appeal

The plaintiff purchased a property for £58,500 which had been valued at £60,000 in a survey which had carelessly failed to detect dry rot. An initial investigation revealed that £6000 worth of work was needed and the property was then worth £50,000 but with a real risk that further rot might be found. Subsequently this transpired to be the case, with repair costs reaching £26,000 and the value of the property prior to that worth £33,200. Damages of £25,300 under this head were awarded.

STAUGHTON LJ (at p. 202):

> With respect to the learned judge, he seems to have overlooked the distinction between the pure subtraction of the cost of reinstatement, i.e. recovering it as a head of damages, from it being perhaps the most reliable, if not the only, method of arriving at a reliable market valuation in the admittedly artificial conditions where the plaintiff in fact would never have bought the house in the first place; and, second, how does one establish a market against which to assess the true value of the house in its defective condition? The distinction between what a surveyor might or might not have said at that time is not relevant as an isolated criterion in reaching that figure. It cannot be taken out of the context of what a reasonably prudent purchaser would have done when receiving the advice from the surveyor.
>
> In this case, notwithstanding the attractive way the matter was put, it is not possible just to look at the figure of £50,000, totally dissociated from the caveat – be careful, because no one can tell how far dry rot has extended until it is actually properly investigated.
>
> [Counsel for Steward] submitted, in my judgment rightly, that the only sort of purchaser who would go ahead and pay the recommended £50,000 on the basis that there was no further undetected invasion of dry rot, would be a speculative builder or somebody who was gambling against the true condition of the house. The basic assessment figure, on all the authorities, is what is the real value of the house tested against a careful inspection and followed by prudent further investigation. It is the necessity for prudent further investigation, even as a hypothetical exercise, which has been overlooked by the learned judge in his judgment that demonstrates the flaw in the argument as it has been put before us by [counsel for Rapley]. At that point of the judgment I agree with the submissions of [counsel for Steward] that it does not fairly record the effect of the authority upon which the judge is relying.

Here the plaintiffs do appear to be gaining the benefit of hindsight. The market value at the date of purchase is difficult to assess because at that time the true extent of the rot was not known. So any figure given at that time would be purely speculative, and potentially highly unreliable. It is to avoid that danger that the later events are taken into account including the actual costs of necessary repairs, giving an accurate figure for the difference in value when it is finally known, albeit that the calculation at a later date incurs the risk for the surveyor that the award of damages will be at a much higher level. This approach would appear to have been endorsed by Lord Templeman in *Smith* v. *Eric S. Bush* and followed by Scott Baker J in *Hipkins* v. *Jack Cotton Partnership* [1989] 45 EG 163.

SYRETT v. CARR AND NEAVE

[1990] 48 EG 118 High Court (Official Referee's Business)

The plaintiff purchased a property for £300,000. A limited survey by the defendants failed to detect beetle infestation and damp. Repairs were estimated to cost £78,000 at 1990 prices but could not be executed at that time due to the plaintiff's financial position. The cost of repair was greater than the diminution in value, but damages were awarded based on the £78,000 cost of repair.

JUDGE BOWSHER QC (at p. 125):

> Following the guideline set out by Megaw LJ in *Dodd Properties (Kent) Ltd* v. *Canterbury City Council* [1980] 1 WLR 433, it seems to me that the appropriate measure of damages to put the plaintiff in the same position as she would have been in if she had not sustained the wrong is the cost of repairs plus the costs ancillary to those repairs plus compensation for the blight on her happiness since October 1988. If she had received a careful rather than a negligent report, she would not have bought this property and she would have bought a property which would have been worth what she paid for it and would not have required large expenditure in repairs and would have given her much enjoyment already.

(and at p. 126, speaking of *Perry* v. *Sidney Phillips*):

> In the same case at p. 1302, Lord Denning MR said:
>
> > 'The general rule of law is that you assess the damages at the date of the breach so as to put the plaintiff in the same position as he would have been in it if the contract had been properly performed. Even if the claim be laid in tort against the surveyor, the damages should be on the same basis.. . .
>
> If the plaintiff is unaware of the breach until two years after the breach took place there may be many reasons for not applying that general rule and, in my judgment, there are cogent reasons for not applying that general rule in this case. Those cogent reasons are that the plaintiff did not have an opportunity of cutting her losses at or about the date of purchase and that she has acted reasonably throughout and that all the expenses which I find she has suffered or is likely to suffer are natural and probable results of the negligent survey. The plaintiff did have an opportunity of cutting her losses two years after purchase of the property in October 1988, but in her circumstances at the time it seems to me to have been reasonable for her to decide not to do so, and in any event there is no evidence before me as to what would have been the financial results of her selling at that time.
>
> Accordingly, to return to the words of Megaw LJ in *Dodd Properties*, I do

not feel that I am 'obliged by some binding authority to arrive at a result which is inconsistent with the fundamental principle.'

I therefore hold that the plaintiff is entitled to damages comprising the cost of repairs together with the ancillary costs associated with those repairs and damages for distress and inconvenience together with return of the fee paid for the report.

The cost of repairs should be assessed as at the date when they are likely to be done, the plaintiff having acted reasonably in delaying the repairs.

Once again the particular facts, especially the known character of the faults and repair work necessary, may justify a departure from the diminution in value principle and, as in *Steward*, damages are awarded with regard to the repair costs. The result is fair to the purchaser, but perhaps over-fair; after all the only result of the negligence of the surveyor is that the purchaser has bought the property, rather than deciding not to do, and this points to the loss of value as the proper measure. If I spend £100,000 on a house now valued at £75,000, I can sell the house and, with an award of £25,000 – the diminution in value – be placed in the position I was in before the negligence; it is not obvious that a repair bill, if higher than £25,000, is of relevance.

Doubt has indeed now been cast on at least this last case and perhaps on others too in the light of an important recent decision of the Court of Appeal.

WATTS v. MORROW

[1991] 4 All ER 937   Court of Appeal

The plaintiffs purchased a large country house after receiving a survey report from the defendant. Subsequently, defects were discovered which cost almost £34,000 to repair; however, evidence suggested that the property had declined in value by only £15,000.

RALPH GIBSON LJ (at p. 950):

The task of the court is to award to the plaintiffs that sum of money which will, so far as possible, put the plaintiff into as good a position as if the contract for the survey had been properly fulfilled: see Denning LJ in *Philips* v. *Ward* [1956] 1 All ER 874 at 875, [1956] 1 WLR 471 at 473. It is important to note that the contract in this case, as in *Philips* v. *Ward*, was the usual contract for the survey of a house for occupation with no special terms beyond the undertaking of the surveyor to use proper care and skill in reporting upon the condition of the house.

The decision in *Philips* v. *Ward* was based upon that principle; in particular, if the contract had been properly performed the plaintiff either would not have bought, in which case he would have avoided any loss, or, after negotiation, he would have paid the reduced price. In the absence of evidence to show that any other or additional recoverable benefit would have been obtained as a result of proper performance, the price will be taken to have been reduced to the market price of the house in its true condition because it cannot be assumed that the vendor would have taken less.

The cost of doing repairs to put right defects negligently not reported may be relevant to the proof of the market price of the house in its true condition: see *Steward* v. *Rapley* [1989] 1 EGLR 159; and the cost of doing repairs and the diminution in value may be shown to be the same. If, however, the cost of repairs would exceed the diminution in value, then the ruling in *Philips* v. *Ward*, where it is applicable, prohibits recovery of the excess because it would give to the plaintiff more than his loss. It would put the plaintiff in the position of recovering damages for breach of a warranty that the condition of the house was correctly described by the surveyor and, in the ordinary case, as here, no such warranty has been given.

It is clear, and it was not argued to the contrary, that the ruling in *Philips* v. *Ward* may be applicable to the case where the buyer has, after purchase, extricated himself from the transaction by selling the property. In the absence of any point on mitigation, the buyer will recover the diminution in value together with costs and expenses thrown away in moving in and out of resale: see Romer LJ in *Philips* v. *Ward* [1956] 1 All ER 874 at 879, [1956] 1 WLR 471 at 478. I will not here try to state the nature or extent of any additional recoverable items of damage.

The damages recoverable where the plaintiff extricates himself from the transaction by resale are not necessarily limited to the diminution in value plus expenses. The consequences of the negligent advice and of the plaintiff entering into the transaction into which he would not have entered if properly advised may be such that the diminution in value rule is not applicable. An example is *Country Personnel (Employment Agency) Ltd* v. *Alan R. Pulver & Co. (a firm)* [1987] 1 All ER 289, [1987] 1 WLR 916, a case of solicitors' negligence, where the plaintiff recovered the capital losses caused by entering into the transaction.

It is also clear, and again there was no argument for the plaintiffs to the contrary, that, if the plaintiff would have bought the house anyway, if correctly advised, the ruling in *Philips* v. *Ward* is applicable: the fact that after purchase he discovers that the unreported defects will cost more than the diminution in value does not entitle him to recover the excess. That is, again, because, if the contract had been performed properly, he would have negotiated and, absent proof of a different outcome, would have done no better than reduction to the market value in true condition.

It was rightly acknowledged for the plaintiffs that proof that the plaintiff, properly advised, would not have bought the property does not by itself cause the diminution in value rule to be inapplicable. It was contended, however, that it becomes inapplicable if it is also proved that it is reasonable for the plaintiff to retain the property and to do the repairs. I cannot accept that submission for the following reasons.

(i) The fact that it is reasonable for the plaintiff to retain the property and to do the repairs seems to me to be irrelevant to determination of the question whether recovery of the cost of repairs is justified in order to put the plaintiff in the position in which he would have been if the contract, i.e. the promise to make a careful report, had been performed. The position is no different from that in *Philips* v. *Ward*: either the plaintiff would have refused to buy or he would have negotiated a reduced price. Recovery of the cost of repairs after having gone into possession, that is to say in effect the acquisition of the house at the price paid less the cost of repairs at the later date of doing those repairs, is not a position into which the plaintiff could have been put as a result of proper performance of the contract. Nor is that cost recoverable as damages for breach of any promise by the defendant because, as stated above, there was no promise that the plaintiff would not incur any such cost.

(ii) In the context of the contract proved in this case, I have difficulty in seeing when or by reference to what principle it would not be reasonable for the purchaser of a house to retain it and to do the repairs. He is free to do as he pleases. He can owe no duty to the surveyor to take any cheaper course. The measure of damages should depend, and in my view does depend, upon proof of the sum needed to put the plaintiff in the position in which he would have been if the contract was properly performed, and a reasonable decision by him to remain in the house and to repair it, upon discovery of the defects, cannot alter that primary sum, which remains the amount by which he was caused to pay more than the value of the house in its condition.

(iii) If the rule were as contended for by the plaintiffs, what limit, if any, could be put on the nature and extent of the repairs of which the plaintiff could recover the cost? [Counsel for Watts] asserted that the cost of repairs awarded in this case was no more than putting the house in the condition in which, on reading the report, they believed the house to be. That, however, contains no relevant standard of reasonableness because, again, the defendant did not warrant that description to be true. To argue that to award damages on that basis is not to enforce a warranty never given but merely to 'reflect the losses which the plaintiffs have incurred' seems to me to be a circular statement.

(iv) I have considered whether the reasonableness of the amount which a plaintiff might recover towards the cost of repairing unreported defects in excess of the diminution in value might be determined by reference to the amount which the plaintiff could recover if he sold the property, i.e. the diminution in value plus any other recoverable losses and expenses. Such a limit was not contended for by [counsel for Watts]. It has the apparent attraction of enabling a plaintiff who chooses to retain the property to recover as much as he would recover if he chose to sell it. It seems to me, however, to be impossible to hold that such is the law in the case of such a contract as was made in this case. The plaintiff must, I think, prove that the loss which he claims to have suffered was caused by the breach of duty proved and he cannot do that by proving what his loss would have been in circumstances which have not happened.

So *Philips* v. *Ward* is reasserted; *Syrett* is disapproved while *Steward*, though a Court of Appeal case, and *Howard*, only remain good law if their facts are thought to be sufficiently unusual to distinguish them from the main-stream of *Philips*, as confirmed by *Watts*.

Another aspect of damages has recently fallen to be considered by the House of Lords, viz whether a lender who lends money on the basis of a careless valuation can recover interest which the borrower was due to have paid from the borrower.

SWINGCASTLE LTD v. GIBSON

[1991] 2 AC 223   House of Lords

The plaintiffs loaned money to a Mr and Mrs Clarke on the strength of a valuation by the defendant. The owners fell into arrears, which attracted a savage interest rate. Finally the plaintiffs took possession of the house and sold it at £6000 less than the valuation.

LORD LOWRY (at p. 238):

> My Lords, it is clear that the lenders ought to have presented their claim on the basis that, if the valuer had advised properly, they would not have lent the money. Where they went wrong was to claim, not only correctly that they had to spend all the money which they did, but incorrectly that the valuer by his negligence deprived them of the interest which they would have received from the borrowers if the borrowers had paid up. The security for the loan was the property but the lenders did not have a further security consisting of a guarantee by the valuer that the borrowers would pay everything, or indeed anything, that was due from them to the lenders at the date, whenever it occurred, on which the loan transaction terminated. The fallacy of the lenders' case is that they have been trying to obtain from the valuer compensation for the borrowers' failure and not the proper damages for the valuer's negligence.

This case again focuses attention on the precise purpose of the work of a surveyor in valuing, and seeks to reflect this in the award of damages.

A particular aspect of the award of damages in surveying cases which has become prominent in recent years is the frequency with which damages for distress and inconvenience are awarded.

PERRY v. SIDNEY PHILLIPS & SONS

[1982] 1 WLR 1297   Court of Appeal

For facts see p. 289 above.

For facts see p. 289 above.

LORD DENNING MR (at p. 1302):

> The second point is as to the distress, worry, inconvenience and all the trouble
> to which Mr Perry was put during the time when he was in the house. [Counsel
> for Sidney Phillips and Sons] sought to say before us that damages ought not
> to be recoverable under this head at all. He referred to the *The Liesbosch* [1933]
> AC 499. In that case Lord Wright said at p. 460 that the loss due to the
> impecuniosity of the plaintiffs was not recoverable. I think that that statement
> must be restricted to the facts of *The Liesbosch*. It is not of general application.
> It is analysed and commented upon in this court in *Dodd Properties (Kent) Ltd
> v. Canterbury City Council* [1980] 1 WLR 433. It is not applicable here. It
> seems to me that Mr Perry is entitled to damages for all the vexation, distress
> and worry which he has been caused by reason of the negligence of the
> surveyor. If a man buys a house – for his own occupation – on the surveyor's
> advice that it is sound – and then finds out that it is in a deplorable condition,
> it is reasonably foreseeable that he will be most upset. He may, as here, not
> have the money to repair it and this will upset him all the more. That too is
> reasonably foreseeable. All this anxiety, worry and distress may nowadays be
> the subject of compensation. Not excessive, but modest compensation.

> Damages under this heading can now be quite substantial. The plaintiff
> in *Syrett* (which admittedly was quite a serious case of its kind) received
> £8000 under this heading which was described by the judge as 'not
> excessive but modest'. However, such an award is by no means invariable.

HAYES v. DODD

[1989] 5 PN 64   Court of Appeal

This was an action against a solicitor who failed to advise the plaintiffs
while they were purchasing premises to be used as a motor workshop that
there were problems concerning the rights of access to the workshop.
Inter alia, damages were awarded for the mental distress suffered by the
plaintiffs at first instance, but this was rejected by the Court of Appeal.

STAUGHTON LJ (at p. 68):

> Hirst J awarded £1,500 to each of the plaintiffs under this head. There can be
> no doubt, and it was accepted in this court that each of them suffered vexation

and anguish over the years to a serious extent, for which the sum awarded was but modest compensation. There is, however, an important question of principle involved.

For my part I would have wished for a rather more elaborate argument than we received on this point, before deciding it, since the law seems to be in some doubt. But I would be most reluctant to impose on Mr and Mrs Hayes, on top of their other misfortunes, two or three days of scholarly argument as to whether and in what circumstances damages can be awarded for mental distress consequent upon breach of contract in a business transaction, possibly at their expense, when the sum involved is only £3000 or roughly three per cent of their total claim. The difficulty is that amost any other case where a plaintiff claims to have suffered mental distress would present a similar problem, that the individual plaintiff ought not to be expected to bear the burden and perhaps also the cost of an elaborate argument. If, as I think, the law needs clarification, it is to be hoped that a case can be found where that will be provided by the House of Lords. Or it may be that the Law Commission can supply it.

Like the Judge, I consider that the English courts should be wary of adopting what he called 'the United States' practice of huge awards'. Damages awarded for negligence or want of skill, whether against professional men or anyone else, must provide fair compensation, but no more than that. And I would not view with enthusiasm the prospect that every shipowner in the Commercial Court, having successfully claimed for unpaid freight or demurrage, would be able to add a claim for mental distress suffered while he was waiting for his money.

In a sense, the wrong done to Mr and Mrs Hayes in this action, for which they seek compensation under this head, lay in the defendant's failure to admit liability at an early state. On 6 July 1983 the defendants acknowledged that there was no right of way, but denied negligence. Had they on that very day admitted liability and tendered a sum on account of damages, or offered interim reparation of some other form, the anxiety of Mr and Mrs Hayes, and their financial problems, could have been very largely relieved. But liability was not admitted until January 1987. I believe that in one or more American States damages are awarded for wrongfully defending an action. But there is no such remedy in this country so far as I am aware.

In *Perry's* case, damages were awarded for the distress, worry, inconvenience and trouble which the plaintiff had suffered while living in the house he bought, owing to the defects which his surveyor had overlooked. Lord Denning, MR at p. 1302 considered that these consequences were reasonably foreseeable. Kerr LJ stated a narrower test, at p. 1307:

'So far as the question of damages for vexation and inconvenience is concerned, it should be noted that the judge was awarded these not for the tension or frustration of a person who is involved in a legal dispute in which the other party refuses to meet its liabilities. If he had done so, it would have been wrong, because such aggravation is experienced by almost all litigants. He has awarded these damages because of the physical consequences of the breach which were all foreseeable at the time. The fact that in such cases damages under this head may be recoverable – if they have been suffered but not otherwise – is supported by the decision of this court in *Hutchinson* v. *Harris* 10 B LR 19.'

I would emphasise the reference to physical consequences of the breach.

I am not convinced that it is enough to ask whether mental distress was reasonably foreseeable as a consequence, or even whether it should reasonably have been contemplated as not unlikely to result from a breach of contract. It seems to me that damages for mental distress in contract are, as a matter of policy, limited to certain classes of case. I would broadly follow the classification provided by Dillon LJ in *Bliss* v. *South-East Thames Regional Health Authority* [1987] ICR 700 at p. 718: 'Where the contract which has been broken was itself a contract to provide peace of mind or freedom from distress'.

It may be that the class is somewhat wider than that. But it should not, in my judgment, include any case where the object of the contract was not comfort or pleasure, or the relief of discomfort, but simply carrying on a commercial activity with a view to profit. So I would disallow the item of damages for anguish and vexation.

This case serves as a reminder that a surveyor in a commercial transaction, on analogy with *Hayes*, would not be liable for mental distress suffered by the parties consequent upon his negligence, because such transactions are not designed to give comfort or pleasure. Equally in a domestic property case, such matters are part of what the plaintiff is seeking and surveyors will continue to be liable for mental distress.

Watts v. *Morrow* stands as important authority here too. The Court of Appeal noted that the plaintiffs' claim under this heading only extended to distress caused by physical consequences of the breach of contract and not more general claims for peace of mind. Accordingly, the trial judge's award of £8000 under this head was reduced to just £750.

In this area, as in others, the general principles of mitigation of damages apply. One case combines this feature with the claim unusually being by the mortgagee not the mortgagor.

LONDON & SOUTH OF ENGLAND BUILDING SOCIETY v. STONE

[1983] 1 WLR 1242   Court of Appeal

The plaintiff instructed the defendant to carry out a valuation survey in respect of a property on security of which a mortgage loan of £11,880 was to be advanced to borrowers. Later the property was found to be very seriously affected by subsidence and major repairs costing £29,000 were necessary. The plaintiff elected not to exercise their right to recover these sums from the borrowers, who in any event were financially stretched.

O'CONNOR LJ (at p. 1256):

> The actual loss to the lenders was £29,000. Russell J has held that it was unreasonable to spend so much money on repairing the house. It is not suggested that the house could have been repaired for less than £11,880. What then is it suggested that the lenders should have done? The judge did not ask himself this question, and as a result did not answer it. Something had to be done, for the evidence was that the house was about to fall down. The borrowers could not afford to put the house into repair; what then should the lenders have done? Should they have called in the loan for breach of contract and repossessed the property? That would have been a pointless exercise, as the house was worthless and indeed a liability, for it either had to be repaired or pulled down and the neighbouring premises shored up.
>
> The truth is that however one looks at this case the lenders have lost the whole of their advance at the very least. This loss has been caused by the negligence of the valuer, for it is quite certain that but for the negligent valuation the lenders would not have embarked on the transaction at all; they would have lent nothing. That loss has not been diminished by the small repayment of capital.
>
> I can see no justification for the suggestion that the lenders were under any duty to the valuer to mitigate this loss by trying to extract money from the borrowers. Let me test the proposition in a simple way: assume that the borrower had agreed to contribute £5000 towards these repairs, could the valuer have claimed credit for this or any part of this sum on the facts of this case? I am quite clear that he could not have done so. I would allow this appeal and enter judgment for the lenders in the full sum of £11,880.
>
> It will be obvious that in my judgment the valuer's cross-appeal asking that more than £3000 should be deducted from the £11,880 must be dismissed.

This decision appears fair and reflects the reality of the building society's position in a typical case.

The surveyor and the building team

At the start of this chapter, reference was made to the separate functions of surveyors and valuers on the one hand and the other groups of potential defendants in construction cases. This is not to say, though, that their various liabilities are entirely unconnected, since changes in one area of law have a ripple effect on others. This area now provides a good example.

The retreat from *Anns*, culminating in its overruling in 1990 has, naturally, reduced the scope of actions framed in negligence against the builder and the local authority in particular, and part of the attack on *Anns* and its successors has been based on the notion that it is wrong for the law to create what in effect has seemed to be a warranty by the builder of the quality of his work. The removal of this 'warranty' has had the effect of focusing more attention, and more litigation, on any other warranties as to the quality of the building that may exist. The work of the surveyor in reviewing the structure of a property and in ascribing a value to it has the effect of producing just such a warranty and it is therefore no surprise that claims against surveyors appear from the evidence of the reported cases, to be increasing in number. The prudent purchaser, perhaps aware of his now limited remedies against the builder and others, will now seek the protection of a survey report and, of course, will be keen to sue the surveyor if he has failed to spot the defects.

Naturally, this increased focus on the role of the surveyor does not automatically lead to greater liability. The plaintiff, as ever, will have to prove that the surveyor has been in breach of his duties and this will not be easy in many latent damage claims or where the defect in an unusual one.

That said, the courts do seem to assist the plaintiff in an action against the surveyor in several respects. As has been noted, the courts seem to expect surveyors to go beyond their strict brief, at least in cases where they should have been put on guard, even if the survey is only a minimal one, as in *Roberts* v. *Hampson*. Add to this the wide liability of the building society surveyor, after *Smith* v. *Eric S. Bush,* and the controls placed there on the use of disclaimers, and a picture begins to emerge where not only does the focus of liability shift towards the surveyor but, in addition, whether coincidentally or not, it becomes relatively easier to sue surveyors now than in the past. Surveyors are now very much in the front line of negligence liability for defective premises.

The connection between the role of surveyors and the activities of others involved in construction work arises in another quite different way too. This time the issue is whether the builder, etc., can escape liability on the ground that the plaintiff had not had the property surveyed, or fully surveyed. Can non-use or inappropriate use of a surveyor have a 'knock-on' effect on the liability of others in the construction field?

SUTHERLAND v. C.R. MATON & SONS LTD

(1976) 240 EG 135   High Court

The plaintiffs purchased a seven-year-old bungalow. They were aware of cracks but assumed that the property was satisfactory since their building society had made an inspection and made the full amount of the requested loan available, although the surveyor's report had said that investigation and remedial works was necessary in respect of the cracks. In spite of this, the builders were held liable to the plaintiffs.

COBB J (at p. 135):

> The chain of responsibility was not broken merely because it might be in a particular builder's contemplation that a purchaser would have a full structural survey carried out. A builder's liability could however end when an examination or survey showing up a fault operated to break the chain of causation. In the present case, there was evidence from an inspector who examined the house on behalf of the Hastings & Thanet Building Society when the plaintiffs were considering buying. The tenor of the inspector's evidence was that not every purchaser could be expected to have an independent, full survey done. He (his Lordship) nevertheless thought that if circumstances arose to put a prospective purchaser on his guard, or if there was a survey which showed up defects, but still the purchaser went ahead and bought, the chain of causation going back to the builder would probably snap.

> The facts were that in 1972, when the plaintiffs were living in Gravesend, they became interested in buying the bungalow. Mr Sutherland noticed a crack in the lounge wall. He spoke to a builder friend who advised him not to worry as it was only settlement. In July the bungalow was inspected – not surveyed – by the building society, and their inspector regarded it as being of fair speculative construction. The building society's report advised the plaintiffs to investigate the cause and make good the crack in the lounge. The building society's inspector said in evidence that the crack could have been due simply to soil settlement, though in fact a full survey would have revealed the cause of the subsequent trouble. No portion of the loan was withheld until the work was done. At the time of their purchase the plaintiffs were not aware of the slope in the kitchen floor, as they visited the bungalow only when it had furniture in it. It was argued for the defendants that the crack in the loung and the slight slope in the kitchen floor should have caused the plaintiffs to have had a full structural survey carried out. If they had, it would have been clear that the clay infill under the concrete slab had subsided. The defect, the defendants submitted, was not latent, and even if the crack did not amount to a patent defect, it was enough to cause any reasonable man to have a survey done. It was clear that Mr Sutherland saw the crack in the lounge, but he knew that the bungalow was only a few years old, he thought that if anything serious was wrong the building society would have withheld some of the amount advanced, and he could not afford a survey. He (his Lordship) accepted that ordinary people relied on building society surveys and did not instruct surveyors. He did not feel it was very helpful to consider what had happened in previous cases. He did not feel that ordinary people would think, seeing a crack, 'My goodness, the concrete slab is collapsing.' He thought the plaintiffs did not act unreasonably in not obtaining a full survey. They knew the bungalow was new, it looked all right, and there was no warning from the

building society. Also the builder friend had advised that there was nothing to worry about. No reasonable man would have acted any differently, as he (Cobb J) saw matters, though possibly an ultra-cautious man might have done, and accordingly the plaintiffs' failure to have a survey carried out did not in his opinion constitute a fault on their part.

Similar views were expressed in *Hone* v. *Benson* (1978) 248 EG 1013 at 1014–5 (see above p. 159) and in *Higgins* v. *Arfon Borough Council* [1975] 1 WLR 525. The position thus appears to be that, at least in the ordinary case, the lack of a private survey, as in *Higgins*, or the failure to act sensibly in the light of the surveyor's report, as in *Sutherland*, will not affect the chain causation leading back to the builder or local authority. Of course different facts can produce different answers: if there has been no building society inspection the plaintiff's case would appear much weaker, as would be the case also if there were special circumstances obvious to the plaintiff that should have the effect of placing him 'on guard'.

An alternative way of looking at the problem is to consider whether it would be appropriate to reduce the amount of any award to the plaintiff by regarding his failure to use a surveyor as building contributory negligence.

This possibility was mooted in *Law* v. *R.J. Haddock Ltd* (1985) 2 EGLR 247 and again it is very much going to be a question of fact, turning on such matters as the presence (or not) of a building society survey, the reputation of the builder, etc.

These views do appear to reflect the reality of the situation. Doubtless for reasons of cost, only a minority of house purchasers bother with a full structural survey of the property. In other words, *de facto* reliance is usually being placed on the building society valuation, or else generally on the reputation of the builder, the recent construction of the building and a range of other such factors.

The Role of Negligence Today

How important is negligence now?

After the detailed analysis of the ramifications of the tort of negligence in preceding chapters, it is appropriate finally to consider the current overall status of the tort as it affects the construction and related industries. It is tempting, but misleading, to take the view that negligence is no longer a matter of concern with the wicked ogre of *Anns* v. *Merton*-based liability having been firmly dealt with by the House of Lords in *Murphy v. Brentwood*. In fact, there remains considerable life in the tort, a considerable potential in it to have a continuing impact on builders and others.

A more accurate representation of the current position may be to characterize negligence as being more narrow in its impact, but also more clearly focused. It is true that the wilder excesses of the tort in recent years have been curtailed. For example, claims for damage to property which merely threatens future injury have been reclassified as economic loss and are accordingly much more difficult to maintain successfully. Likewise, the threat of widespread liability being placed on local authorities in spite of their secondary role in the construction process has now been greatly diminished.

Care should be taken to remember, however, where negligence still has an important role to play. It seems likely that in those cases where buildings do cause actual injury to people that the builder, almost certainly, and the local authority, probably, will still be liable. The law's renewed insistence on a relationship of 'close and direct proximity' between the parties to exist before a duty of care comes into being means that fewer parties may be found liable in negligence; this is not all good news for those involved with construction, however, since it simply means that liability will fall fully on the remaining parties who are in such a close and direct relationship. Those parties will now bear 100 per cent of the liability and will be unable to easily extract contributions from the others involved. The unsuccessful attempt of the architects and engineers in

Peabody v. *Parkinson* to ensnare the local authority in the web of liability is an excellent example of this; the decision of the House of Lords to exempt the local authority from liability effectively increased the burden on the architects and engineers.

It should also not be ignored that claims for economic loss have far from disappeared from the negligence scene. Professional men are very much at risk from the ongoing and undoubted liability that exists for careless misstatements and surveyors, as a result, are, as we have seen, now bearing a major share of negligence liability. The controversial decision in *Junior Books* v. *Veitchi*, creating a duty to avoid performing negligent acts which cause economic loss, remains with us. Its precise scope and impact will remain to be determined by future litigants, but it is clear that if it is to be of any importance in any sector, it will be in the construction world, where the close relationships between employer, contractor and (nominated) sub-contractor and the generally high degree of pre-planning that goes on between the employer and all the members of the building team creates just the circumstances of reliance and close proximity that appear to be needed for *Junior Books* liability to arise.

Negligence v. contract

It is in these types of claim that the tort of negligence appears to come closest to overlapping with (or, as some would say, usurping) the role of contract liability as the main route for allocating liability between the parties involved. Is this a problem? Arguably, it is not. Much of the havoc wrought by the onset of negligence liability in the construction arena arose simply because it was new and unexpected, and therefore parties had not taken precautions to avoid or control their liabilities. As an example, the hapless architect in *Eames* (p. 192, above) was uninsured, faced bankruptcy as a result of the ruling against him and had to represent himself, unsuccessfully, in the court hearing. Now, the picture is entirely different.

The fact that the presence of negligence is now well-known means that its consequences can be controlled. Appropriate insurance cover, for example against liability for economic loss, can be arranged. Advantage can be taken of the background of contract liability in a typical construction case to allocate liabilities in accordance with the intention of the parties; indeed within a commercial relationship, the parties will normally be free to exclude or restrict liability by means of exclusion clauses and/or disclaimers of liability.

Those who fear tort's usurpation of contract's functions centre their fears on the fact that claims in respect of pure economic losses amount to protection in tort for the purchaser of property in respect of defects of quality, a protection not afforded by the contract of purchasers of property with its traditional approach of *caveat emptor* – let the buyer beware. Against such fears, several comments may be made. Firstly, claims for economic loss, whether based on acts or statements, are only available in close relationships of proximity akin to a contractual relationship in almost all respects apart from the absence of consideration. Secondly, such a view represents a fundamental misunderstanding of the respective roles of tort and contract. The words of Lord Salmon in *Anns* v. *Merton*, [1978] AC 728 at p. 767, are as pertinent now as when they were made.

> 'If the foundations do give way, there is no warranty by the council which has inspected them that they are sound. The council is responsible only if it has exercised its powers to inspect; and the defects in the foundations would have been detected by an inspection carried out with reasonable care and skill. It seems to me to be manifestly fair that any damage caused by negligence should be borne by those responsible for the negligence rather than by the innocents who suffer from it.'

The only occasion when negligence may provide even the resemblance of a warranty as to quality to future purchasers of a property is when the builder, or whoever, has fallen below the standard of care of his peers i.e. has been careless. Obligations in contract, often to achieve a given result ('the building will be habitable') go much further than this tortious duty merely to take reasonable care. The obligations created by the tort of negligence have rightly been described as a bottom line – an obligation to take the level of care that would be expected from one's peers that is triggered off only when that very basic level is not attained and creates a building that is actually damaged i.e. is in need of repair. Contracts, on the other hand, can provide for higher standards of performance and higher levels of quality, and the law of tort, far from clashing with this, provides an underpinning to it. It is for this reason that it has been argued (at p. 78, above) that tortious actions should normally be available as between parties to a contract.

Negligence and consumer protection

Although it is clear that negligence retains an important role in the area of construction law, the very fact that it does not provide a general warranty of quality to subsequent purchasers, coupled with the various other restrictions placed on its sue in recent years, means that some purchasers of defective properties will go without any legal remedy. The typical subsequent purchaser of a dwelling-house with ailing foundations only has a contract with his vendor. This will embody the *caveat emptor* principle so will provide no relief for the purchaser. (Things may well be different in a commercial context where the purchaser may well nowadays insist on warranties and bonds to ensure quality.) The domestic purchaser immediately then has to have recourse to the tort of negligence and it now provides a very limited answer to his or her problems. If the building collapses and causes injury, then a remedy will be available as some consolation. If it merely threatens to collapse, as is more usual, now that this is categorized as economic loss the typical domestic purchaser will be without a remedy not being in the necessary especially close relationship of proximity with the members of the building team.

The opportunities for recourse in such a case may best be described as 'patchy'. If the purchaser's (or his building society's) surveyor has been careless in not noticing the defect, then an obvious defendant swims into view, but in a case of slow deterioriation of the property, it may be difficult to show that the surveyor was in breach of duty. Other possible actions, in appropriate circumstances, will lie in respect of materials and products if they turn out to be defective. However, if it is the actual building and/or design work that turns out to be defective, the purchaser's only remedy in negligence will depend on the possibility (no more) of bringing an action based on today's version of the 'complex structure' theory or on the threat the defective building poses to its neighbours. Both these are uncertain remedies and, even if they can be used, represent a somewhat quirky response to the purchaser's problems, giving, at best, an erratic remedy, contingent on chance outside factors.

More law reform?

Is this, then, a case for major reform of the law? Certainly, it is unfair to examine liability in negligence in isolation from the newly-important Defective Premises Act 1972 and the NHBC scheme, and this latter scheme provides some protection but only against major structural defects and for a limited period of time (10 years). So claimants, such as *Murphy* himself, can still remain remedy-less. What is particularly unfortunate, it is suggested, is that the House of Lords have reached their decision to deprive such plaintiffs of a remedy not generally, by looking at their needs, insurance provision and respective position vis-à-vis the builder, but rather by the highly technical adjustment of the legal definitions of physical harm and economic loss. It therefore would seem to be an appropriate area for law reform agencies, and ultimately the legislature, to consider, in due course.

Until such time as any reform occurs, the tort of negligence is assured of an ongoing role in construction litigation. In commercial cases, it will continue to supplement the contractual provisions and provide for limited liabilities beyond the framework and time limits of the contract. The 'consumer' of domestic property will also continue to need a negligence remedy in many cases, and so will have to explore either the various aspects of the present law of negligence that clearly are of assistance or else those which have been left open by the courts for now. Further litigation will in due course doubtless explore their scope and applicability.

Table of Cases

BLOCK CAPITALS indicate case extracts which appear in the text.
Italic page numbers indicate extracts from the case or statute.

Table of Statutes

Index